Advanced Theory and Application of Magnetic Actuators

Advanced Theory and Application of Magnetic Actuators

Editors

Suyuan Yu
Jin Zhou
Feng Sun
Ming Zhang

MDPI

Basel • Beijing • Wuhan • Barcelona • Belgrade • Novi Sad • Cluj • Manchester

Editors
Suyuan Yu
Tsinghua University
Beijing
China

Jin Zhou
Nanjing University of
Aeronautics and Astronautics
Nanjing
China

Feng Sun
Shenyang University of
Technology
Shenyang
China

Ming Zhang
Shenyang University of
Technology
Shenyang
China

Editorial Office
MDPI AG
Grosspeteranlage 5
4052 Basel, Switzerland

This is a reprint of articles from the Special Issue published online in the open access journal *Actuators* (ISSN 2076-0825) (available at: https://www.mdpi.com/journal/actuators/special_issues/ATAMA).

For citation purposes, cite each article independently as indicated on the article page online and as indicated below:

Lastname, A.A.; Lastname, B.B. Article Title. *Journal Name* **Year**, *Volume Number*, Page Range.

ISBN 978-3-7258-1753-5 (Hbk)
ISBN 978-3-7258-1754-2 (PDF)
doi.org/10.3390/books978-3-7258-1754-2

Cover image courtesy of Feng SUN

Contents

actuators

MDPI

Article

Dynamic and Thermal Investigations of the Forward Dry-Friction Whirl/Whip of a Vertical Rotor-AMB System during Touchdowns

Zilin Li [1], Mindong Lyu [2], Guojun Yang [3], Jingjing Zhao [3], Yuming Wang [1] and Zixi Wang [1,*]

1 State Key Laboratory of Tribology, Tsinghua University, Beijing 100084, China
2 Institute of Magnetic Levitation and Electromagnetic Propulsion, China Aerospace Science & Industry Corp., Beijing 100143, China
3 Institute of Nuclear and New Energy Technology, Tsinghua University, Beijing 100084, China
* Correspondence: zxwang@tsinghua.edu.cn; Tel.: +86-13521026173

Abstract: When an active magnetic bearing (AMB) rotor drops, it impacts the touchdown bearing (TDB) and produces friction on its surface. The vertical AMB rotor has no stable support in the radial direction, and the rotor exhibits a violent whirl motion in the gap of the TDB. In this study, a complete dynamic and thermal model of the AMB-rotor-TDB system was established, and the complete drop process was simulated. When the rotor dropped, it obtained stable support after several bounces on the thrust surface of the TDB inner ring in the axial direction. In the radial direction, the rotor entered whirl motion after the initial collisions. There is a natural whirl frequency so that the drop forward whirl is divided into the dry-friction whirl and whip states. The contact force and heat generation of the TDB were monitored in the simulation and had different performancs in the two states. Both the initial collisions and the stabilized whirl motions were studied to evaluate the reliability of the TDB. Finally, a series of drop tests were performed, and the experimental results were in good agreement with the simulation.

Keywords: active magnetic bearing; touchdown bearing; rotor dynamics; dry-friction whirl/whip

Citation: Li, Z.; Lyu, M.; Yang, G.; Zhao, J.; Wang, Y.; Wang, Z. Dynamic and Thermal Investigations of the Forward Dry-Friction Whirl/Whip of a Vertical Rotor-AMB System during Touchdowns. *Actuators* **2022**, *11*, 291. https://doi.org/10.3390/act11100291

Academic Editor: Takeshi Mizuno

Received: 14 September 2022
Accepted: 10 October 2022
Published: 11 October 2022

1. Introduction

When an active magnetic bearing (AMB) rotor drops to touchdown bearing (TDB), the rotor cannot get stable support. Therefore, the TDB will be subject to violent collisions, friction, and heat generation, which can cause failure and damage to the TDB in severe cases [1]. Although the drop dynamic response at the initial stage is highly complex, the international standard ISO14839-4 [2] characterizes the rotor drop dynamic response to evaluate the impact and friction forces on the TDB to determine its reliability. Lyu [3,4] proposed a drop trajectory identification method based on the instantaneous frequency transient response. The online condition monitoring implemented in the system was mainly used for vibration suppression and rotor re-suspension [5]. Some external means, such as the installation of squeeze film dampers between the TDB and the bearing support in a study by Kang [6], can also effectively suppress the transient rotor drop vibration response.

Ishii [7] performed early research on the transient response of the AMB–rotor drop process and established an analytical model of the Jeffcott rotor and TDB to study the influence of nonlinear contact on the transient vibration of the bearing. Kärkkäinen [8] established a flexible rotor model based on Timoshenko theory and a TDB model based on a detailed ball bearing model. The Hunt–Crossley contact model [9], which evaluates the collision between the rotor and the TDB, was also widely used in subsequent drop studies. Studies have shown that the dynamic response of rotor drop is affected by many factors, such as the stiffness damping coefficient [10], friction coefficient [11], rotor unbalance [12], and rotor installation misalignment [13]. In addition to the drop dynamic responses of the rotor and the TDB, the heat generation and temperature rise of the TDB are other important monitoring parameters when studying the reliability of the system. The maximum contact force generally occurs in the initial impact of the drop, and the temperature increases continuously with the friction torque as the drop decelerates. Gupta [14] established a dynamic, thermal-coupled model of rolling bearings. The heat in the bearing is mainly caused by internal frictional resistance; note that hybrid ceramic ball bearings generate less heat than steel ball bearings. For the drop process, the friction between the rotor and the inner ring contact surface also needs to be considered in the thermal model. Sun [15] established an effective thermal model of the drop process, including a one-dimensional radial heat transfer grid, and found that the thermal growth at the rotor–TDB interface is the highest, indicating that the contact friction heating power between the rotor and the TDB is higher than that in the TDB.

Most of the above studies and models are based on horizontally oriented rotors, but there are some vertically oriented rotors in industrial applications [16,17]. There is no difference between the two rotor types in the control level of AMB. However, in a drop study, the direction of gravity greatly affected the dynamic response. Hawkins [18] conducted a drop test of a vertical rotor, and it took 2.7 h for the rotor to decelerate from 36,000 r/min to a complete stop with an unassisted spin down from its full speed. Since the rotor is radially unconstrained, friction in the thrust surface brings more constant wear and heating. Although Zhao [19] modeled and simulated the drop dynamics and thermal behaviors of the vertical rotor, the frictional heat generation of the contact surface in the thrust surface was not considered in the model.

In addition, some of our experiments (Figure 1) show that TDB damage does not occur at the drop moment; instead, the TDB is damaged by the exposure to high temperatures during the drop period. Therefore, it is necessary to monitor the entire drop process, especially regarding temperature and stress evaluations. At present, the research in the field mainly focuses on the contact force and temperature changes in the drop transient response, and there are few accurate simulation studies on the full drop process. In this paper, the corresponding thermal model is established based on Wilkes's research [20,21] on the dynamic response of the vertical rotor dry-friction whirl/dry-friction whip. The whole drop process is simulated and analyzed, particularly the thermal growth difference between the rolling friction of the rotor–TDB dry-friction whirl and the sliding friction of the rotor–TDB dry-friction whip. The simulation results are verified by experiments.

Figure 1. Burn and wear scars on the sleeve after severe TDB failure.

2. Modeling

2.1. Friction between the Rotor and TDB

The types of contact in the system need to be defined before establishing a dynamic model. There are three types of contact between the rotor and TDB: axial surface contact, radial line contact, and internal bearing contact. The frictional forces for all three types of contact are calculated by the Coulomb friction model. For the first two forms of contact, the Hunt–Crossley collision contact model [9] is used, which can accurately describe the collision contact force and puncture depth [22]:

$$F_r(t) = Kx^{\beta}(t)\left(1 + \frac{3}{2}\lambda\dot{x}(t)\right) \qquad when\ x(t) > 0 \tag{1}$$

The contact stiffness is given by the Hertz contact model:

$$K = \frac{4}{3\left(\frac{1-v_1^2}{E_1} + \frac{1-v_2^2}{E_2}\right)}\left(\frac{r_1 r_2}{r_1 + r_2}\right)^{\frac{1}{2}} \tag{2}$$

Among the first two contact types, the collision friction calculation for the axial contact surface is more complicated (Figure 2). To calculate the collision friction on the entire surface, the relative motion state of the rotor and inner ring of the TDB must be solved. First, the friction for the contact point A is calculated as follows:

$$\vec{v}_{A,rb}(r,\varphi) = \left[\dot{\rho} + \left(\dot{\theta}_r - \dot{\theta}_b\right) r\sin\varphi\right]\vec{i} + \left[\left(\dot{\alpha} - \dot{\theta}_r\right)\rho - \left(\dot{\theta}_r - \dot{\theta}_b\right) r\cos\varphi\right]\vec{j} \tag{3}$$

Then, the friction force and friction torque are calculated by integrating the whole surface as follows:

$$F_{a,i} = -P_a\mu_a \cdot \int_{R_1}^{R_2}\int_{0}^{2\pi} \frac{\left[\dot{\rho} + \left(\dot{\theta}_r - \dot{\theta}_b\right) r\sin\varphi\right]\vec{i} \cdot r d\varphi dr}{\left\{\left[\dot{\rho} + \left(\dot{\theta}_r - \dot{\theta}_b\right) r\sin\varphi\right]^2 + \left[\left(\dot{\alpha} - \dot{\theta}_r\right)\rho - \left(\dot{\theta}_r - \dot{\theta}_b\right) r\cos\varphi\right]^2\right\}^{\frac{1}{2}}} \tag{4}$$

$$F_{a,j} = -P_a\mu_a \cdot \int_{R_1}^{R_2}\int_{0}^{2\pi} \frac{\left[\left(\dot{\alpha} - \dot{\theta}_r\right)\rho - \left(\dot{\theta}_r - \dot{\theta}_b\right) r\cos\varphi\right]\vec{j} \cdot r d\varphi dr}{\left\{\left[\dot{\rho} + \left(\dot{\theta}_r - \dot{\theta}_b\right) r\sin\varphi\right]^2 + \left[\left(\dot{\alpha} - \dot{\theta}_r\right)\rho - \left(\dot{\theta}_r - \dot{\theta}_b\right) r\cos\varphi\right]^2\right\}^{\frac{1}{2}}} \tag{5}$$

$$M_a = \frac{2F_a\mu_a\left(R_2^3 - R_1^3\right)}{3\left(R_2^2 - R_1^2\right)} \tag{6}$$

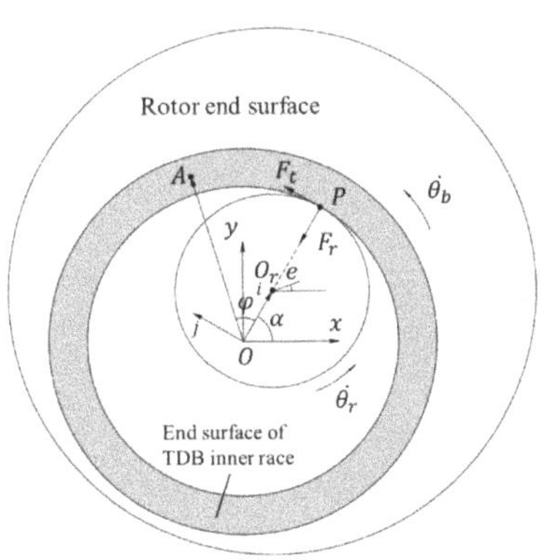

Figure 2. Contact model of the rotor and TDB inner ring.

2.2. *Dynamics of the Vertical Rotor–TDB System*

TDBs can be described by some classical models of rolling bearings [23]. Since it is difficult to design a lubrication system, the TDB is usually in the state of no lubrication or grease lubrication, so the most important thing that needs clarification is the friction torque inside the bearing, which is directly related to the energy loss and temperature rise of the TDB. A high bearing temperature can lead to lubricant deterioration, wear increase, and even bearing damage.

The internal friction torque of the bearing is calculated by Palmgren's empirical formula [24]:

$$M_b = M_0 + M_1 \tag{7}$$

$$M_0 = \begin{cases} 160 \times 10^{-7} f_0 d_m^3, & v\dot{\theta}_b \le 2000 \\ f_0 \left(v\dot{\theta}_b\right)^{\frac{2}{3}} d_m^3 \times 10^{-7}, & v\dot{\theta}_b > 2000 \end{cases} \tag{8}$$

$$M_1 = f_1 P_1 d_m \tag{9}$$

where M_0 represents the friction torque related to the bearing type, the bearing rotational speed, and the lubricant properties, $N \cdot mm$. M_1 represents the friction and wear related to elastic hysteresis and local differential sliding, $N \cdot mm$.

In addition, there is a spin friction torque that affects the heat generation of the bearing at high speeds, which is calculated for the inner and outer ring and each rolling body as follows:

$$M_{si} = \frac{3}{8} \mu_i F_i a_i L_i(k)\ ; M_{so} = \frac{3}{8} \mu_o F_o a_o L_o(k) \tag{10}$$

$$M_{ij} = \frac{d_b}{z \cdot d_i} M\ ; M_{oj} = \frac{d_b}{z \cdot d_o} M \tag{11}$$

By considering the vertical magnetic bearing rotor as the modeling object, the rigid rotor dynamics model is established in Figure 3. According to the Lagrange equation, the dynamic equation of the rotor falling in three directions is established as follows:

$$\begin{cases} m_r \left[\ddot{x}_r - e\sin(\theta_r)\ddot{\theta}_r - e\cos(\theta_r)\dot{\theta}_r^2 \right] = -(F_r - F_{a,i})\cos\alpha - \left(F_t + F_{a,j}\right)\sin\alpha \\ m_r \left[\ddot{y}_r + e\cos(\theta_r)\ddot{\theta}_r - e\sin(\theta_r)\dot{\theta}_r^2 \right] = -(F_r - F_{a,i})\sin\alpha + \left(F_t + F_{a,j}\right)\cos\alpha \\ m_r \ddot{z} = -m_r g + F_{a,z} \\ m_r e\left[-\sin(\theta_r)\ddot{x}_r + \cos(\theta_r)\ddot{y}_r\right] + \left(m_r e^2 + J_r\right)\ddot{\theta}_r = F_r e\sin(\alpha - \theta_r) + F_t(r - e\cos(\alpha - \theta_r)) - M_a \end{cases} \tag{12}$$

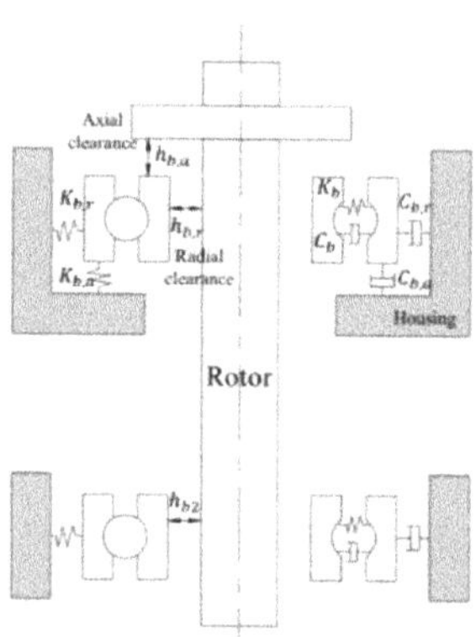

Figure 3. Contact model of the rotor and TDB inner ring.

The dynamic equation of the TDB inner ring can be written as follows:

$$\boldsymbol{M}_b \ddot{\boldsymbol{x}} = \boldsymbol{F}_c \tag{13}$$

where $\boldsymbol{M}_b = [m_b\ m_b\ J_b]$ and $\boldsymbol{x} = [x_b\ y_b\ \theta_b]^T$, which are the mass and displacement matrices of the TDB inner ring, respectively. $\boldsymbol{F}_c$ represents the contact force and friction matrix on the TDB inner ring, and it is defined as follows:

$$\boldsymbol{F}_c = \begin{bmatrix} (F_r - F_{a,i})\cos\alpha + (F_t + F_{a,j})\sin\alpha - F_{nb}\cos\alpha_b + (M/r_s)\sin\alpha_b \\ (F_r - F_{a,i})\sin\alpha - (F_t + F_{a,j})\cos\alpha - F_{nb}\sin\alpha_b - (M/r_s)\cos\alpha_b \\ -F_t r_i - M_b + M_a \end{bmatrix} \tag{14}$$

2.3. Thermal Model of the Rotor–TDB System

During rotor drop, heat is generated by the frictional forces between the rotor and the TDB inner ring and by the frictional torque inside the TDB. This heat causes a temperature rise in the bearing, which in turn changes the bearing dynamics and performance. Therefore, the thermal characterization of the drop is also a crucial aspect of the rotor–TDB system reliability assessment.

The heat generation mainly comes from the contact friction defined in the previous section on dynamics. The node heat exchanges comprise the heat conduction and convection of each component and the air, respectively. The temperature node model and the heat transfer grid model are established as shown in Figure 4. The heat transfer capability of each part is determined by the contact thermal resistance. The contact thermal resistance of each part is calculated as follows:

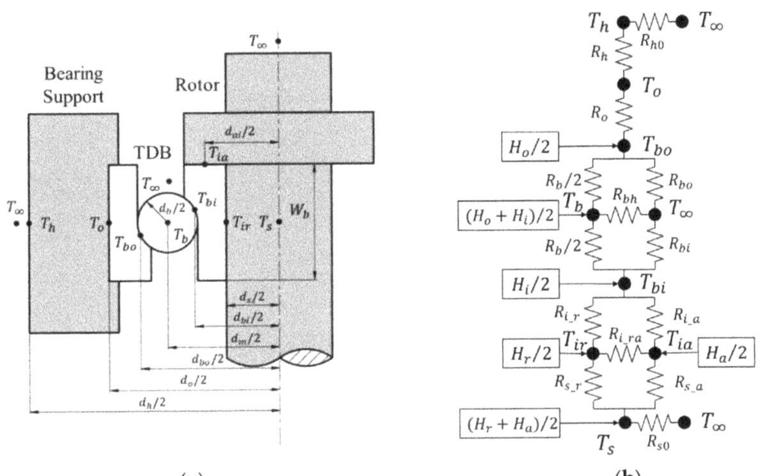

Figure 4. Thermal model of the rotor–TDB system: (**a**) Temperature node model; (**b**) Heat transfer grid model.

Rolling element:

$$R_b = \frac{2z}{k_b \pi d_b};\ R_{bh} = \frac{z}{h_b \pi d_b^2} \tag{15}$$

TDB inner ring:

$$R_{bi} = \frac{z}{h_i \pi d_{bi} W_b};\ R_{i_r} = \frac{\ln(d_{bi}/d_s)}{2k_i \pi W_b};\ R_{i_a} = \frac{2W_b}{k_i \pi (d_{bi}^2 - d_s^2)};R_{i_{ra}} = \frac{2}{k_i \pi \sqrt{(d_{ai} - d_s)^2 + W_b^2}} \tag{16}$$

TDB outer ring:

$$R_{bo} = \frac{z}{h_o \pi d_{bo} W_b};\ R_o = \frac{\ln(d_o/d_{bo})}{2k_o \pi W_b} \tag{17}$$

Rotor:

$$R_{s_r} = \frac{1}{k_s \pi W_b};\ R_{s_a} = \frac{2}{k_s \pi d_{ai}};R_{s0} = \frac{4L_s}{k_s \pi d_s^2} + \frac{4}{h_s \pi d_s^2} \tag{18}$$

TDB support:

$$R_h = \frac{\ln(d_h/d_o)}{2k_h \pi W};\ R_{h0} = \frac{1}{h_h \pi d_h W} \tag{19}$$

Assuming that the heat flux is uniformly distributed radially, the heat transfer mesh of the system can be established, and the temperature of each node can be solved with a first-order differential equation:

$$mC_p \Delta T = \Delta Q \tag{20}$$

Overall, the calculation process can be summarized by Figure 5. The axial dynamic response, radial dynamic response, and thermal response of the system are calculated in sequence. Among them, the spatial position relationship of the rotor and the TDB causes the axial and radial dynamics to influence each other. The temperature change caused by the dynamics also affects the dynamic response because of the viscosity–temperature characteristics of the lubricant.

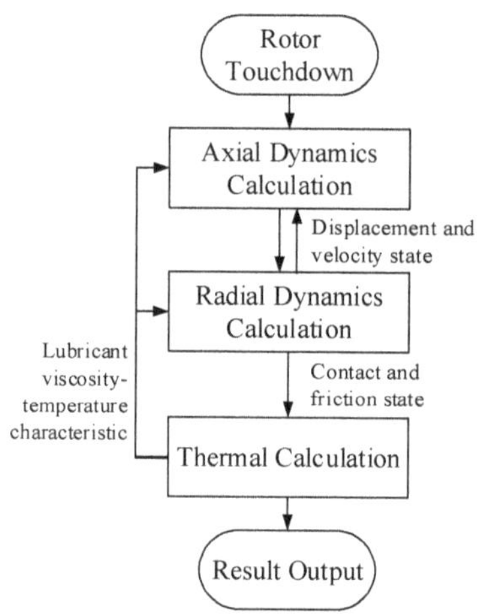

Figure 5. Contact model of the rotor and TDB inner ring.

3. Rotor Touchdown Experiments

To verify the validity of the model, an AMB–rotor–TDB system experimental bench with a rigid rotor is built, which is capable of horizontal and vertical arrangements and is suitable for various research needs. The mechanical structure is modularized, relying on four optical axes to ensure coaxial positioning accuracy. It mainly consists of the rotor, high-speed asynchronous motor, radial AMB, axial AMB, and TDBs (see Figure 6). Two sets of TDBs support the rotor; each set of TDB uses a pair of angular contact ball bearings with face-to-face preloaded. The upper TDB bears both axial and radial loads, while the lower TDB only bears radial loads. The axial and radial clearances between the rotor and the TDB are both 0.2 mm. The electrical control section mainly consists of eddy current displacement sensors, a photoelectric speed sensor, analog filter, controller, and power amplifier. The monitoring part consists of a CAN (Controller Area Network) communication card, acquisition card, thermal imager, and host computer, which can record and display key information, such as the rotor trajectory, rotational speed, control current, and PID control parameters. At present, this system has achieved a maximum speed of 36,000 r/min in suspension tests. The relevant parameters of the experimental bench are shown in Table 1.

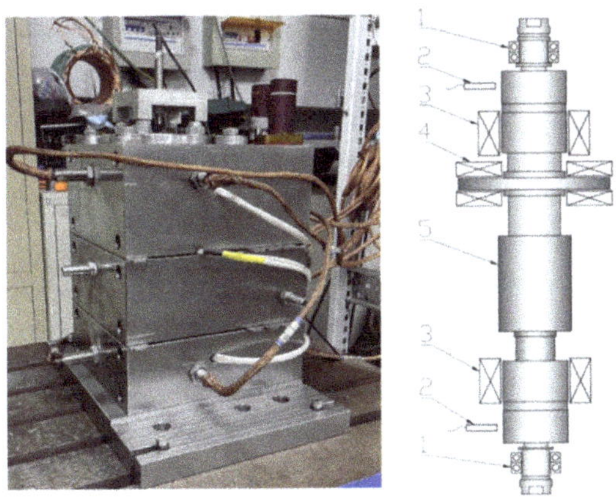

Figure 6. AMB–rotor–TDB drop test bench: 1—TDBs, 2—eddy current sensors, 3—radial AMBs, 4—axial AMB, 5—rotor.

Table 1. Parameters of the AMB–rotor–TDB system.

Parameters	Value
Rotor mass	2.5 kg
Rotor length	293 mm
Rotor moment of inertia	8957.32×10^{-6} kg·m^2
Imbalance eccentricity	1.5×10^{-5} m
Radial magnetic bearings distance	155 mm
Two TDBs' distance	253 mm
Radial AMB clearance	0.3 mm
Axial AMB clearance	0.4 mm
TDB clearance	0.2 mm
Contact friction coefficient	0.12
Motor power	4 kW
Bias current	2 A
Radial AMB current stiffness	81.268 N/A
Radial AMB displacement stiffness	-5.864×10^5 N/m
Axial AMB current stiffness	50.26 N/A
Axial AMB displacement stiffness	-2.720×10^5 N/m
Maximum speed	36,000 r/min

The test protocol is as follows: start the AMB to levitate the rotor when the rotor is stationary; use the motor to accelerate the rotor to the set speed; turn off the motor and AMB simultaneously to let the rotor drop freely without applying additional braking force; use the eddy current sensor and photoelectric speed sensor to monitor the rotor trajectory and rotational speed; and use the thermal imager to monitor the temperature change of TDB.

4. Result and Discussion

4.1. Simulation Results

A drop simulation model was performed using the method described in Section 2. The parameters of the experimental bench were substituted into the simulation model calculations. Three cases of rotor drop speed—5000, 10,000, and 20,000 r/min—were used for analysis. The drop trajectory, rotational speed, contact force, and temperature variation were solved for each drop case. The simulations not only focused on the dynamic and thermal responses of the rotor and TDB at the beginning of the drop but also on the entire drop process as the rotor decelerated to its stop.

The simulation results of the rotor dropping to the TDB at 5000 r/min are shown in Figure 7. From the drop trajectory in Figure 7a,b, the rotor initially experienced several collisions before achieving axial stable support. Radially, it entered stable whirl motion, of which the direction was obtained from the rotor speed in Figure 7c,d. Figure 7c,d displays the rotor speed, TDB inner ring speed, and rotor whirl motion. The inner ring accelerated to a speed equal to the rotor soon after the drop, while the whirl speed was initially negative (backward whirl) and then gradually became a forward whirl. Then, the whirl speed increased to the same speed as the rotor, and the two synchronously decelerated to zero velocity. From the force diagram in Figure 7e,f, the maximum contact force occurred in the first few collisions. The maximum axial contact force was 542 N, and the radial maximum contact force was 322 N. According to the temperature change in Figure 7g,h, the most obvious heat generation was in the rolling elements with the fastest heat dissipation. The highest temperature, which was 2.4 °C higher than the ambient temperature, appeared within the first 3 s of the drop.

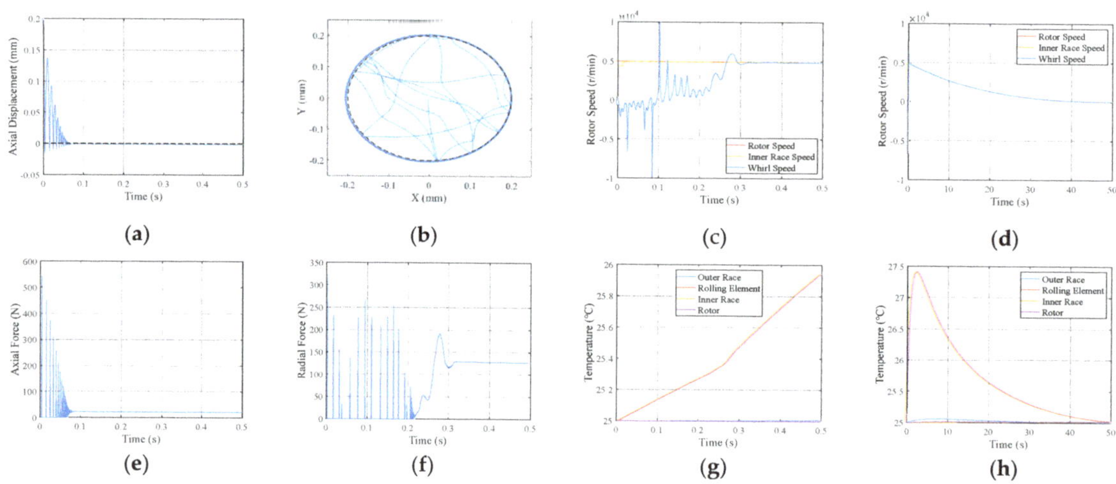

Figure 7. Numerical simulation results of the rotor touchdown at 5000 r/min: (**a**) Axial displacement, (**b**) Radial trajectory, (**c**) Rotor speed change in the first 0.5 s, (**d**) Rotor speed change for the whole process, (**e**) Axial contact force of the TDB, (**f**) Radial contact force of the TDB, (**g**) Temperature change in the first 0.5 s, (**h**) Temperature change for the whole process.

The results of increasing the drop speed to 10,000 r/min are shown in Figure 8. The difference of the drop trajectory in Figure 8a,b was not obvious. The most obvious change was in the whirl speed in Figure 8c,d, where the whirl speed could not reach the rotor speed after acceleration; however, the sub-synchronous forward whirl occurred at a fixed speed range (4480–8000 r/min with an average of 6240 r/min) until the rotor decelerated to the whirl speed, and then the rotor and TDB inner ring transitioned to a synchronous whirl and then decelerated to zero velocity. The contact force and temperature rise of the rotor significantly increased as there was an increase in both factors during the transition from sub-synchronous whirl to synchronous whirl. The radial maximum contact force was 716 N. The maximum temperature rise of the rolling element was 9.2 °C.

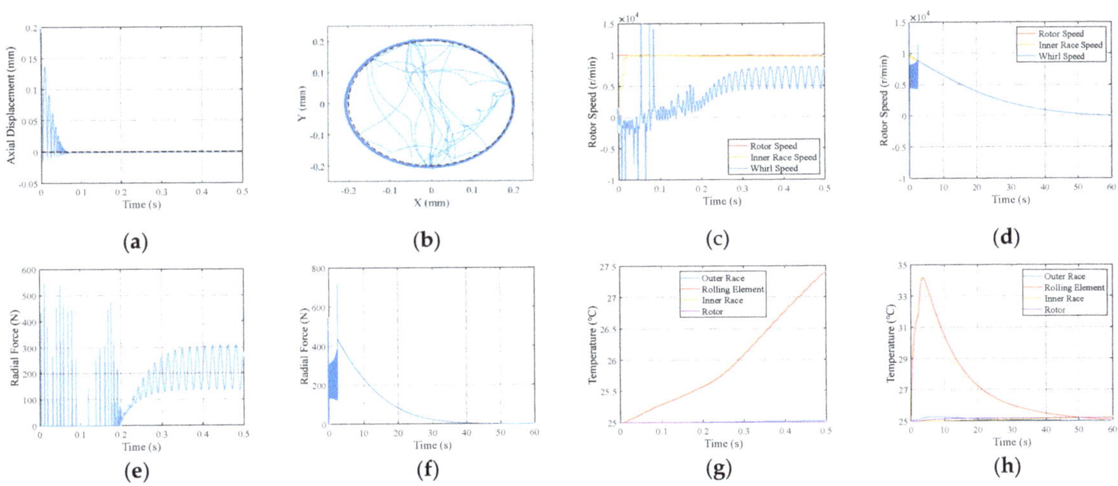

Figure 8. Numerical simulation results of the rotor touchdown at 10,000 r/min: (**a**) Axial displacement, (**b**) Radial trajectory, (**c**) Rotor speed change in the first 0.5 s, (**d**) Rotor speed change for the whole process, (**e**) Radial contact force of the TDB in first 0.5 s, (**f**) Radial contact force of the TDB for the whole process, (**g**) Temperature change in first 0.5 s, (**h**) Temperature change for the whole process.

The speed was further increased to 20,000 rpm, and the simulation results are shown in Figure 9. The change in whirl speed showed that the whirl speed remained at 6240 r/min and did not increase with increasing rotor speed, unlike when the speed was 10,000 rpm. In addition, the contact force and heat generation level at the beginning of the drop further increased as the rotor speed increased. The radial maximum contact force was 1155 N. The maximum temperature rise of the rolling element was 10.5 °C. After entering the stable forward whirl motion, the contact force and heat generation did not change much compared to the lower speed, where the contact force was approximately 200 N.

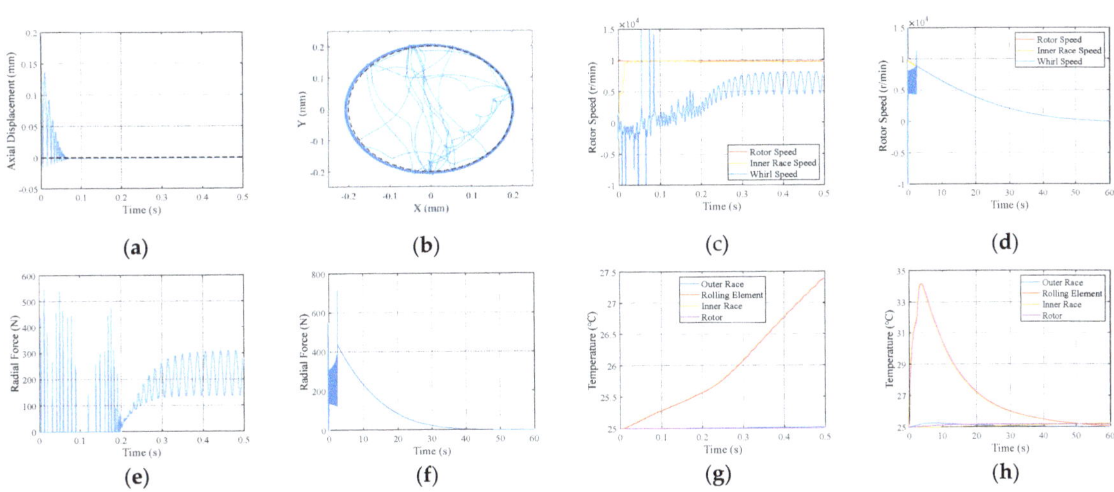

Figure 9. Numerical simulation results of the rotor touchdown at 5000 r/min: (**a**) Axial displacement, (**b**) Radial trajectory, (**c**) Rotor speed change in the first 0.5 s, (**d**) Rotor speed change for the whole process, (**e**) Radial contact force of the TDB in the first 0.5 s, (**f**) Radial contact force of the TDB for the whole process, (**g**) Temperature change in the first 0.5 s, (**h**) Temperature change for the whole process.

4.2. Model Analysis

The axial ring surface and tangential contact friction drive the rotor to whirl. The whirl direction depends on the direction of the friction force F_t. In the early stages of the drop, the inner ring is stationary or operating at a low speed, and the F_t direction is opposite to the rotor rotation direction, so the rotor exhibits a backward whirl. After the axial surface causes the inner ring to accelerate, the inner ring will maintain the same speed as the rotor. The total friction direction is the same as the rotor rotation direction, so the rotor will begin to forward whirl after stabilization.

Although the initial collision process is unpredictable, the stable forward whirl state had a certain regularity, which is consistent with the dry-friction whirl and dry-friction whip states expressed by Wilkes [20,21]. These states are defined as follows.

1. **Dry-friction whirl (Dry whirl)**: the rotor rolls on the surface of the stator without slipping, and the precession frequency is governed by the radius-to-clearance ratio at the contact location;
2. **Dry-friction whip (Dry whip)**: the rotor slides continuously on the surface of the stator, and the precession frequency is controlled by the combined natural frequency of the rotor–stator system.

In the dry-friction whirl state, since the diameters of the rotor and the TDB inner ring are approximately equal, the speeds of the rotor and inner ring are also approximately equal. However, the natural whirl frequency under dry-friction whip is not equal to the natural frequency of the rotor at stable-support [16], which is derived and analyzed below.

It is assumed that after entering a stable forward whirl state, the rotor performs whirl motion with a fixed radius, and the model does not consider the change in the rotational speed that occurs over a short time. Therefore, Equation (12) is simplified and expressed in polar coordinates as follows:

$$m_r\left[\rho_0\ddot{\alpha}ie^{i\alpha} + \rho_0\dot{\alpha}^2e^{i\alpha} + e_0\Omega^2e^{i\Omega t}\right] = K\rho_0e^{i\alpha} + \left(\mu K\rho_0 + F_{a,j}\right)ie^{i\alpha} \tag{21}$$

Equation (21) is a second-order differential equation for the angular displacement, so there is a special solution of this equation—the natural whirl frequency. In addition, there is a time-dependent imbalance parameter $e_0\Omega^2e^{i\Omega t}$, which explains why the simulated natural whirl frequency is not a fixed value but fluctuates within a certain range.

According to the above analysis, the stable whirl speed at each speed level was solved by simulation, and the variation spectrum of whirl speed with the initial drop speed is shown in Figure 10. The switching between the two states of dry-friction whirl and dry-friction whip was distinguished by an obvious critical speed, which was jointly determined by the natural whirl frequency and its fluctuation range.

The two stable forward whirl states were also different in force and heat generation. The dry-friction whirl occurred at a lower speed than the dry-friction whip; therefore, the dry-friction whirl had a lower collision contact force at the beginning of the drop. After entering the dry-friction whip state, the contact force did not continue to rise with increasing drop speed. In the dry-friction whirl state, the rotor rotated synchronously with the TDB inner ring, and there was almost no sliding friction between them. Therefore, the overall heat generation was low, mainly from the internal heat of the bearing. After entering the dry-friction whip state, the sliding friction between the rotor and the inner ring of the TDB became the dominant factor in the heat generation of TDB, so the heat generation was significantly increased compared that of the dry-friction whirl state. With the energy release and deceleration of the rotor, the high-speed rotor also changed from the dry-friction whip state to the dry-friction whirl state when it decelerated to zero velocity. The force and temperature simulation of the entire process helped to accurately assess the reliability of the TDB.

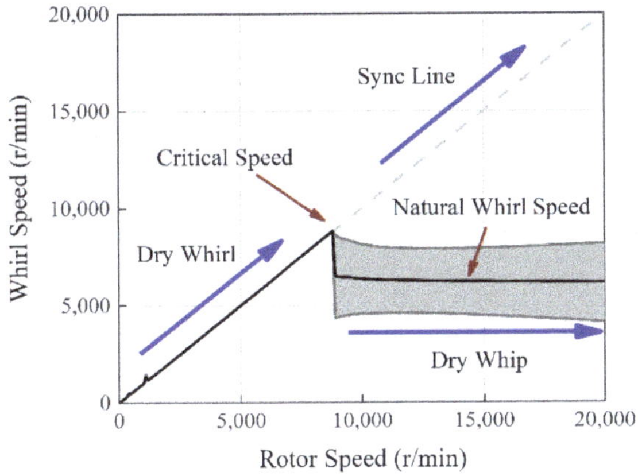

Figure 10. Whirl speed variation spectrum with initial touchdown speed.

4.3. Experimental Verification

After arrangement of the experimental table vertically, a series of drop tests were conducted by setting the rotational speed to 5000 rpm and 20,000 rpm, and the results were obtained, as shown in Figures 11–13. Figure 11 shows the rotor drop trajectory and temperature measured in the experiment, and similar results were obtained at different speed settings.

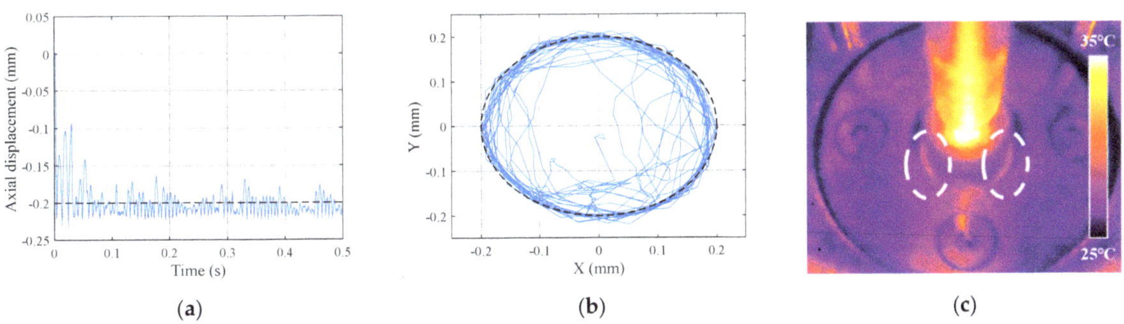

Figure 11. Experimentally measured rotor drop process: (**a**) Axial displacement, (**b**) Radial trajectory, and (**c**) Temperature change (The circle shows the position of TDB).

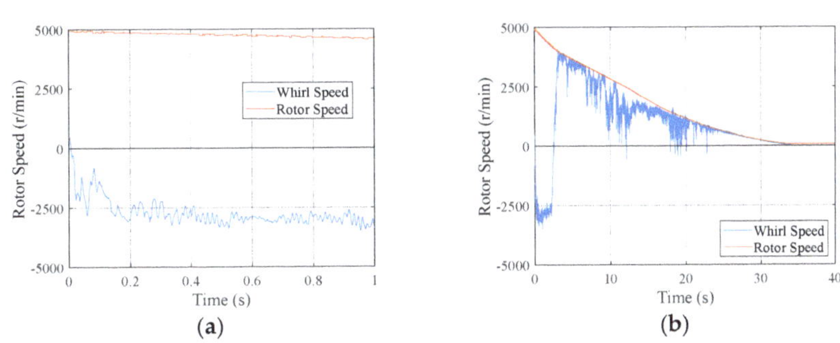

Figure 12. Rotor speed and whirl speed measured in the 5000 r/min drop test: (**a**) First second, (**b**) Whole process.

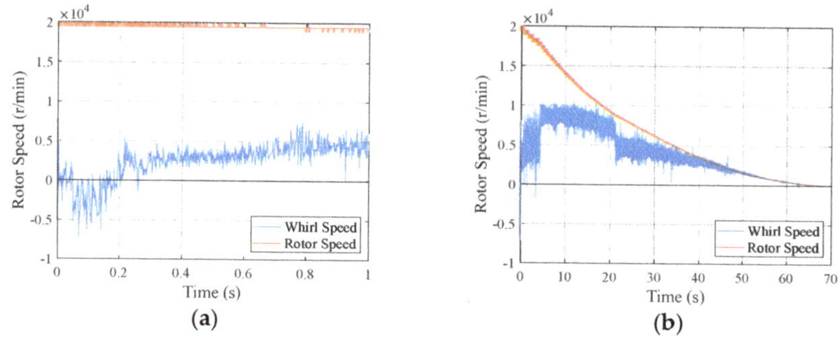

Figure 13. Rotor speed and whirl speed measured in the 20,000 r/min drop test: (**a**) First second, (**b**) Whole process.

At a 5000 r/min drop in Figure 12, the rotor quickly entered a full-rub whirl state. Compared to the simulation results, the rotor experienced a longer period of backward whirl and entered a forward dry-friction whirl state, then decelerated to zero velocity synchronously as in the simulation. At a 20,000 r/min drop in Figure 13, the rotor moved more quickly from the backward whirl into the forward whirl than at the 5000 r/min drop. Then, the rotor remained at a whirl speed of 6174 r/min, resulting in a dry-friction whip state. After the speed was greatly reduced, the whirl speed also gradually decreased synchronously with the rotor, changing from dry-friction whip to dry-friction whirl. The maximum temperature of the rolling body measured by the thermal imaging camera was 1.5 °C at 5000 r/min and 9.5 °C at 20,000 r/min above the ambient temperature, which were slightly lower than the simulation results, 2.4 °C at 5000 r/min and 10.5 °C at 20,000 r/min.

Based on a comparison of the simulation and experimental results from Figures 7, 9 and 11–13, the results are in good agreement, especially the occurrence of each state in the whole rotor drop process, which was accurately predicted. In addition, comparing the deceleration of the rotor in Figure 14 proved the computational validity of the simulation model throughout the whole drop process, indicating that the simulation model could accurately evaluate the process of drop energy conversion, which basically met the research needs to study the contact force and heat generation.

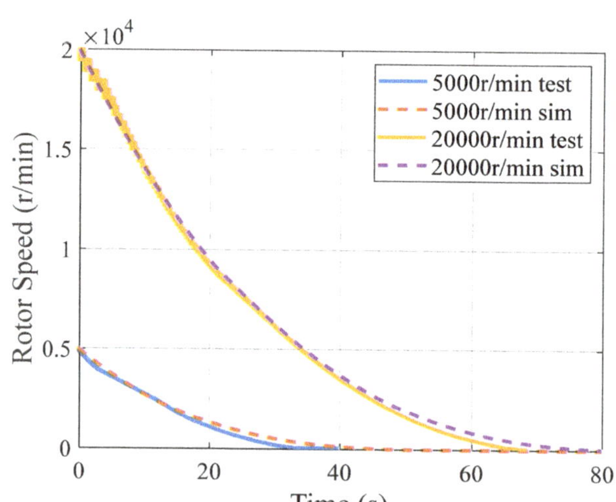

Figure 14. Comparison of rotor speed changes between simulation and experiment.

5. Conclusions

When a vertical rotor drops on the TDB, after several bounces in the axial direction, the rotor obtains a stable support. The friction of the contact surface drives the TDB inner ring to accelerate to the same rotational speed as the rotor. In the radial direction, the rotor enters whirl motion after the initial collisions. The whirl direction depends on the direction of friction, so the system first enters the backward whirl state and then stabilizes in the forward whirl state. There is a natural whirl frequency (rotational speed) and a critical speed, so the drop forward whirl is divided into the dry-friction whirl and whip states; the former occurs at low speeds, and the latter occurs at high speeds.

The dry-friction whip has a higher TDB contact force and temperature rise than the dry-friction whirl, which is due to the rotor speed and friction state. The contact force at the beginning of the drop increases with the drop speed. The contact force of the dry-friction whirl also increases with the drop speed, and the contact force of the dry-friction whip is not affected by the drop speed. Nonetheless, the significant temperature rise caused by sliding friction in dry-friction whip exposes the TDB to potential wear failure.

The above conclusions are drawn through the simulation of the dynamic and thermal coupled model. Compared with models in other studies, this model can simulate the whole process of drop and deceleration. A five-degree-of-freedom AMB–rotor–TDB system experimental bench was established, and a series of drop tests were performed. The drop speed was set to 5000 r/min and 20,000 r/min, and the results showed that the model and the experiment were in good agreement when considering the drop trajectory, rotation speed, and temperature rise, especially for the simulation of the rotor deceleration process, which had good repeatability.

Author Contributions: Methodology, writing—original draft preparation, Z.L.; conceptualization, M.L.; data curation, validation, G.Y.; software, J.Z.; resources, supervision, Y.W.; project administration, writing—review and editing, Z.W.; All authors have read and agreed to the published version of the manuscript.

Funding: This research was funded by the National Key R&D Program of China, grant No. 2018YFB2 000102 and the National Natural Science Foundation of China, grant No.52005467.

Conflicts of Interest: The authors declare no conflict of interest.

Nomenclature

a	contact ellipse long semiaxis
d_b, d_i, d_o	diameter of ball, inner ring, outer ring
d_m	$= (d_b + d_i)/2$
E	modulus of elasticity
f_0	surface friction media parameter
$F_{a,z}$, F_a, P_a,	contact force, frictional force, pressure on the axial end surface
F_r, F_t	radial impact force and tangential frictional force between rotor and TDB inner ring
$h_{b,a}$, $h_{b,r}$	axial and radial clearance of rotor and TDB
H	heat generation
k	conduction thermal conductivity coefficient
K	contact stiffness
$L(k)$	the second kind of completely elliptic integrals
m_r, J_r, e	rotor mass, inertia, and unbalance
M_a	friction torque on the axial end surface
M_b	friction torque of TDB
r	radius of curvature in the Hertz model
R	thermal resistance
R_1, R_2	inner and outer radius of TDB inner ring
T	temperature

v	Poisson's ratio
$\vec{v}_{A,rb}, r, \varphi$	relative sliding velocity, radius, and phase angle at point A
z	number of balls
β	the nonlinear exponent determined from material and geometric properties of the local region of the contacting bodies
λ	contact coefficient of Hunt-Crossley model, valued 0.08-0.32 s/m
θ_b, θ_r	rotational angle of TDB inner ring and rotor
μ_a, μ_r	axial and radial friction coefficient between rotor and TDB inner ring
ρ, α	polar coordinate representation of rotor center
Subscripts	
a	axial direction
b	ball
h	housing
i	Inner race
o	outer race
r	radial direction
s	rotor

References

1. Schweitzer, G. Safety and Reliability Aspects for Active Magnetic Bearing Applications—A Survey. *Proc. Inst. Mech. Eng. Part I J. Syst. Control Eng.* **2005**, *219*, 383–392. [CrossRef]
2. *ISO 14839-4*; Mechanical Vibration—Vibration of Rotating Machinery Equipped with Active Magnetic Bearings Part 4: Technical Guidelines. International Organization for Standardization: Geneva, Switzerland, 2012.
3. Lyu, M.; Liu, T.; Wang, Z.; Yan, S.; Jia, X.; Wang, Y. Orbit Response Recognition during Touchdowns by Instantaneous Frequency in Active Magnetic Bearings. *J. Vib. Acoust. Trans. ASME* **2018**, *140*, 1–11. [CrossRef]
4. Lyu, M.; Wang, Z.; Liu, T.; Jia, X.; Wang, Y. Frequency Analysis of the Orbit Responses of Active Magnetic Bearings in Touchdown Using Hilbert Transform. *Int. J. Struct. Stab. Dyn.* **2017**, *17*, 17500869. [CrossRef]
5. Lyu, M.; Liu, T.; Wang, Z.; Yan, S.; Jia, X.; Wang, Y. A Control Method of the Rotor Re-Levitation for Different Orbit Responses during Touchdowns in Active Magnetic Bearings. *Mech. Syst. Signal Process.* **2018**, *105*, 241–260. [CrossRef]
6. Kang, X.; Palazzolo, A.; Zhong, W. Auxiliary Bearing Squeeze Film Dampers for Magnetic Bearing Supported Rotors. *Tribol. Int.* **2020**, *146*, 106181. [CrossRef]
7. Ishii, T.; Kirk, R.G. Transient Response Technique Applied to Active Magnetic Bearing Machinery during Rotor Drop. *ASME J. Vib. Acoust.* **1996**, *118*, 154–163. [CrossRef]
8. Kärkkäinen, A.; Sopanen, J.; Mikkola, A. Dynamic Simulation of a Flexible Rotor during Drop on Retainer Bearings. *J. Sound Vib.* **2007**, *306*, 601–617. [CrossRef]
9. Hunt, K.; Crossley, E. Coefficient of Restitution Interpreted as Damping in Vibroimpact to Cite This Version: HAL Id: Hal-01333795 Coefficient of Restitution Interpreted as Damping in Vibroimpact. *J. Appl. Mech. Am. Soc. Mech. Eng.* **1975**, *42*, 440–445. [CrossRef]
10. Jarroux, C.; Dufour, R.; Mahfoud, J.; Defoy, B.; Alban, T.; Delgado, A. Touchdown Bearing Models for Rotor-AMB Systems. *J. Sound Vib.* **2019**, *440*, 51–69. [CrossRef]
11. Neisi, N.; Sikanen, E.; Heikkinen, J.E.; Sopanen, J. Stress Analysis of a Touchdown Bearing Having an Artificial Crack. *Proc. ASME Des. Eng. Tech. Conf.* **2017**, *8*, 1–9. [CrossRef]
12. Fonseca, C.A.; Santos, I.F.; Weber, H.I. Influence of Unbalance Levels on Nonlinear Dynamics of a Rotor-Backup Rolling Bearing System. *J. Sound Vib.* **2017**, *394*, 482–496. [CrossRef]
13. Halminen, O.; Kärkkäinen, A.; Sopanen, J.; Mikkola, A. Active Magnetic Bearing-Supported Rotor with Misaligned Cageless Backup Bearings: A Dropdown Event Simulation Model. *Mech. Syst. Signal Process.* **2015**, *50–51*, 692–705. [CrossRef]
14. Gupta, P.K.; Taketa, J.I.; Price, C.M. Thermal Interactions in Rolling Bearings. *Proc. Inst. Mech. Eng. Part J J. Eng. Tribol.* **2020**, *234*, 1233–1253. [CrossRef]
15. Sun, G. Auxiliary Bearing Life Prediction Using Hertzian Contact Bearing Model. *J. Vib. Acoust.* **2006**, *128*, 203. [CrossRef]
16. Caprio, M.T.; Murphy, B.T.; Herbst, J.D. Spin Commissioning and Drop Tests of a 130 KW-Hr Composite Flywheel. In Proceedings of the 9th International Symposium on Magnetic Bearings, Lexington, KY, USA, 3–6 August 2004.
17. Reid, C.M.; Miller, T.B.; Hoberecht, M.A.; Loyselle, P.L.; Taylor, L.M.; Farmer, S.C.; Jansen, R.H. History of Electrochemical and Energy Storage Technology Development at NASA Glenn Research Center. *J. Aerosp. Eng.* **2013**, *26*, 361–371. [CrossRef]
18. Hawkins, L.; McMullen, P.; Vuong, V. Development and Testing of the Backup Bearing System for an AMB Energy Storage Flywheel. In Proceedings of the Turbo Expo: Power for Land, Sea, and Air, ASMEDC, Montreal, QC, Canada, 14–17 May 2007; pp. 1055–1062.
19. Zhao, Y.; Yang, G.; Shi, Z.; Zhao, L. Thermal Analysis and Simulation of Auxiliary Bearings and Its Application in the High Temperature Reactor-10. *J. Tribol.* **2016**, *138*, 011102. [CrossRef]
20. Wilkes, J.C.; Childs, D.W.; Dyck, B.J.; Phillips, S.G. The Numerical and Experimental Characteristics of Multimode Dry-Friction Whip and Whirl. *J. Eng. Gas Turbines Power* **2010**, *132*, 052503. [CrossRef]

21. Wilkes, J.; Moore, J.; Ransom, D.; Vannini, G. An Improved Catcher Bearing Model and an Explanation of the Forward Whirl/Whip Phenomenon Observed in Active Magnetic Bearing Transient Drop Experiments. *J. Eng. Gas Turbines Power* **2014**, *136*, 1–11. [CrossRef]
22. Jacobs, D.A.; Waldron, K.J. Modeling Inelastic Collisions with the Hunt-Crossley Model Using the Energetic Coefficient of Restitution. *J. Comput. Nonlinear Dyn.* **2015**, *10*, 021001. [CrossRef]
23. Halminen, O.; Aceituno, J.F.; Escalona, J.L.; Sopanen, J.; Mikkola, A. Models for Dynamic Analysis of Backup Ball Bearings of an AMB-System. *Mech. Syst. Signal Process.* **2017**, *95*, 324–344. [CrossRef]
24. Takabi, J.; Khonsari, M.M. On the Thermally-Induced Failure of Rolling Element Bearings. *Tribol. Int.* **2015**, *94*, 661–674. [CrossRef]

actuators

MDPI

Article

Analysis of Trajectory Tracking Characteristics of a Magnetically Driven Oil-Free Scroll Compressor

Ce Shi [1], Feng Sun [1,*], Fangchao Xu [1], Junjie Jin [1], Ling Tong [1], Qing Zhou [1] and Koichi Oka [2]

1 School of Mechanical Engineering, Shenyang University of Technology, Shenyang 110801, China
2 School of Systems Engineering, Kochi University of Technology, Kochi 782-8502, Japan
* Correspondence: sunfeng@sut.edu.cn; Tel.: +86-189-0404-0100

Abstract: The conventional scroll compressor cannot run oil-free because of wear and tear and lubrication problems during operation due to some parts, such as anti-rotation devices. The magnetic drive oil-free scroll compressor (MDOFSC) uses a contactless drive method to avoid this drawback. In order to solve the swing problem of the orbiting scroll during the operation of the MDOFSC, decentralized control and centralized control are used to study the trajectory tracking characteristics. Firstly, the structure and working principle of the MDOFSC are introduced, and the system's magnetic circuit and differential control principle are analyzed. Then, the dynamic model of the MDOFSC under the condition of non-compressed gas is established, and the coordinate matrix decoupling method is used to analyze the relationship between the degree of freedom of the system and the measurement distance of the displacement sensor. Finally, the system is simulated and experimentally studied under centralized PID control, and the experimental comparison study between decentralized control and centralized control is conducted. The results show that centralized control dramatically improves the trajectory control ability of the system.

Keywords: oil-free scroll compressor; magnetic drive; centralized control; matrix decoupling; PID control; experimental study

Citation: Shi, C.; Sun, F.; Xu, F.; Jin, J.; Tong, L.; Zhou, Q.; Oka, K. Analysis of Trajectory Tracking Characteristics of a Magnetically Driven Oil-Free Scroll Compressor. *Actuators* **2022**, *11*, 312. https://doi.org/10.3390/act11110312

Academic Editor: Katsushi Furutani

Received: 9 September 2022
Accepted: 25 October 2022
Published: 27 October 2022

1. Introduction

Scroll compressors have many unique advantages, such as a simple structure, low noise, high efficiency, and good reliability. As the application of scroll compressors extends to medicine, food, and other fields requiring clean compressed gas, developing an oil-free scroll compressor has essential practical significance [1,2].

Li proposed a sealing mechanism model to achieve sealing of scroll tooth axial clearance by installing self-lubricating material sealing strips and springs in the sealing groove opened on the tooth end face [3]. In existing oil-free scroll compressors, the first type uses self-lubricating bearings instead of oil-lubricated bearings; the second type coats a layer of self-lubricating material on the surface of the anti-rotation device and the contact surface of the moving and stationary scroll. Although the two methods achieve the goal of being oil-free, there are some problems with friction and heating. Sun proposed an oil-free scroll compressor with a solenoid instead of an anti-rotation device and no wearing parts. Only modeling and simulation were performed in the article, and no experimental study was conducted [4,5]. Magnetic levitation technology, with its fast response, high accuracy, and clean and pollution-free features, provides new solutions for several industries and has been widely used under the intensive research of many scholars, with typical applications such as magnetic levitation bearings [6,7], magnetic levitation trains [8], and magnetic levitation planar motors [9]. Zhao proposed a new magnetic levitation system with two rotating magnets that can achieve a zero-power horizontal levitation state under eccentric loading [10]. Soni investigated the dynamical behavior and stability of energy-efficient electromagnetic levitation with time-lag rotors.

The control formulas were transformed to the frequency domain, the time lag values of the feedback signals were modeled using the Pade approximation, and stability analysis was performed by analyzing the system poles [11]. Zhang X provided a six-degree-of-freedom levitation platform. This study manipulated the levitation carrier completely without a tether in a large area and calculated the wrench matrix quickly and accurately using the magnetic node method and the Lorentz force law [12]. Graphene levitation and magnetic field orientation control were studied in the literature [13]. Torques induced by antimagnetic forces in all three spatial directions were used to predict stability conditions for millimeter-scale graphite plates of different shapes. Article [14] introduces the design principle, and the initial model of radial-axial magnetic bearing proposes a multi-objective optimization method of bearing and provides the analytical expression of the equivalent magnetic circuit, verified by finite element analysis and experiments. For the problem that the magnetic saturation of the core reduces the maximum output force of the magnetic bearing, an improved core design criterion for the cross-polar magnetic bearings is proposed in the paper [15]. The three-dimensional finite element method and experiments also successfully verified the optimization method. The method can provide a reference for designing and selecting active magnetic bearings. In order to predict the performance and guide the design of permanent magnet-biased three-degree-of-freedom magnetic levitation bearings, Yun proposed an accurate analytical model including the eddy current effect and leakage effect. The stiffness was investigated by static and transient finite element methods (fem). Finally, a prototype was designed and fabricated, and the validity of the analysis was verified using the finite element results and experimental results [16]. In the articles [17], the dynamics of a six-degree-of-freedom bearingless linear motor system were modeled using a set-sum modeling approach. The validity of the established model was verified by comparing the time-domain simulation results with the experimental results.

The model can be applied to time-domain simulation, real-time control system development, and various system analyses. The article [18] proposes the integration of magnetic bearings into a tubular linear actuator (TLA), resulting in a new self-bearing (bearingless) TLA. The system is derived from a standard TLA by changing its stator geometry. The working principle is illustrated using the finite element method, and critical design aspects are investigated. Sun proposes a three-degree-of-freedom permanent magnetic levitation platform for cleanroom conveying systems, which is supported by four magnetic poles and uses a variable flux path mechanism. Based on the analysis of the dynamics, the magnetic levitation platform's mathematical model and state space are established. A double closed-loop controller for stable levitation and motion control is designed, and the control system is decoupled using a coordinate transformation matrix. Finally, simulations and experiments were conducted to achieve stable levitation and three-degree-of-freedom motion control with remarkable positioning accuracy [19–21].

Magnetic levitation has been widely used in the biomedical industry due to its advantages of cleanliness and non-contact. Ke proposed a planar Litz coil sandwiched between two ferrite substrates optimized design method that wirelessly powers a novel mechanical artificial anal sphincter system for treating severe fecal incontinence [22]. Zhou designed a novel implantable puborectal-like artificial anal sphincter (PAAS) that replicates rectal perception with a low risk of ischemic necrosis. A pressure sensor embedded in the PAAS was used to determine the relationship between the stool mass and pressure and to develop a stool mass estimation model based on in vitro studies [23]. Articles [24–26] applied magnetic levitation technology to studies on artificial hearts to optimize the improvement of magnetic pumps. Forrai investigated the system identification and control of a nonlinear electromagnetic actuator that could be used in an artificial heart actuator [27]. Haisler described a 3D culture method, the magnetic levitation method (MLM), in which cells are combined with magnetic nanoparticle assemblies overnight to provide them with magnetic properties. When resuspended in the medium, an external magnetic field causes the cells to levitate and concentrate at the gas–liquid interface, where they aggregate to form larger 3D cultures [28]. Parfenov proposed the rapid creation of 3D scaffolds using the magnetic levitation of calcium phosphate particles. Label-free suspension assembly was achieved in the presence of gadolinium salts by using a custom-designed magnetic system that allowed the suspension of calcium phosphate particles. The chemical conversion of tricalcium phosphate to octacalcium phosphate under magnetic levitation conditions in a non-uniform magnetic field was also demonstrated [29]. Yaman developed a magnetization-rate-based protein detection scheme using a low-cost, miniaturized magnetic levitation device consisting of two opposing magnets to generate a magnetic field gradient, a glass capillary channel to retain the sample, and two lateral mirrors to monitor the interior of the channel. The method involves using polymeric microspheres as a mobile analysis surface and magnetic nanoparticles as markers. The assay is achieved by trapping the target protein in the polymer microspheres [30].

The MDOFSC proposed in this paper has the advantages of no oil and stable operation at a low-pressure ratio. It can be used in the medical environment to provide oxygen for patients and improve cardiac extracorporeal membrane oxygenation. Based on the electromagnetic drive principle, this paper proposes the MDOFSC, the relationship between the nonlinear magnetic model and the working air-gap length is analyzed, and the decoupling control model of MDOFSC is established by applying the coordinate transformation strategy. With the goal of trajectory control, a centralized controller was designed and experimentally verified.

2. Description of the System

2.1. Structure and Principle

The structure of the MDOFSC is shown in Figure 1, which is mainly composed of orbiting scroll, static scroll, electromagnet, armature, magnetic ring, sensor, etc. The working principle of the MDOFSC is that three groups of electromagnets attract the armature and orbital scrolls to achieve orbital motion and compress the gas. In order to make the MDOFSC work with a specific stiffness, each group of electromagnets is controlled differentially. The MDOFSC is equipped with six electromagnets inside. The open-loop electromagnetic drive system is unstable, and difficult to control the running trajectory. In order to realize the closed-loop control of the running track of MDOFSC, each group of electromagnetic units is executed by a servo system composed of a controller, a sensor, and a power amplifier. The sensor in the Y direction measures the position of the armature and the two sensors in the X direction measure the position and rotation angle of the armature. Through real-time detection and control, the distance between the two X directions is kept straight and equal to ensure the translational movement of the orbiting scroll, which replaces the anti-rotation mechanism, reducing friction and realizing oil-free operation. As shown in Figure 1b, four sets of permanent magnet rings are mounted in four bracket recesses, forming symmetrically distributed mutually exclusive upper and lower sections to support the suspension of the orbiting scroll.

2.2. Magnetic Circuit Analysis

Figure 2 shows the magnetic flux density distribution of the electromagnet and armature. The electromagnetic drive system is composed of three groups of electromagnets and armatures. The air-gap distance between the armature and the electromagnet shown in the figure is 1.5 mm, the number of coil turns is $N_x = 235$, two coils in the Y direction are connected in series, $N_y = 260$, and the current is 2A. The magnetic flux that leaked into the air cannot be ignored, and the leakage coefficient varying with the air-gap distance needs to be introduced. As seen in Figure 2, the magnetic flux density of the armature is almost zero. It can be seen from Figure 3 that the flux path starts from the north pole and returns to the south pole through the core. In order to form a specific stiffness when controlling the operation of the orbiting scroll, the system adopts the method of differential control. When the orbiting scroll is stationary, the two electromagnets in the up and down direction of Y are input with bias current and pull the orbiting scroll mutually. When the positive signal in the direction is given, the upper electromagnet current increases, and the lower electromagnet current decreases.

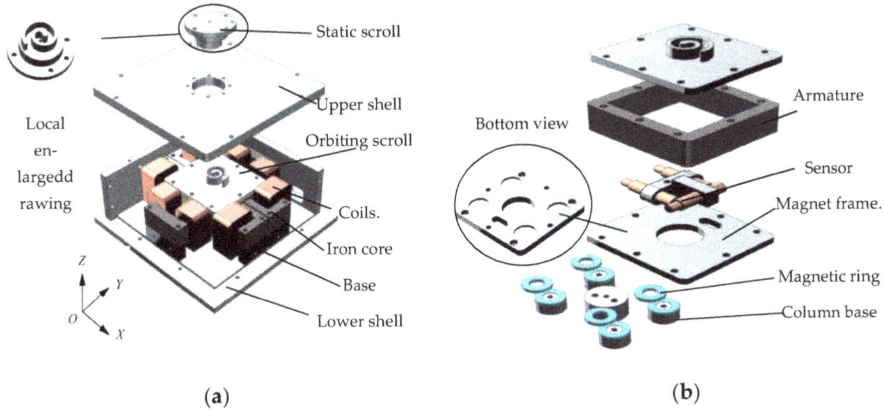

Figure 1. Structure of MDOFSC (**a**) Overall structure; (**b**) Suspension mechanism.

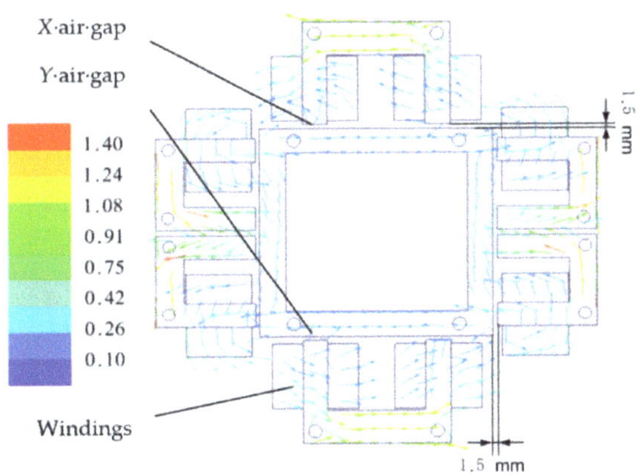

Figure 2. Structure of MDOFSC Magnetic flux density distribution on the electromagnet. Current = 2 A, Air gap x,y = 1.5 mm, and N_x = 235, N_y = 260.

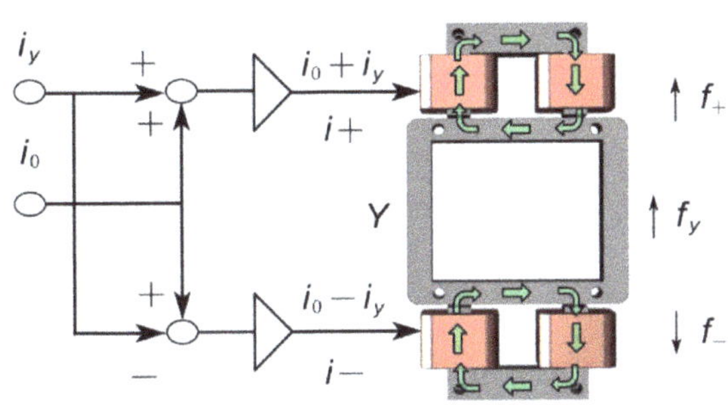

Figure 3. Magnetic flux path through the U-core electromagnet.

2.3. System Model

The distance measured by the sensor needs to be converted into the air gap between the electromagnet and the armature, and the relationship is shown in Formula (1). As shown in Figure 4, when the displacement of the MDOFSC in the X direction is x, the displacement in the Y direction is y, and the rotation angle of the orbiting scroll around the Z-axis is θ. The transformation relationship between the change of air gap between the three groups of magnetic poles and armature and the three degrees of freedom of the platform is shown in Formula (2). After linear approximation, the coordinate transformation relationship in matrix form can be obtained, as shown in Formula (3). As shown in Formula (4), solve the pseudo-inverse matrix of the coordinate transformation matrix, and the coordinate inverse transformation relationship of the system can be obtained, as shown in Formula (5).

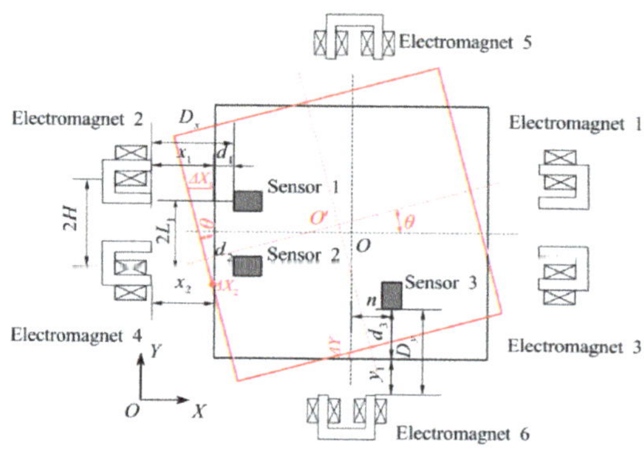

Figure 4. Relationship between the working air gap and sensor measurement distance.

As shown in Figure 4 and Formula (1): D_x, and D_y are the distances between the sensor and the corresponding end face of the electromagnet, which are fixed values; d_1, d_2, and d_3 are the distances between the sensor and the armature; x_1, x_2, and x_3 are the air-gap distances between the end face of the electromagnet and the armature. x, y, and θ are the three degrees of freedom of the suspended solids. $2L_1$ is the distance between two sensors in the X direction.

$$\begin{bmatrix} x_1 & x_2 & y_1 \end{bmatrix}^T = \begin{bmatrix} D_x & D_x & D_y \end{bmatrix}^T - \begin{bmatrix} d_1 & d_2 & d_3 \end{bmatrix}^T, \tag{1}$$

$$\begin{cases} x_1 = x + y \cdot \tan\theta - H \cdot \sin\theta \\ x_2 = x + y \cdot \tan\theta + H \cdot \sin\theta \\ y_1 = -x \cdot \tan\theta + y \end{cases}, \tag{2}$$

$$\begin{bmatrix} x_1 & x_2 & y_1 \end{bmatrix}^T = N_1 \begin{bmatrix} x & y & \theta \end{bmatrix}^T, \tag{3}$$

$$N_1 N_2 = E_{3\times 3}, \tag{4}$$

$$\begin{bmatrix} x & y & \theta \end{bmatrix}^T = N_2 \begin{bmatrix} x_1 & x_2 & y_1 \end{bmatrix}^T, \tag{5}$$

where N_1 is the coordinate transformation matrix, and N_2 is the inverse coordinate transformation matrix.

$$N_1 = \begin{bmatrix} 1 & 1 & -H \\ 1 & 1 & H \\ -1 & 1 & 0 \end{bmatrix}, \quad N_2 = \begin{bmatrix} 1/4 & 1/4 & -1/2 \\ 1/4 & 1/4 & 1/2 \\ -1/(2H) & 1/(2H) & 0 \end{bmatrix},$$

As shown in Figure 5, $2H$ is the center distance between Electromagnet 2 and Electromagnet 4; $2L$ is the distance from the center of the magnetic spring composed of an annular permanent magnet and a cylindrical permanent magnet to the center of the actuating platform.

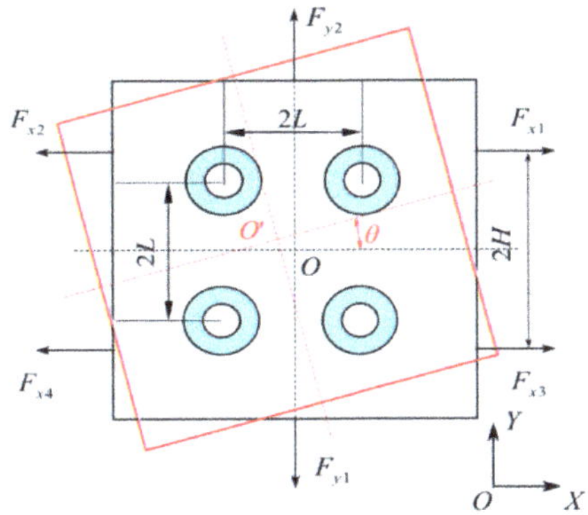

Figure 5. Force and motion of the platform.

In order to analyze the feasibility of the MDOFSC, the control system needs to be simulated. Therefore, the mathematical model describing the controlled object should be established according to the dynamic characteristics of the controlled object. In order to simplify mathematical modeling, this paper assumes that the following complex elements can be ignored. (1) The mass of the suspension platform of the MDOFSC is evenly distributed, and the center of gravity overlaps the geometric center of the platform. (2) The suspended platform is an ideal rigid body; the deformation is negligible. (3) The dimensional accuracy is sufficient, and the magnetic characteristics of each magnetic pole are the same. (4) It is difficult to derive the friction formula between the orbiting scroll and the static scroll, supplemented by increasing the system damping. The force and motion of the platform are shown in Figure 4. Then, the differential formula is established through the force and motion analysis of each degree of freedom, as shown in Formula (6).

$$\begin{cases} m\ddot{x} = F_{qx1} + F_{qx2} + 4F_m - c\dot{x} \\ m\ddot{y} = F_{qy} + 4F_m - c\dot{y} \\ J\ddot{\theta} = (2(F_{qx2} - F_{qx1})H + 8F_m L)\theta - c\dot{\theta} \end{cases}, \tag{6}$$

$$F_v = k_w \frac{i_2}{(d_w + \lambda_w)^2};\ v = x_1,\ x_2,\ x_3,\ x_4,\ y_1,\ y_2;\ w = x,\ y \tag{7}$$

$$F_w = k_w\left(\left(\frac{i_0 + i_w}{d_0 - d_w - \lambda_w}\right)^2 - \left(\frac{i_0 - i_w}{d_0 + d_w + \lambda_w}\right)^2\right), \tag{8}$$

where m is the mass of suspended solids, F_{qx1}, F_{qx1}, and F_{qy} are three groups of differential electromagnetic forces in X_1, X_2, and Y directions, respectively, and F_m is the magnetic force of a single group of permanent magnetic rings. The derivation process of the magnetic force of permanent magnetic rings is shown in the paper [5]. J is the rotational inertia of the suspended solids, and F_v is the magnetic force of a single electromagnet. i_2 is the input current of the coil, d is the air-gap length between the electromagnet and the magnetized target, F_w is the differential electromagnetic force of a single group of electromagnets, λ_w is the compensation constant for the air gap, k_w is the constant of magnetic force, i_0 is the bias current, i_w is the control current, d_0 is the air-gap distance at the balance position, and d_w is the displacement of the platform. The structure of the electromagnetic system determines the above two parameters.

$$F_m = k_m\delta, \tag{9}$$

$$\Delta F = k_{iw}(i_0 + i_w) + k_{dw}(d_0 + d_w);\ w = x, y \tag{10}$$

where K_m is the stiffness coefficient of the permanent magnetic force, δ. It is the horizontal displacement change between a pair of permanent magnetic rings and the linearization result of a single group of differential electromagnetic forces. K_{iw} is the current stiffness after linearizing a single group of differential electromagnetic forces, and k_{dw} is the displacement stiffness after linearizing a single group of differential electromagnetic forces. k_{iw} and k_{dw} are current and displacement stiffness coefficients, respectively.

According to the system dynamics formula, the system state space formula is established as follows:

$$\begin{cases} \dot{X} = AX + BU \\ Y = CX \end{cases}, \tag{11}$$

where,

$$X = \begin{bmatrix} x & \dot{x} & \theta & \dot{\theta} & y & \dot{y} \end{bmatrix}^T, \tag{12}$$

$$Y = \begin{bmatrix} y_1 & y_2 & y_3 \end{bmatrix}^T, \tag{13}$$

$$U = \begin{bmatrix} i_1 & i_2 & i_3 \end{bmatrix}^T, \tag{14}$$

$$A = \begin{bmatrix} 0 & 1 & 0 & 0 & 0 & 0 \\ \frac{2k_x+4k_m}{m} & \frac{-c}{m} & 0 & 0 & 0 & 0 \\ 0 & 0 & 0 & 1 & 0 & 0 \\ 0 & 0 & \frac{2H^2k_x+4L^2k_m}{J} & \frac{-c}{J} & 0 & 0 \\ 0 & 0 & 0 & 0 & 0 & 1 \\ 0 & 0 & 0 & 0 & \frac{k_y+4k_m}{m} & \frac{-c}{m} \end{bmatrix}$$

$$B = \begin{bmatrix} 0 & 0 & 0 \\ \frac{k_{ix}}{m} & \frac{k_{ix}}{m} & 0 \\ 0 & 0 & 0 \\ \frac{-k_{ix}}{J} & \frac{k_{ix}}{J} & 0 \\ 0 & 0 & 0 \\ 0 & 0 & \frac{k_{iy}}{m} \end{bmatrix}$$

$$C = \begin{bmatrix} 1 & 0 & 0 & 0 & 0 & 0 \\ 0 & 0 & 1 & 0 & 0 & 0 \\ 0 & 0 & 0 & 0 & 1 & 0 \end{bmatrix}$$

3. Controller Design and System Simulation

3.1. Motion Controller Design

To solve the deviation problem caused by the different magnetic characteristics of each magnetic pole of the MDOFSC under a decentralized control, a three-degree freedom centralized control strategy is designed to realize the direct control of the freedom of the MDOFSC. There is a coupling between the input and output of the system. The system can be decoupled into three single input and single output systems by Formulas (1) and (5), and PID controller. The integration link can correct the control current of each magnetic pole in real time, compensate for the influence of the difference in magnetic characteristics, and correct the inclination of the orbiting scroll. Figure 6 shows the principle of the controller, and the expression is written as follows:

$$\begin{array}{c} e_{du} = \text{Ref } u - u(t);\ u = x, y, \theta, \\ i_u(t) = k_{pu} \cdot e_{du}(t) + k_{iu} \int e_{du}(t)\mathrm{d}t + k_{du} \cdot \frac{\mathrm{d}e_{du}(t)}{\mathrm{d}t}, \end{array} \tag{15}$$

where e_{du} is the input error signal, Ref u is the reference input of three degrees of freedom, $u(t)$ is the actual degree of freedom of feedback, $i_u(t)$ is the input current. The k_{pu}, $k_{iu,}$ and k_{du} are the proportional gain, integral gain, and differential gain of the PID controller, respectively.

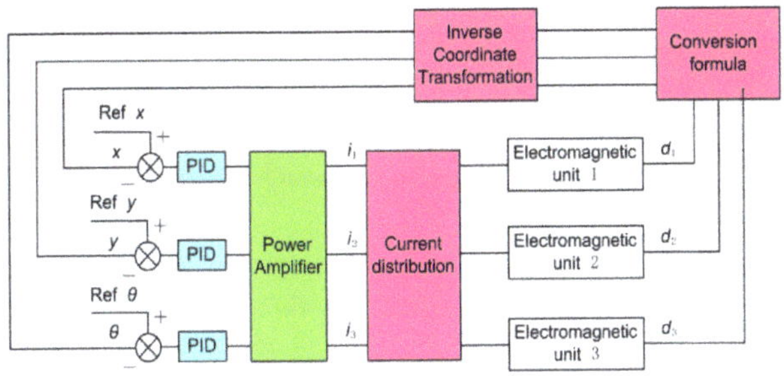

Figure 6. Block diagram of the control system.

3.2. Control System Simulation

The feasibility of the double closed-loop controller is verified by numerical simulation using Simulink software. Table 1 shows the parameters. In this program, the simulation lasts for 0.5 s, and at 0 s, each electromagnetic unit inputs a signal, respectively. Figure 7 shows the simulation results (track tracking states and corresponding control currents in the *X* and *Y* directions). The control parameters in the X direction are P = 100, I = 2, and D = 10. During the movement in the *X* direction, a negative sinusoidal signal is input in the X direction at 0 s, and the controller quickly inputs a negative current to the electromagnetic unit. This system simulation does not consider the friction between the orbiting scroll and the static scroll, so the system stability is poor, and a larger D is required to enhance the system damping and control effect. However, the differential effect amplifies noise disturbances, causing the control current to oscillate rapidly. The threshold of the control current is ±10 A, and the starting current of track tracking reaches the peak. The peak time of track tracking is 0.053 s, and the maximum overshoot is 3%. The control parameters in the *Y* direction are P = 30, I = 2, and D = 20. In the trajectory motion in the *Y* direction, a cosine signal is an input in the *Y* direction at 0 s, at which time the controller quickly outputs the maximum negative current. The larger D value makes the system have the effect of starting overshoot and strengthening control, which can effectively reduce the dynamic deviation of the control process. There is no overshooting in the trajectory tracking, and the tracking effect is good.

Table 1. System parameters.

Description	Parameter	Quantity
Moment of inertia	J	0.0015097 kg·m^2
Mass of suspended solids	m	1.7 kg
Current stiffness coefficient in the *X* direction	k_{ix}	150
Displacement stiffness coefficient in the *X* direction	k_{dx}	−75
Current stiffness coefficient in the *Y* direction	k_{iy}	180
Displacement stiffness coefficient in the *Y* direction	k_{dy}	−57
Magnetic stiffness coefficient of a magnetic ring group	k_m	4.544 N/mm
Damping of the system	c	0.015 N/(mm/s)
Center distance between two electromagnets in the *X* direction	H	19.5 mm
Center distance of adjacent magnetic rings	L	21 mm

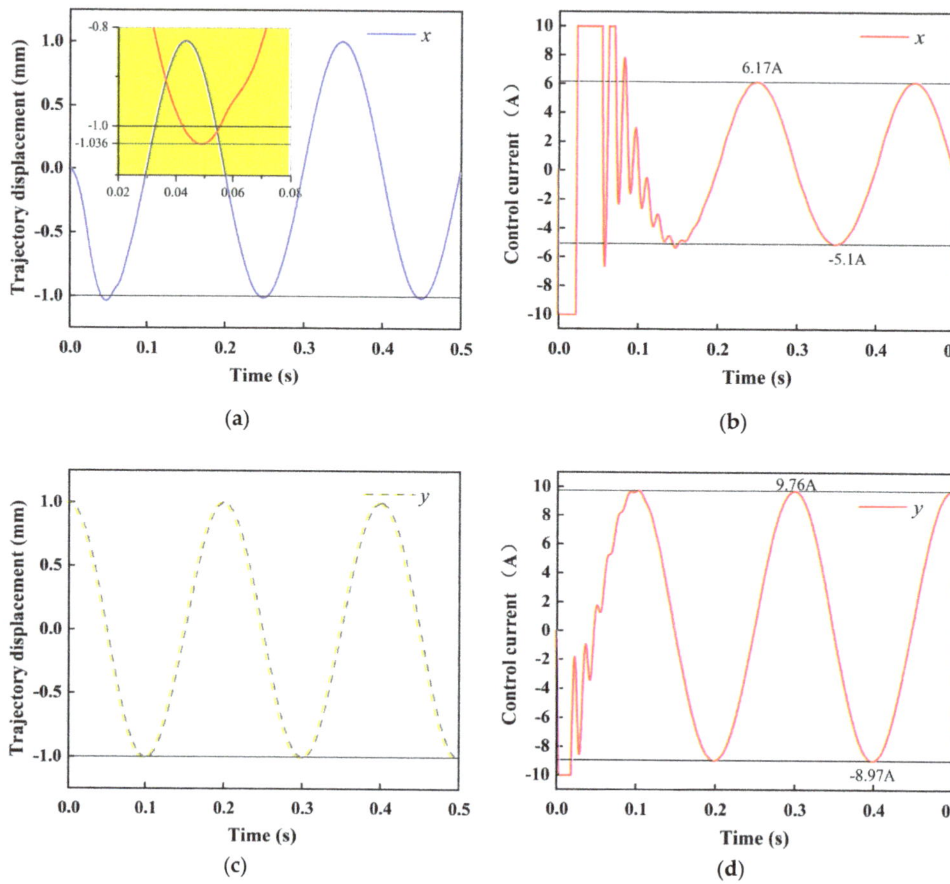

Figure 7. Simulation results of trajectory tracking. (**a**) Trajectory in *X* direction; (**b**) Control current in *X* direction; (**c**) Trajectory in *Y* direction; (**d**) Control current in *Y* direction.

4. Experiment and Analysis

4.1. Experimental System

Figure 8 shows the experimental system of the MDOFSC, mainly including the prototype, hardware equipment, and control system. The control system is based on the MicroLabBox produced by the dSPACE company. MATLAB and dSPACE software kits are installed on the upper computer. The power amplifier adopts the current control mode. Pu-05 eddy current displacement sensor of the AEC company is used for air-gap detection, with a range of 0–2 mm and a resolution of 0.5 μm. The analog output voltage range is −5 V to 5 V.

4.2. Experimental Result

Trajectory tracking experimental parameters: P = 30, I = 2, D = 0.01. The parameters of PID come from many experiments. The appropriate parameters are determined by the step experiment's positioning effect and the trajectory tracking stability. Figure 8 shows the trajectory tracking experiment of the MDOFSC in the *X* direction. The period is 0.2 s, and the motion track is a circle with a radius of 0.5 mm. It can be seen from Figure 9a that at 0 s, a negative sinusoidal trajectory in the *X*-axis direction is input. Under PID decentralized control, the track difference between x_1 and x_2 increases gradually from the track center to the track peak, the maximum time difference reaches 0.148 mm, and the maximum relative error rate is 29.6%. As shown in Figure 9b, the trajectory tracking effect is better. In the whole tracking process, the maximum error of x_1 and x_2 is 0.024 mm, and the maximum relative error rate is 4.8%. The results show that the centralized control system has a good tracking effect.

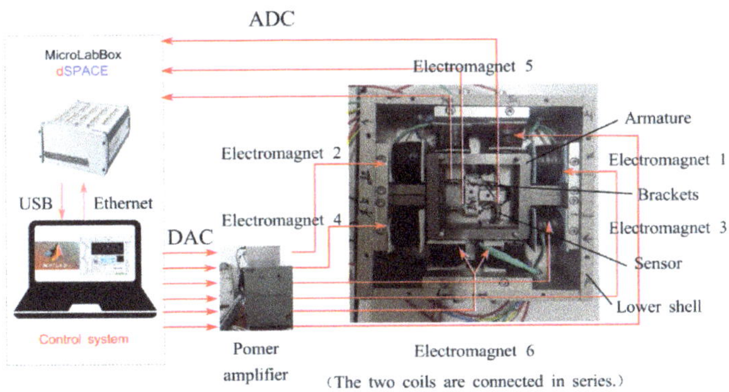

Figure 8. Control system hardware.

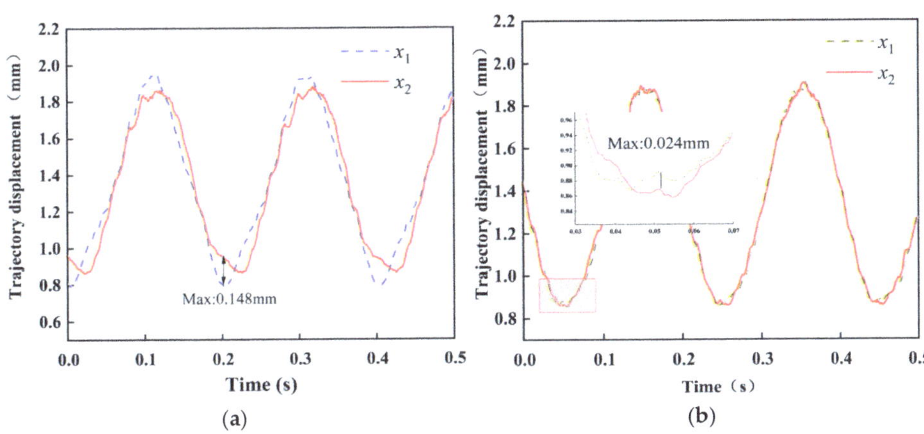

Figure 9. Trajectory tracking experiment. (**a**) *X*-direction trajectory tracking under decentralized control; (**b**) *X*-direction trajectory tracking under centralized control.

Figure 10 shows the tracking track in the *Y* direction. The cosine displacement signal is input at 0 s. The track tracking is not complete at the second and third peaks, which is quite different from the cosine signal. The reasons are as follows: 1. After the trajectory signal is given, the static scroll changes from static to motion and vibrates due to inertia and magnetic ring interference. It is not easy to control the system. From the fourth peak, the tracking effect is better; 2. PID control is more suitable for linear systems. At the trajectory's peak, the magnetic unit's nonlinearity is strong, and the control effect is poor.

Figure 11 shows the current control diagram during track tracking. Figure 8 shows that coils 1, 2, 3, and 4 are in the *X* direction. When the time is 0, the trajectory signal begins to input, and the orbiting scroll makes a negative sinusoidal motion along the *X* direction. At this time, the currents of coil 2 and coil 4 are 4.38 A and 6.77 A, which provides the system with high response-ability and provides acceleration. At this time, the current of coil 3 is 0 because the magnetic ring will provide suction to fight against the current of coil 2,4, and the permanent magnet is a passive magnetic force, which is challenging to maintain the stable movement in the *X* direction, so the current of coil 1 constantly changes. At 0 s, the current of coil 1 starts to rise from 0, which can not only cooperate with the permanent magnetic ring to control the torsion of the orbiting scroll but also pull with coil 2,4 to form a specific stiffness.

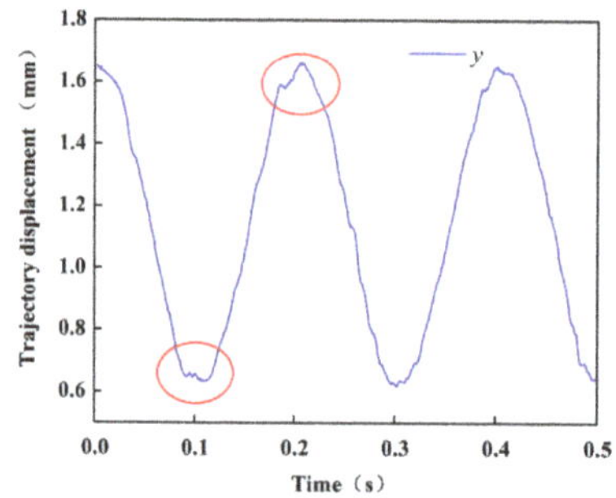

Figure 10. Trajectory tracking in *Y*-direction under centralized control.

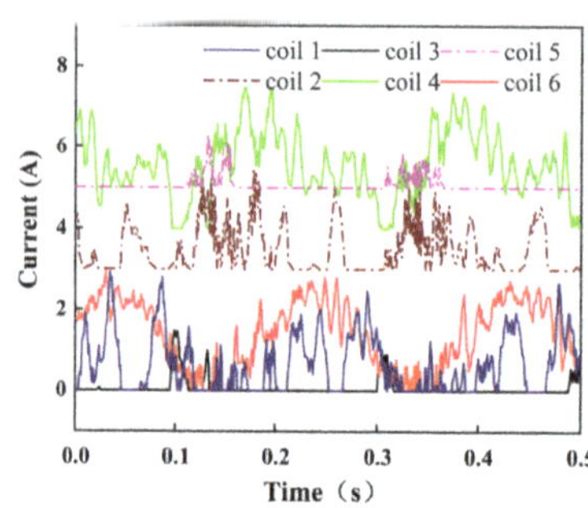

Figure 11. Control current.

Figure 12 shows the trajectory tracking of MDOFSC under decentralized control and centralized control, respectively. Under decentralized control, the circles fitted by the X_1 and X_2 direction trajectories and the Y direction trajectories have poor coincidence. While under centralized control, x_1 and x_2, respectively, coincide with the circular trajectory formed by the trajectory in the Y direction. Since the trajectories of x_1 and x_2 are controlled separately under a decentralized control, the value of x_1 and x_2 is not always equal due to the hysteresis of position control. By adopting the centralized control method, the three degrees of freedom of the system are directly controlled, and the control rotation angle is always zero. The effect of track tracking is improved, and the coincidence degree of tracks is higher.

An experimental prototype of MDOFSC was built to verify the trajectory tracking experiment. The experimental results show that under a decentralized control, the maximum difference between the synchronization trajectories of x_1 and x_2 is 0.148 mm, and the maximum relative error rate is 29.6%. Through centralized control after decoupling, the maximum difference between the synchronization trajectories of x_1 and x_2 is 0.024 mm, and the maximum relative error rate is 4.8%. The results show that through the decoupled centralized control, the system has a self-tuning characteristic, which can ensure that the trajectories of x_1 and x_2 directions remain constant at all times, reduce the left and right wobble of the orbiting scroll, and reduce the wobble and air leakage during the actual operation of the MDOFSC, which is of great significance for further research on the MDOFSC.

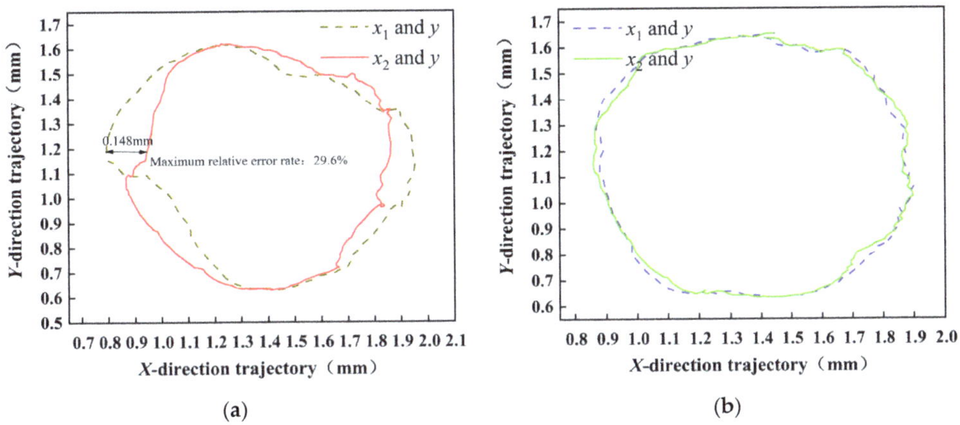

Figure 12. Trajectory tracking diagram. (**a**) Trajectory tracking under decentralized control; (**b**) Trajectory tracking under centralized control.

5. Conclusions

This paper introduces the structure and working principle of MDOFSC. Then, the magnetic circuit of the electromagnetic unit is analyzed through finite element simulation, and the mathematical model of MDOFSC no-load operation is established. Finally, the relationship between the air-gap distance and the degree of freedom is decoupled, and the decoupling controller is designed. The trajectory tracking characteristics of MDOFSC are studied through simulation and experiment. Experimental results: Under decentralized control, the maximum difference between x_1 and x_2 is 0.148 mm, and the maximum relative error rate is 29.6%. Through the centralized control after decoupling, the maximum difference between x_1 and x_2 is 0.024 mm, and the maximum relative error rate is 4.8%. The experimental results show that the system has self-tuning characteristics through decoupling centralized control, which can ensure that the trajectories of x_1 and x_2 directions remain unchanged, reduce the left and right swing of the orbiting scroll, and improve the tracking effect of the system. MDOFSC has good trajectory tracking characteristics.

Author Contributions: Conceptualization, F.S., C.S. and F.X.; methodology, F.S., C.S., Q.Z. and J.J.; software, F.S., C.S. and F.X.; validation, F.S., L.T., Q.Z. and J.J.; formal analysis, F.S., F.X., Q.Z. and J.J.; investigation, F.S., C.S., F.X. and K.O.; resources, F.S., K.O. and F.X.; data curation, Q.Z. and L.T.; writing—originaldraft preparation, F.S., C.S., K.O. and L.T.; writing—review and editing, F.S., C.S. and F.X.; visualization, F.S., C.S., F.X. and J.J.; supervision, F.S. and F.X. project administration, F.S.; funding acquisition, F.S. All authors have read and agreed to the published version of the manuscript.

Funding: This research is supported by the National Natural Science Fund of China (No. 52005345, No. 52005344), National Key Research and Development Project (No. 2020YFC2006701), China Scholarship Council (No. 202108210125), Scientific Research Fund Project of Liaoning Provincial Department of Education (No. LFGD2020002), Major Project of the Ministry of Science and Technology of Liaoning Province (No. 2022JH1/10400027).

Conflicts of Interest: The author declares no conflict of interest.

References

1. Li, B.R.; Zhang, X.; Yang, G.Y. Numerical simulation of interior flow field for oil-free scroll vacuum pump. *Vacuum* **2016**, *53*, 7–10.
2. Cardone, M.; Gargiulo, B. Numerical Simulation and Experimental Validation of an Oil Free Scroll Compressor. *Energies* **2020**, *13*, 5863. [CrossRef]
3. Li, H.S.; Chen, Y.H.; Wu, K.B.; Zhang, X.W.; Wang, J.S. Structural Design on Tooth Seal of Oil-Free Scroll Compressor. *Lubr. Eng.* **2015**, *40*, 13–16.

4. Shi, C.; Sun, F.; Dou, R.; Ren, H.; Li, Q.; Xu, F.; Zhang, X. Modeling and simulation analysis of oil-free scroll compressor driven by magnetic force. *Int. J. Appl. Electromagn. Mech.* **2020**, *64*, 1269–1278. [CrossRef]
5. Shi, C.; Xu, F.C.; Sun, F. Mechanical Properties of Scroll Compressor with Permanent Magnetic Compliance Mechanism. *J. Southwest Jiaotong Univ.* **2022**, *57*, 597–603.
6. Wang, X.N.; Hu, Y.F.; Wu, H.C.; Dong, R. Research of bearing capacity model of large-air-gap hybrid magnetic bearings. *J. Mech. Eng.* **2015**, *51*, 153–160. [CrossRef]
7. Deng, Z. Research and Development Status of High Temperature Superconducting Magnetic Bearings. *Trans. China Electrotech. Soc.* **2009**, *24*, 1–8.
8. Li, H.; Deng, Z.; Huang, H.; Liao, H.; Yuan, Y.; Zhang, W. Experiments and Simulations of the secondary suspension system to improve the dynamic characteristics of HTS maglev. *IEEE Trans. Appl. Supercond.* **2021**, *31*, 1–8. [CrossRef]
9. Zhu, H.Y.; Teo, D.; Pang, C.K. Magnetically levitated parallel actuated dual-stage (Maglev-PAD) system for six-axis precision positioning. *IEEE/ASME Trans. Mechatron.* **2019**, *24*, 1829–1838. [CrossRef]
10. Zhao, C.; Oka, K.; Sun, F.; Harada, A.; Jin, J.; Zhang, M. Design of Zero-Power Control Strategy with Resisting Tilt of Hybrid Magnetic Levitation System. *IEEE Trans. Ind. Electron.* **2022**, *69*, 11. [CrossRef]
11. Soni, T.; Dutt, J.K.; Das, A.S. Dynamic Behavior and Stability of Energy Efficient Electro-Magnetic Suspension of Rotors Involving Time Delay. *Energy* **2021**, *231*, 120906. [CrossRef]
12. Zhang, X.; Trakarnchaiyo, C.; Zhang, H.; Khamesee, M.B. MagTable: A tabletop system for 6-DOF large range and completely contactless operation using magnetic levitation. *Mechatronics* **2021**, *77*, 102600. [CrossRef]
13. Chao, N.; Feng, L.; Wang, Z.M.; Bao, J.; Hu, J. Graphene levitation and orientation control using a magnetic field. *J. Appl. Phys.* **2018**, *123*, 044302.
14. Han, B.C.; Qin, J.; Yuan, Q. Multi-objective Optimization of a Combined Radial-Axial Magnetic Bearing for Magnetically Suspended Compressor. *IEEE Trans. Ind. Electron.* **2016**, *63*, 2284–2293. [CrossRef]
15. Xu, S.; Sun, J.; Wang, Z. Improved design and analysis of a radial magnetic bearing with paired-pole or alternating-pole configurations. *IET Electr. Power Appl.* **2022**, *16*, 382–393. [CrossRef]
16. Yun, L.; Sun, J.; Han, B. Modeling and design of 3-DOF Magnetic Bearing for High-Speed Motor Including Eddy-Current Effects and Leakage Effects. *IEEE Trans. Ind. Electron.* **2016**, *63*, 3656–3665.
17. Hosseinzadeh, R.; Martin, F.; Hinkkanen, M. A Dynamic Model for Six-Degree-of-Freedom Bearingless Linear Motor Systems. *IEEE Trans. Ind. Appl.* **2021**, *57*, 6921–6930. [CrossRef]
18. Miric, S.; Kuttel, P.; Tuysuz, A.; Kolar, J.W. Design and Experimental Analysis of a New Magnetically Levitated Tubular Linear Actuator. *IEEE Trans. Ind. Electron.* **2019**, *66*, 4816–4825. [CrossRef]
19. Zhao, C.; Sun, F.; Pei, W.Z.; Jin, J.J.; Xu, F.C.; Zhang, X.Y. Independent Cascade Control Method for Permanent Magnetic Levitation Platform. *J. Southwest Jiaotong Univ.* **2022**, *57*, 618–626.
20. Sun, F.; Pei, W.Z.; Jin, J.J.; Zhao, C.; Xu, F.C.; Zhang, M. Research on Floating Control Method for Permanent Magnetic Levitation Platform Using a Variable Flux Path Mechanism. *J. Southwest Jiaotong Univ.* **2022**, *57*, 531–539.
21. Sun, F.; Pei, W.Z.; Zhao, C.; Jin, J.; Xu, F.; Zhang, X. Permanent Maglev Platform Using a Variable Flux Path Mechanism: Stable Levitation and Motion Control. *IEEE Trans. Magn.* **2022**, *58*, 7. [CrossRef]
22. Ke, L.; Yan, G.; Yan, S.; Wang, Z.; Li, X. Optimal Design of Litz Wire Coils With Sandwich Structure Wirelessly Powering an Artificial Anal Sphincter System. *Artif. Organs* **2015**, *39*, 615–626. [CrossRef] [PubMed]
23. Zhou, Z.; Yan, G.; Wang, Z.; Jiang, P.; Hua, F.; Yao, S.; Ding, Z. Design and Evaluation of Puborectalis ...ike Artificial Anal Sphincter That Replicates Rectal Perception. *Artif. Organs* **2020**, *44*, E300–E312. [CrossRef] [PubMed]
24. Luciani, G.B.; Hoxha, S.; Torre, S.; Rungatscher, A.; Menon, T.; Barozzi, L.; Faggian, G. Improved Outcome of Cardiac Extracorporeal Membrane Oxygenation in Infants and Children Using Magnetic Levitation Centrifugal Pumps. *Artif. Organs* **2016**, *40*, 27–33. [CrossRef]
25. Liu, Y.; Wang, S.; Qin, Y.; Guo, H.; Li, Z.; Li, Y.; Tang, J.; Ma, Z.; Liu, J. 5 nm-scale surface evenness movement measurement method based on the electron spin in diamond. *Laser Phys. Lett.* **2021**, *18*, 015202. [CrossRef]
26. Yuhki, A.; Nogawa, M.; Takatani, S. Development of a compact, sealless, tripod supported, magnetically driven centrifugal blood pump. *Artif. Organs* **2015**, *24*, 501–505. [CrossRef]
27. Forrai, A.; Ueda, T.; Yumura, T. Electromagnetic Actuator Control: A Linear Parameter-Varying (LPV) Approach. *IEEE Trans. Ind. Electron.* **2007**, *54*, 1430–1441. [CrossRef]
28. Haisler, W.L.; Timm, D.M.; Gage, J.A.; Tseng, H.; Killian, T.C.; Souza, G.R. Three-dimensional cell culturing by magnetic levitation. *Nat. Protoc.* **2013**, *8*, 1940–1949. [CrossRef]
29. Parfenov, V.A.; Mironov, V.A.; Koudan, E.V.; Nezhurina, E.K.; Karalkin, P.A.; Pereira, F.D.; Petrov, S.V.; Krokhmal, A.A.; Aydemir, T.; Vakhrushev, I.V.; et al. Fabrication of calcium phosphate 3D scaffolds for bone repair using magnetic levitational assembly. *Sci. Rep.* **2020**, *10*, 4013. [CrossRef]
30. Yaman, S.; Tekin, H.C. Magnetic Susceptibility-Based Protein Detection Using Magnetic Levitation. *Anal. Chem.* **2020**, *92*, 12556. [CrossRef]

actuators

MDPI

Article

Unbalance Vibration Suppression of Maglev High-Speed Motor Based on the Least-Mean-Square

Huachun Wu [1,2,3], Mengying Yu [1,*], Chunsheng Song [1,2,3] and Nianxian Wang [4]

1 School of Mechanical and Electronic Engineering, Wuhan University of Technology, Wuhan 430070, China
2 Shenzhen Research Institute, Wuhan University of Technology, Shenzhen 518057, China
3 Hubei Provincial Engineering Technology Research Center for Magnetic Suspension, Wuhan 430070, China
4 School of Machinery and Automation, Wuhan University of Science and Technology, Wuhan 430081, China
* Correspondence: 318705@whut.edu.cn

Abstract: The harmonic response caused by unbalanced excitation vibration for the high-speed rotating machinery will reduce the control accuracy and stability of the maglev high-speed motor, and limit the increase of its speed. When the active magnetic bearing is used to solve the unbalanced vibration, it will increase additional electromagnetic force and energy consumption, sometimes leading to the saturation of the power amplifier, and will transfer to the bearing foundation, causing the foundation to vibrate. In this paper, we analyzed periodic unbalance excitation force and the principle of rotor unbalanced vibration suppression, and the unbalance vibration model of the maglev rotor is derived. The Least-Mean-Square (LMS) algorithm is introduced into the PID control, an unbalance vibration control strategy based on real-time filtering compensation of rotor displacement signal is proposed, the vibration is eliminated by filtering the synchronous frequency and harmonic signal of the input of the PID control. The experimental results show that the proposed method can improve the maglev rotor's rotation accuracy, reduce the magnetic bearing's maximum control current, and decrease the vibration of the supporting foundation.

Keywords: unbalanced vibration; LMS; maglev motor; vibration control; foundation vibration

Citation: Wu, H.; Yu, M.; Song, C.; Wang, N. Unbalance Vibration Suppression of Maglev High-Speed Motor Based on the Least-Mean-Square. *Actuators* **2022**, *11*, 348. https://doi.org/10.3390/act11120348

Academic Editors: Suyuan Yu, Jin Zhou, Feng Sun and Ming Zhang

Received: 21 October 2022
Accepted: 24 November 2022
Published: 26 November 2022

1. Introduction

The motor is the power source of industrial equipment and plays an important role in the whole industrial field. Maglev high-speed motors are characterized by high speed, no transmission, and small inertia. Compared with traditional motors, high-speed motors have higher power density, smaller volume, smaller noise, faster dynamic response, and so on. With the wide application of maglev technology in the field of bearings and trains [1–3], researchers have made many attempts to apply maglev technology to high-speed motors. But, there is a certain imbalance in the rotor due to material, machining accuracy, and working deformation. The unbalanced force caused by the unbalance rotor causes the unbalanced vibration of the rotor, which not only affects the operation accuracy of the rotor, but also reduces the reliability of the system. At present, the unbalanced vibration compensation for maglev rotors can be divided into two categories according to their compensation methods. One is the minimum displacement compensation [4]. That is, based on the displacement of the rotor, an inverse electromagnetic force is applied to rotate it around the geometric center. The other is the minimum force compensation, which is to make the rotor rotate around its inertia axis to minimize the inertia force of the rotor. The minimum displacement compensation can effectively reduce the unbalanced displacement of the rotor, but the amplitude and phase of the compensation signal are usually calculated more complexly. The principle of minimum force compensation is to eliminate the control of unbalanced vibration and make it rotate freely around the inertial axis, which can effectively reduce the vibration force and improve the rotation accuracy [5].

R. Herzog et al. used the generalized narrow band notch filter to control the unbalanced vibration of the maglev rotor and used the sensitivity matrix at a certain speed to calculate the notch filter parameters at that speed [6]. K. Y. Lum et al. proposed to compensate for the rotor's unbalance vibration in the whole speed range by the unbalance parameters identified at a certain speed [7]. J. Setiawan et al. used bias current excitation to identify unbalanced forces, allowing the rotor to rotate around a geometric axis with high precision [8]. A. Matras et al. applied the adaptive disturbance rejection method to the maglev rotor system and achieved an obvious control effect at a constant frequency [9]. Chiacchiarini et al. applied an iterative learning algorithm to imbalance compensation, compared the effects of using a forgetting factor and non-causal low-pass filter, and found that the latter is better [10]. F. Betschon et al. obtained the rotor unbalance parameters by gaining scheduling adaptive control, and carried out automatic balance [11]. Zheng Shiqiang et al. combined the feedforward control strategy with the adaptive notch filter to compensate for the displacement stiffness force component and the current stiffness force component in the unbalanced vibration force, respectively, to achieve automatic balance control [12]. K. Nonami et al. proposed an unbalanced vibration control method based on an attempt-adjustment method [13]. A. Rundell [14] and A. Mohamed et al. [15] used sliding mode control and the Q-parameterization method to realize automatic balance control, respectively. Han et al. designed a sliding mode variable structure disturbance observer to compensate for the unbalanced disturbance force and torque at the synchronous frequency [16]. Zhang Yue et al. proposed an adaptive odd repetitive control that does not need the speed sensors to suppress the active magnetic bearing odd harmonic currents [17,18].

The above literature proposes some solutions for the unbalanced vibration suppression of the maglev rotor. Although some results have been achieved, there are problems such as complex algorithms, slow convergence, the need for an accurate model of the control object, and poor compensation effect. LMS algorithm is an adaptive filtering algorithm, which is widely used in the field of signal filtering because of its simple structure and fast convergence speed [19,20]. According to the principle of minimum inertia force compensation, the key is to eliminate the controller's control of unbalanced vibration and to achieve the purpose of rotor rotation around the inertial spindle. In this regard, this paper uses the LMS algorithm to filter the unbalanced signal and its harmonics at the input end of the PID controller in real-time. It can reduce the amplitude of the control current of the magnetic bearing coil, reduce its constraint on the rotor inertia force, and reduce the vibration transmitted from the rotor to the base. The unbalanced vibration suppression of the maglev rotor is realized and better compensation effect and convergence speed are obtained.

2. Mechanism of Rotor Unbalance of Maglev Motor

Take the rotor centroid section and establish a coordinate system as shown in Figure 1, with the stator center O as the origin. The coordinate axes are two mutually perpendicular electromagnetic force directions x and y, and the stator coordinate system xOy is established. Taking the rotation center O_r of the rotor as the origin, the rotor coordinate system $x\prime O\prime y\prime$ is established. When the rotor coordinate system is in the initial state, the coordinate axis direction is consistent with the stator coordinate system. The rotor coordinate system takes the rotor rotation center O_r as the origin, and the speed is the rotor speed ω as the rotor rotates synchronously. $\overline{O_cO_g} = e_1$ is the centroid distance. $\overline{O_rO_g} = e$ is the distance from the rotation axis to the center of mass. Assuming that the rotor section is a standard circle, the coordinates x_c, y_c of the centroid $O_c(x_c, y_c)$ in the stator coordinate system are expressed as the displacement deviation of the rotor in the x-axis direction and the y-axis direction, respectively.

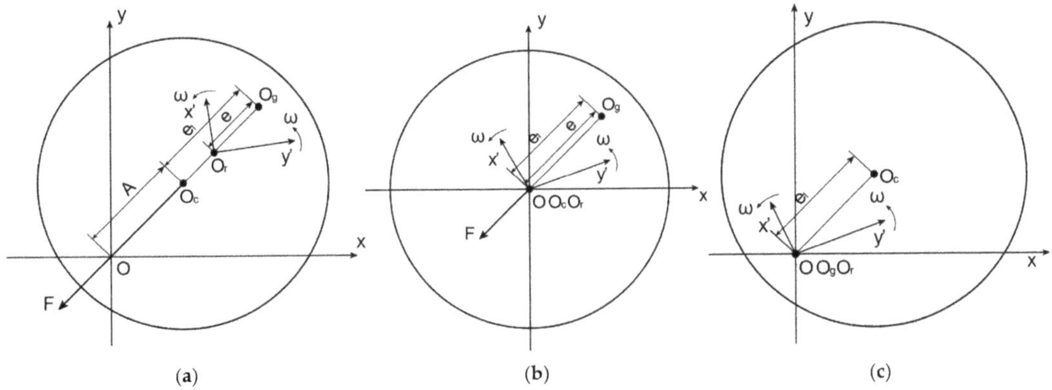

Figure 1. Three cases of rotor unbalance. (**a**) When the rotor rotation center is between the centroid centroids; (**b**) Rotor rotation center at the centroid; (**c**) When rotor rotation center is in the center of mass.

Due to the relatively small support stiffness of the magnetic bearing, when the rotor rotates at high speed, it will rotate around an axis between the geometric center O_c and the centroid O_g. Because the inertia spindle does not coincide with the rotation shaft, the rotor will produce force vibration [21]. Since the geometric center of the rotor does not coincide with the rotation center, the rotor will produce radial whirl and radial unbalanced displacement vibration [21].

When the rotor rotates around the rotation center and is between the centroid and the centroid, as shown in Figure 1a, the differential equation of motion of the centroid in the x direction is:

$$m\ddot{x}_c = f_u + F_x \tag{1}$$

Thereinto, m is the rotor mass, f_u is the unbalanced excitation force generated by the eccentricity of the rotor, and F_x is the electromagnetic force generated by the magnetic bearing to control the rotor displacement. There is:

$$m\ddot{x}_c = me\omega^2 \sin \omega t + k_i i + k_x x_c \tag{2}$$

Thereinto, k_i and k_x are the current stiffness and displacement stiffness of the magnetic bearing. And i is the control current.

It can be seen that the resultant force of the rotor is the resultant force of the inertial force and the electromagnetic force of the magnetic bearing. Because the magnetic bearing control is the displacement deviation control, that is, the rotor inertial force is constrained by the dynamic electromagnetic force, the rotor rotates around its geometric center and $x_c \to 0$ realizes active control of the rotor. To make the rotor rotate around its geometric center, the minimum displacement unbalance control is realized, that is $x_c = 0$ in Equation (2). At this time, the rotor rotation center and the centroid weight are merged at the stator centroid O, as shown in Figure 1b. It can be seen from the figure that at this time $e = e_1$, the unbalanced force reaches the maximum value. From Equation (2), the unbalanced force is completely offset by the $k_i i$ term.

When the rotor rotates at high speed, due to the existence of mass eccentricity of the rotor, to make the rotor rotate around its geometric center and reduce x_c, the inertia force f_u of the rotor increases. To meet the requirements of formula (2), the periodic control current i must increase, and the electromagnetic force generated by the periodic control current will be transmitted to the supporting foundation so that the foundation will vibrate. At the same time, when the rotor rotates, the magnetic bearing base is subject to periodic excitation interference, reducing the stability of the system [22].

When the rotor rotates at high speed, because the unbalanced force of the rotor points to the centroid of the rotor, the inertial force makes the rotor rotate automatically close to its centroid, which is the self-centering effect [23]. The displacement deviation control of the magnetic bearing forces the rotor to rotate around its geometric center as much as possible. If the control of the magnetic bearing on the unbalanced vibration is eliminated in some way, the rotor will automatically rotate around its center of inertia. Minimizing both the inertial force and the electromagnetic force, i.e., the minimum inertial force control. As shown in Figure 1c. At this time, the rotor centroid coincides with the rotation center, that is $e = 0$, the unbalanced inertia force $f_u = 0$, and the rotor displacement vibration amplitude $A = e_1$. The rotor has only unbalanced displacement, but no unbalanced force. The minimum inertia force control is realized.

From the above analysis, it can be seen that reducing the periodic control current of the magnetic bearing can make the rotor more inclined to its inertial axis rotation, effectively reducing the electromagnetic force and its influence on the supporting foundation.

3. Unbalance Vibration Control of Maglev Motor Based on LMS

3.1. Unbalance Vibration Control Model of Maglev Motor Rotor

The schematic diagram of the maglev rotor control system with filter compensation is shown in Figure 2. Thereinto, $F_d(t)$ is the unbalanced force of the rotor, which causes the rotor to produce an unbalanced displacement $X(t)$. Because the actual position of the rotor is inconsistent with the center position, the controller input will produce a periodic deviation signal. The controller produces the corresponding control signal according to the deviation signal. And the periodic electromagnetic force is transmitted to the base reversely, which will also produce the periodic base vibration. From the above analysis, it can be seen that to achieve the rotor unbalanced vibration suppression, the appropriate filter can be selected to filter out the signal of the synchronous frequency as the speed at the input of the controller. The LMS algorithm can effectively filter or extract the sine wave of the specified frequency in the signal, so it is introduced into the rotor control.

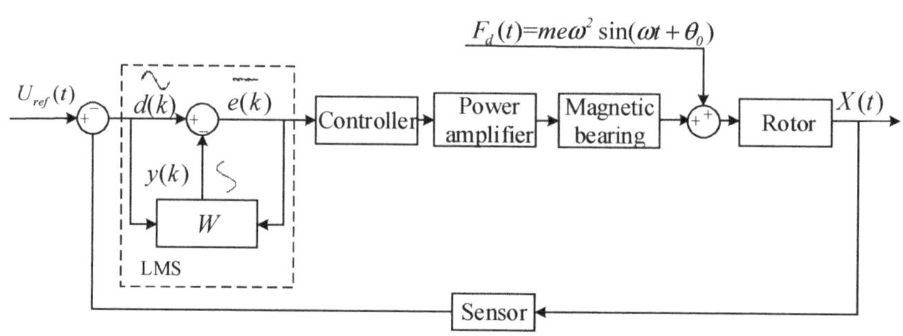

Figure 2. Principle of unbalanced vibration suppression control algorithm for maglev rotor.

The rotor unbalanced displacement signal is sinusoidal in the form of the following:

$$d(t) = A\sin(\omega_0 t + \varphi) \tag{3}$$

In the formula, A is the amplitude function of the rotor displacement signal concerning time t, ω_0 is the rotor rotation angle frequency, and φ is the displacement phase. Expand Equation (3) as follows:

$$d(t) = A\sin\varphi\cos(\omega_0 t) + A\cos\varphi\sin(\omega_0 t) \tag{4}$$

Let $A\sin\varphi = w_1(t)$, $A\cos\varphi = w_2(t)$, then:

$$d(t) = w_1(t)\cos(\omega_0 t) + w_2(t)\sin(\omega_0 t) \tag{5}$$

The discretizations of pair (5) are:

$$d(kT) = w_1(kT)\cos(\omega_0 kT) + w_2(kT)\sin(\omega_0 kT) \tag{6}$$

Thereinto, $k \in [0, +\infty)$. T is the sampling period of the digital controller.

From the above analysis, the discrete LMS algorithm shown in Figure 3 can be constructed. In Figure 3, $X = \begin{bmatrix} \sin(\omega_0 kT) & \cos(\omega_0 kT) \end{bmatrix}^T$ is defined as the input vector of the algorithm and $W(kT) = \begin{bmatrix} w_1(kT) & w_2(kT) \end{bmatrix}^T$ is the weight vector. $d(kT)$ is the expected signal, that is, the rotor displacement signal in the AMB system. $y(kT)$ is the algorithm following signal, that is, the signal obtained by real-time amplitude and phase tracking of the signal with the synchronous frequency as the speed in the rotor displacement. $e(kT)$ is the algorithm error signal and ω_0 is the filter angular frequency.

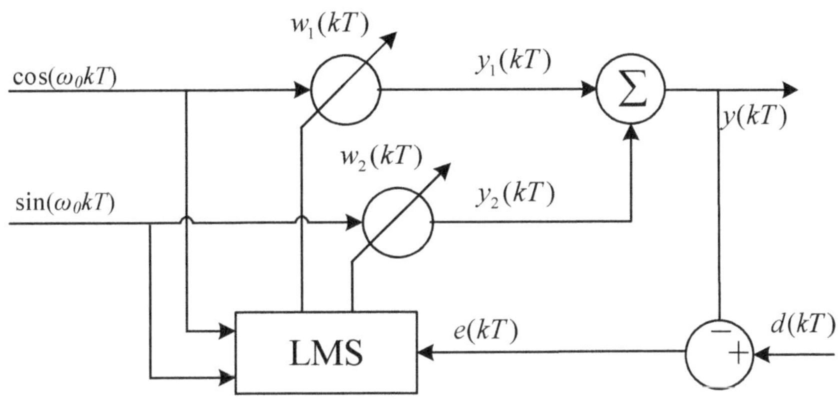

Figure 3. LMS adaptive filtering principle.

The principle of the LMS algorithm is to minimize the objective function by using the gradient random descent method. The specific definition is that the weight vector is updated in a given proportion along the negative direction of the gradient estimation of the error performance surface during iteration. The objective function here is defined as:

$$\delta(kT) \overset{define}{=} E\left[e^2(kT)\right] \tag{7}$$

Among them:

$$e(kT) = d(kT) - y(kT) = d(kT) - X^T(kT)W(kT) \tag{8}$$

$$y(kT) = y_1(kT)w_1(kT) + y_2(kT)w_2(kT) \tag{9}$$

The joint (7) and (8) expansions are:

$$\delta(kT) = E(d^2(kT) + W^T(kT)RW(kT) - 2H^T W(kT)) \tag{10}$$

where $R \overset{define}{=} E\left[X(kT)X^T(kT)\right]$, $H \overset{define}{=} E[X(kT)d(kT)]$. R and H are autocorrelation matrix and cross-correlation matrix functions, respectively. The gradient function of the attainable objective in (10) is:

$$\nabla\delta(kT) \overset{define}{=} \frac{\partial\delta(kT)}{\partial W(kT)} = -2H + 2RW(kT) \tag{11}$$

The R and H functions in (11) can be obtained only when the characteristics of the input signal and the desired signal are fully estimated, which leads to an increase in the amount of calculation. The gradient of the objective function can be replaced by the gradient of the mean square error of the $\nabla e^2(kT)$ estimation, which is:

$$\nabla e^2(kT) = \frac{\partial e^T(kT)}{\partial W(kT)} = -2e(kT)X(kT) \tag{12}$$

The iterative update formula of the weight vector can be obtained from Equation (12):

$$W((k+1)T) = W(kT) - \frac{\mu}{2}\nabla e^2(kT) = W(kT) + \mu e(kT)X(kT) \tag{13}$$

In Equation (13), μ is the iteration step length, which satisfies the following condition:

$$0 < \mu < \frac{1}{\lambda_{\max}} \tag{14}$$

where $\lambda_{\max}$ is the maximum eigenvalue of R in Equation (14). In practical applications, μ should be as small as possible to meet the requirements of high precision, but too small μ will lead to slow convergence.

From Equations (6), (9) and (13) and Reference [24], the closed-loop transfer function from $e(kT)$ to $y(kT)$ is:

$$G(z) = \frac{e(z)}{d(z)} = \frac{z^2 - 2\cos(2\pi f_0 T_s)z + 1}{z^2 + (\mu - 2)\cos(2\pi f_0 T_s)z + (1 - \mu)} \tag{15}$$

3.2. LMS Filtering Performance Analysis

The amplitude-frequency characteristics, phase-frequency characteristics, and signal extraction effect of Formula (15) are simulated and studied. The sampling period $T_s = 0.00005$ s, the filtering frequency $f_0 = 100$ Hz, and the step size $\mu = 0.003$ are selected. The transfer function Bode diagram is shown in Figure 4.

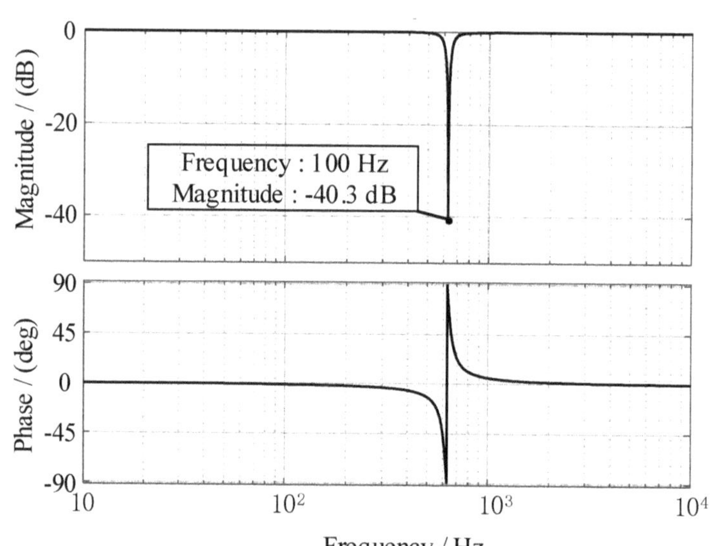

Figure 4. Standard LMS algorithm closed-loop transfer function bode diagram.

According to Figure 4, the output response of the closed-loop transfer function of the LMS algorithm is −40.3 dB at the filtering frequency of 100 Hz, and the LMS algorithm can filter the signal at 100 Hz.

The maglev rotor is set to stably rotate at 6000 r/min. According to the above filter parameter setting, the rotor displacement signal and LMS real-time extraction signal are obtained as shown in Figure 5.

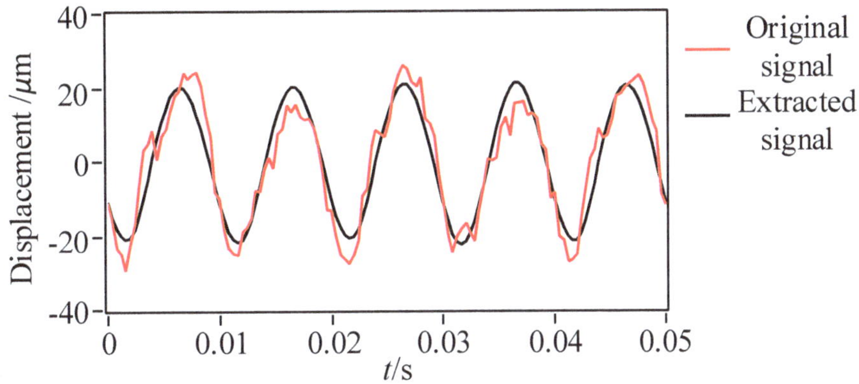

Figure 5. LMS algorithm real-time extraction effect.

4. Experimental Verification

4.1. Experimental Installation

To verify the effectiveness of the algorithm in this paper, an experimental platform of maglev high-speed motor is built, and the rotor online unbalanced compensation experiment is carried out. The experimental platform is shown in Figure 6. The mechanical structure of the experimental platform includes a motor, a rotor, left and right magnetic bearings, a counterweight plate, and so on. The rotor is driven by the right-end motor, and the left and right magnetic bearings support the rotor. The electronic control part of the platform is mainly composed of a dSPACE1007 controller, power amplifier, eddy current displacement sensor, acceleration sensor, data acquisition instrument, and so on. There are two eddy current displacement sensors placed at 90 degrees near the left and right magnetic bearings, which can collect four displacement signals of the rotor. The signal is input into the controller for analysis, and the control voltage is obtained. The control voltage is amplified by the power amplifier to the control current to supply the magnetic bearing to realize the active control of the rotor. At the same time, two acceleration sensors are installed on the supporting base to detect the vibration transmitted by the magnetic bearing to the supporting foundation.

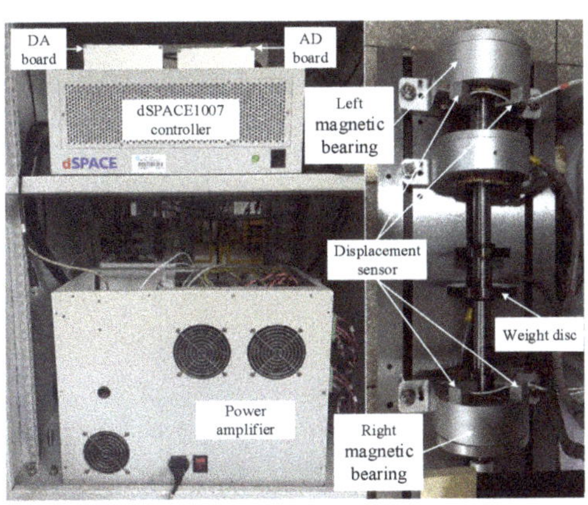

Figure 6. Experimental platform.

4.2. Experimental Result

Firstly, the digital PID control algorithm is used to control the stable suspension of the rotor. The controller control period is Ts = 0.00004 s, and then the LMS algorithm is added to suppress the unbalanced vibration. Figure 7 shows the displacement and current diagram in the x direction of the left end when the rotor is statically suspended. The motor speed is set to 6000 r/min. The real-time displacement of the rotor is measured by the eddy current displacement sensor, and the output current of the controller is collected. Figure 8 shows the changes in rotor displacement and current when the LMS algorithm is used to compensate. The comparison of the time domain and frequency domain before and after LMS compensation is shown in Figures 9 and 10.

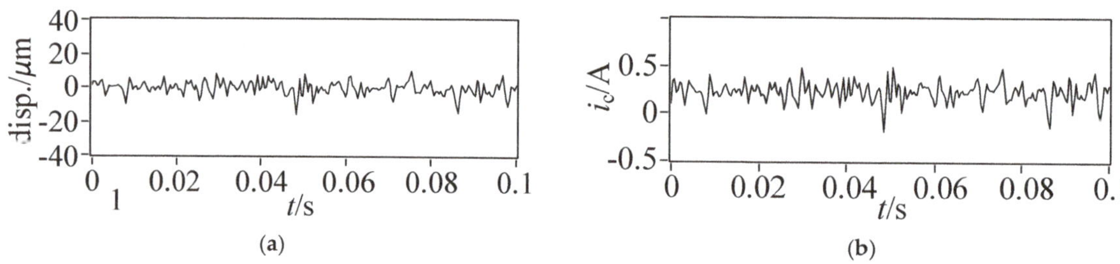

Figure 7. Rotor static suspension displacement and control current. (**a**) displacement diagram; (**b**) current graph.

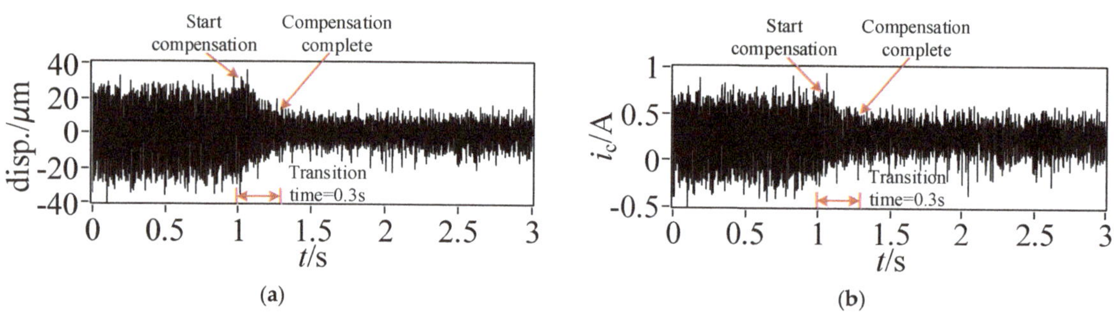

Figure 8. LMS compensation process. (**a**) displacement diagram; (**b**) current graph.

It can be seen from Figures 9 and 10 that before compensation, due to the unbalanced vibration of the rotor, to compensate for the vibration, the controller outputs the control current in real-time. From the spectrum diagram, it can be seen that the main frequency components in the rotor displacement and the magnetic bearing control current are in synchronous frequency with the speed. That is, the main factor affecting the rotor vibration is unbalanced vibration, and the main component of the magnetic bearing control output is also the unbalanced control current with the synchronous frequency. After compensation, the co-frequency control current of the rotor is reduced by 92%, close to 0. And the co-frequency vibration displacement of the rotor is reduced by 87%, reduced to 3 μm. This is due to the reduction of the co-frequency control current, which makes the rotor rotate closer to its inertial spindle. The unbalanced inertia force is reduced, and the rotor amplitude is reduced. At this time, the rotor unbalance force is the smallest, but there is a small unbalance displacement whose amplitude is close to the distance between the rotor centroid and the centroid.

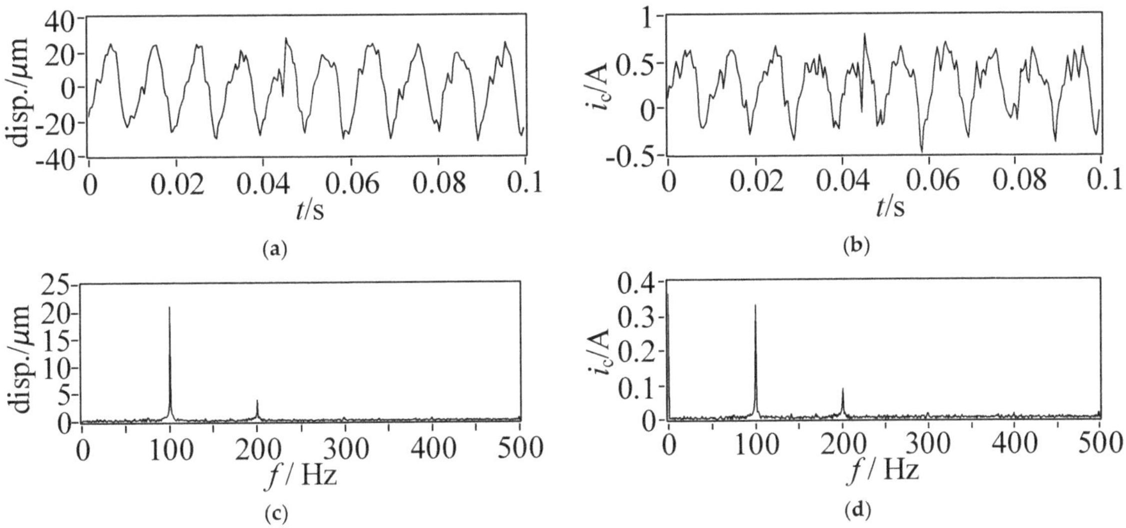

Figure 9. The amplitude-frequency diagram of displacement and current before rotor LMS compensation. (**a**) Time domain spectrum of displacement; (**b**) Time domain spectrum of current; (**c**) Frequency spectrum of displacement; (**d**) Frequency spectrum of current.

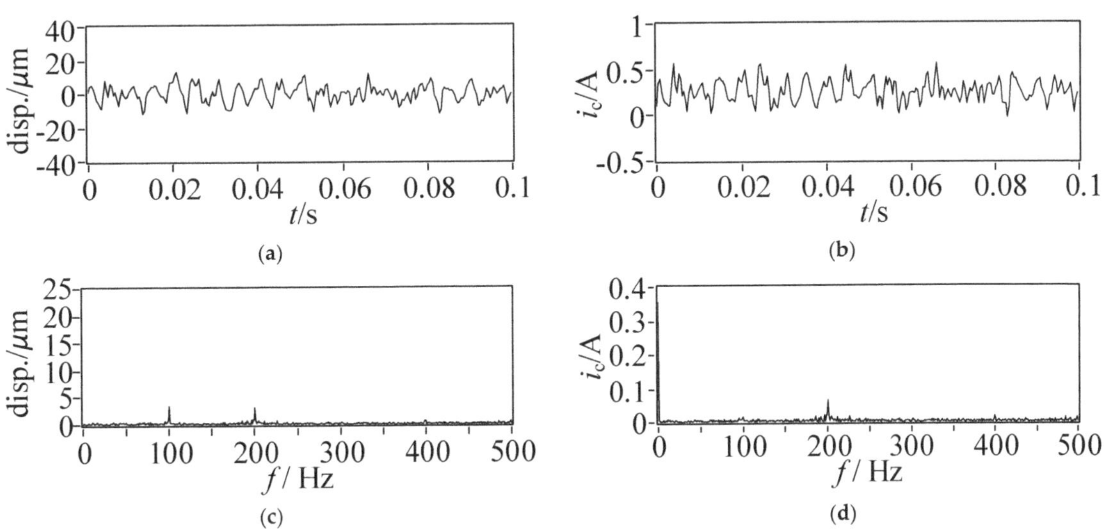

Figure 10. The amplitude-frequency diagram of displacement and current after rotor LMS compensation. (**a**) Time domain spectrum of displacement; (**b**) Time domain spectrum of current; (**c**) Frequency spectrum of displacement; (**d**) Frequency spectrum of current.

It can be seen from Figure 10 that the vibration and control current with the synchronous frequency as rotational speed are reduced after LMS compensation, and the main component of the vibration and control current is its frequency doubling component. To better control the vibration of the maglev rotor and reduce unbalance components in the control current of the magnetic bearing. In this paper, a harmonic control algorithm based on LMS is proposed to compensate for the harmonic vibration while compensating for the synchronous frequency vibration of the speed. In the control algorithm shown in

Figure 2, based on compensating the fundamental frequency vibration of the rotor, the main harmonic vibration component frequency (200 Hz here) is used as the input frequency. The harmonic vibration is also filtered and compensated according to the control method shown in Figure 2. The compensation result is shown in Figure 11. From Figure 11, it can be seen that after further harmonic compensation, the harmonic vibration amplitude of the rotor does not change much, the harmonic control current is reduced by 64% and the amplitude of the synchronous frequency of the speed and its harmonic components in the current spectrum is reduced close to 0. There is only a maximum peak of 0.36 A at 0 Hz, which is a constant current provided by the magnetic bearing to overcome the rotor gravity and keep the rotor in a stable suspension state.

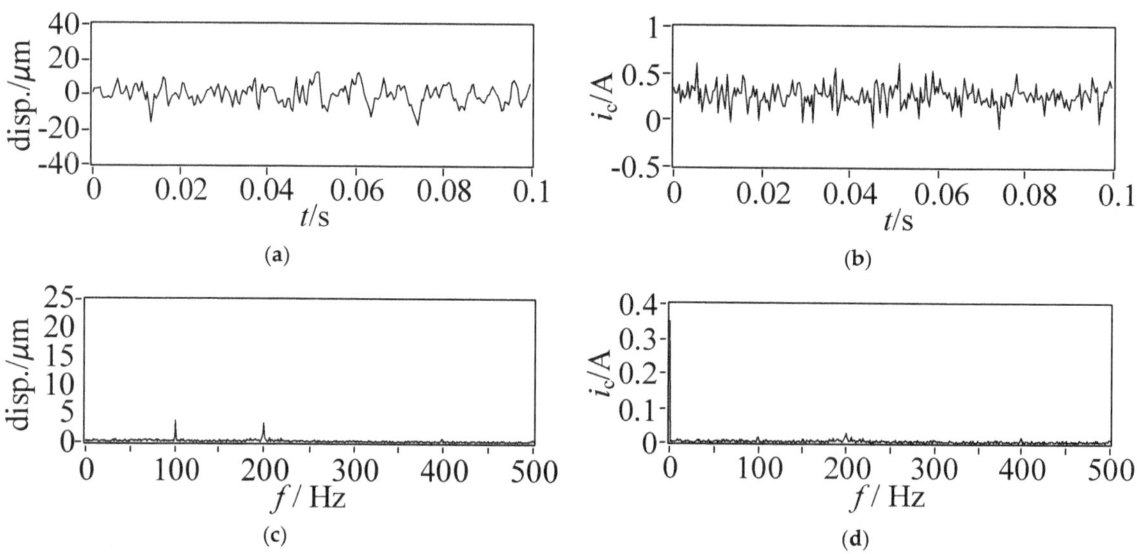

Figure 11. The amplitude-frequency diagram of displacement and current after rotor LMS harmonic compensation. (**a**) Time domain spectrum of displacement; (**b**) Time domain spectrum of current; (**c**) Frequency spectrum of displacement; (**d**) Frequency spectrum of current.

The data with a time of 1 s were randomly taken, and the peak-to-peak mean was obtained. The results are shown in Table 1. Compared with the uncompensated, the peak-to-peak average value of the rotor displacement vibration is reduced by 67.4% and the peak-to-peak average value of the control current is reduced by 31% after using the fundamental frequency and harmonic vibration control algorithm based on LMS in this paper. Compared with the displacement and control current in static suspension, they only increased by 32.6% and 6.5%, respectively.

Table 1. Rotor Displacement and Current Peak-Peak Mean Value.

	Peak-Peak Mean Value	
	Displacement (μm)	**Current (A)**
static levitation	9.32	0.46
100 Hz rotation, no compensation	38.77	0.71
100 Hz rotation, after compensation	12.63	0.49

Through the acceleration sensor installed on the base, and we measured acceleration signals of rotor static suspension before and after LMS control. Figure 12 shows the base acceleration spectrum when the rotor is statically suspended. Figure 13 shows the base acceleration spectrum of the before and after LMS compensation.

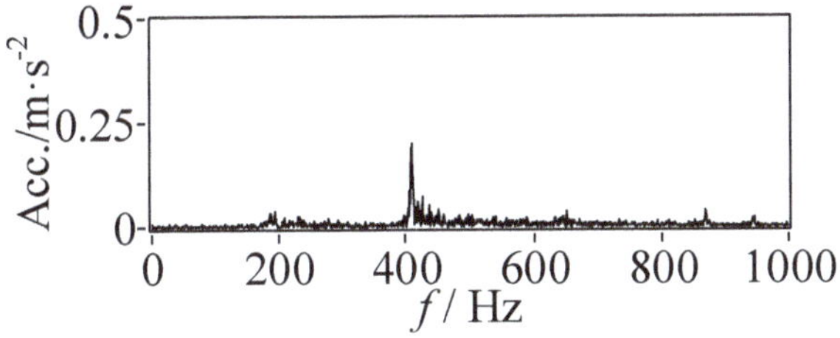

Figure 12. Base acceleration spectrum of the rotor in static suspension.

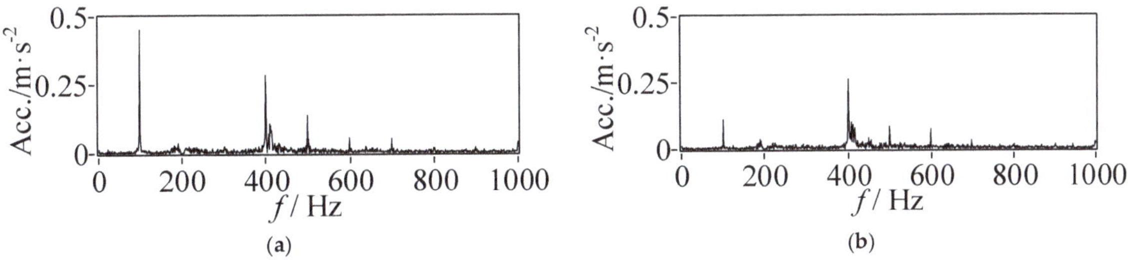

Figure 13. Base acceleration spectrum before and after rotor LMS compensation. (**a**) Compensation for acceleration spectrum of the front base; (**b**) Acceleration spectrum of the base after compensation.

From Figure 12, it can be seen that when the rotor is statically suspended, the base mainly has a vibration of about 400 Hz, which is the working frequency of the magnetic bearing. The frequency is mainly determined by the control parameters and the output characteristics of the hardware equipment such as the power amplifier, which does not belong to the research scope of this paper. The following mainly studies the base vibration caused by rotor imbalance.

It can be seen from Figure 13a that before LMS compensation, the maximum peak of the base acceleration signal is located at the synchronous frequency of 100 Hz. Since the magnetic bearing has no mechanical contact with the rotor, the base vibration is mainly generated by the control reaction force of the magnetic bearing to the rotor, which is the control electromagnetic force generated by the active suppression of the rotor unbalanced vibration by the magnetic bearing. Figure 13b is the base acceleration signal after LMS compensation. After the LMS algorithm compensation in this paper, the co-frequency vibration amplitude of the base speed is attenuated by 80.4%. Figure 14 shows the time-frequency spectrum of the base acceleration. It can be seen that before the compensation, the base acceleration has an obvious 100 Hz co-frequency component. When the compensation starts, the frequency component decreases rapidly and maintains a low vibration level. When the compensation stops, the 100 Hz component of the base acceleration is restored to the vibration level before compensation.

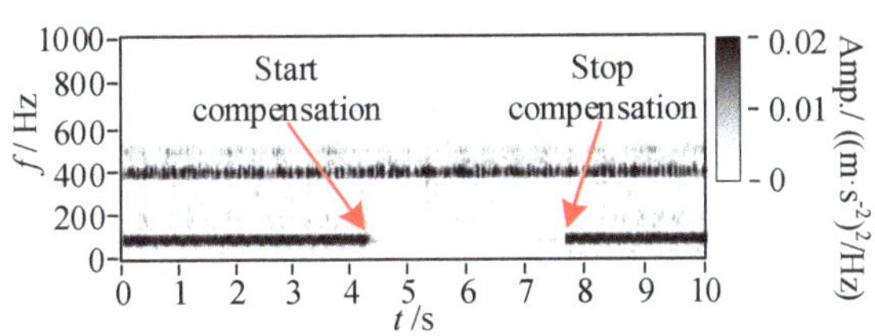

Figure 14. Time-frequency spectrum of base acceleration.

5. Conclusions

The unbalance vibration suppression of maglev high-speed motor based on the Least-Mean-Square is investigated in this paper. The Least-Mean-Square control method is adopted to actively control the vibrations caused by the unbalanced excitation vibration. The numerical simulation and experimental research are carried out. From the present investigations, the following key observations can be obtained:

(1) The LMS algorithm can effectively filter out the sine wave with a specified frequency in the signal, which is introduced into the unbalance compensation process of the maglev rotor to realize the unbalance vibration suppression of the maglev rotor.

(2) Using the LMS algorithm control in this paper, after filtering the unbalanced signal of the synchronous frequency of 100 Hz, the synchronous frequency displacement of the maglev rotor and the synchronous frequency control current of the magnetic bearing is reduced by 87% and 92%, respectively. Further filtering the harmonic signal of the maglev rotor can effectively reduce the harmonic control current of the magnetic bearing. Further reduction of unbalanced control of rotors by magnetic bearings to better realize the unbalanced vibration control of the rotor.

(3) The final compensation results show that the peak-to-peak mean value of rotor displacement decreases by 67.4% and the peak-to-peak mean value of control current decreases by 31% after using the LMS-based fundamental frequency and harmonic vibration control algorithm. Compared with the displacement and control current in static suspension, they only increase by 32.6% and 6.5%, respectively. At the same time, the vibration amplitude of the synchronous frequency with the base speed is attenuated by 80.4%.

The experimental results show that the algorithm can realize the rotor's unbalanced vibration control, make the rotor rotate around its inertial axis, reduce the rotor's unbalanced displacement vibration and forced vibration, reduce the control current of the magnetic bearing, reduce the power consumption of the control system, and decrease the vibration of the magnetic bearing to the supporting foundation.

Author Contributions: H.W.: conception of the study, propose theory and method, supervisor; M.Y.: literature search, figures, and data collection; C.S.: manuscript preparation and writing; N.W.: data interpretation and analysis. All authors have read and agreed to the published version of the manuscript.

Funding: This research was funded by Shenzhen Science and Technology Plan Project (Grant No. JCYJ20190809150603586) and National Natural Science Foundation (Grant No. 51975427).

Institutional Review Board Statement: Not applicable.

Informed Consent Statement: Not applicable.

Data Availability Statement: Not applicable.

Conflicts of Interest: The authors declare no conflict of interest.

References

1. Sun, F.; Jin, J.J.; Oka, K. *Permanent Magnetic Suspension System-Principle, Model, Simulation and Experiment*; Beijing Science Press: Beijing, China, 2018.
2. Zhao, C.; Sun, F.; Jin, J.J.; Tang, H.J.; Xu, F.; Li, Q.; Oka, K. Analysis of quasi-zero power characteristic for a permanent magnetic levitation system with a variable flux path control mechanism. *IEEE-ASME Trans. Mechatron.* **2021**, *26*, 1416–1420. [CrossRef]
3. Zhou, R.; Yan, M.Y.; Sun, F.; Jin, J.; Li, Q.; Xu, F.; Zhang, M.; Zhang, X.; Nakano, K. Experimental validations of a magnetic energy-harvesting suspension and its potential application for self-powered sensing. *Energy* **2022**, *239*, 122205. [CrossRef]
4. Xu, Y.; Wu, H.; Guan, X. Unbalance Suppression for AMB Rotor System Using APF-SRF Algorithm. *Shock. Vib.* **2020**, 1–10. [CrossRef]
5. Mohamed, M.E.; Bonello, P. The efficient inclusion of rotation-induced inertia effects in a shaft-blisk assembly model using zero-speed modes. *J. Sound Vib.* **2020**, *479*, 115357. [CrossRef]
6. Herzog, R.; Buhler, P.; Gahler, C.; Larsonneur, R. Unbalance compensation using generalized notch filters in the multivariable feedback of magnetic bearings. *IEEE Trans. Control. Syst. Technol.* **1996**, *4*, 580–586. [CrossRef]

7. Lum, K.Y.; Coppola, V.T.; Bernstein, D.S. Adaptive auto-centering control for an active magnetic bearing supporting a rotor with unknown mass imbalance. *IEEE Trans. Control. Syst. Technol.* **1996**, *4*, 587–597.
8. Setiawan, J.; Mukherjee, R.; Maslen, E. Synchronous Disturbance Compensation in Active Magnetic Bearings Using Bias Current Excitation. In Proceedings of the IEEE/ASME International Conference on Advanced Intelligent Mechatronics, Como, Italy, 8–12 July 2001; pp. 707–712.
9. Matras, A.; Flowers, G.; Fuentes, R.; Balas, M.; Fausz, J. Suppression of persistent rotor vibrations using adaptive techniques. *J. Vib. Acoust.* **2003**, *128*, 682–689. [CrossRef]
10. Chiacchiarini, H.; Mandolesi, P. Unbalance compensation for active magnetic bearings using ILC. In Proceedings of the IEEE International Conference on Control Applications, Mexico City, Mexico, 7 September 2001; pp. 58–63.
11. Betschon, F.; Knospe, C.R. Reducing magnetic bearing currents via gain scheduled adaptive control. *IEEE/ASME Trans. Mechatron.* **2001**, *6*, 437–443. [CrossRef]
12. Zheng, S.; Feng, R. Feedforward compensation control of rotor imbalance for high-speed magnetically suspended centrifugal compressors using a novel adaptive notch filter. *J. Sound Vib.* **2016**, *366*, 1–14. [CrossRef]
13. Nonami, K.; Qifu, F.; Ueyama, H. Unbalance vibration control of magnetic bearing systems using an adaptive algorithm with disturbance frequency estimation. *JSME Int. J.* **2008**, *41*, 220–226. [CrossRef]
14. Rundell, A.; Drakunov, S.; Decarlo, R. A sliding mode observer and controller for stabilization of rotational motion of a vertical shaft magnetic bearing. *IEEE Trans. Control. Syst. Technol.* **1996**, *4*, 598–608. [CrossRef] [PubMed]
15. Mohamed, A.; Ilene, B. Imbalance compensation and automatic balancing in magnetic bearing systems using the Q-parameterization theory. In Proceedings of the American Control Conference, Baltimore, MD, USA, 30 June–2 July 1994; pp. 2952–2957.
16. Han, B.; Cui, H.; Tang, E. Vibration suppression of magnetic bearing based on sliding mode disturbance observer. *Opt. Precis. Eng.* **2012**, *20*, 563–570.
17. Zhang, Y.; Zhou, J.; Han, X.; Zhou, Y. Adaptive odd repetitive control for magnetically suspended rotor harmonic currents suppression. *J. Vib. Control.* **2022**, *89*, 510–526. [CrossRef]
18. Xu, Y.; Jiang, Q.; Yang, K.; Zhou, J.; Guo, Q. A novel ultra-high-resolution inclination sensor based on diamagnetic levitation. *Sens. Actuators A Phys.* **2022**, *343*, 113686. [CrossRef]
19. Ashkezari-Toussi, S.; Sabzevari, V.R. Early arrhythmia prediction based on Hurst index and ECG prediction using robust LMS adaptive filter. *Signal Image Video Process.* **2021**, *15*, 1813–1820. [CrossRef]
20. Nishikawa, K. The LMS-Type Adaptive Filter Based on the Gaussian Model for Controlling the Variances of Coefficients. *IEICE Trans. Fundam. Electron. Commun. Comput. Sci.* **2020**, *E103.A*, 1494–1502. [CrossRef]
21. Zhang, H.; JLiu Zhu, R.; Chen, H.; Yuan, H. Nonlinear adaptive harmonics vibration control for an active magnetic bearing system with rotor unbalance and sensor runout. *IEEE Sens. J.* **2021**, *21*, 12245–12254. [CrossRef]
22. Soni, T.; Dutt, J.K.; Das, A.S. Parametric Stability Analysis of Active Magnetic Bearing Supported Rotor System With a Novel Control Law Subject to Periodic Base Motion. *IEEE Trans. Ind. Electron.* **2020**, *67*, 1160–1170. [CrossRef]
23. Wang, M.; Han, Q.; Wen, B.; Zhang, H.; Guan, T. Modal characteristics and unbalanced responses of fan rotor system with flexible support structures in aero-engine. *Proc. Inst. Mech. Eng. Part G J. Aerosp. Eng.* **2017**, *231*, 1686–1705. [CrossRef]
24. Gao, H.; Song, L. Modified LMS algorithm applied to maglev flywheel's vibration compensation problem. *Int. J. Appl. Electromagn. Mech.* **2017**, *53*, 359–369. [CrossRef]

actuators MDPI

Article

High Precision Magnetic Levitation Actuator for Micro-EDM

Boran Luan [1], Xiaoyou Zhang [1,2,*], Fangchao Xu [1], Guang Yang [1,3], Junjie Jin [1], Chengcheng Xu [1], Feng Sun [1] and Koichi Oka [4]

[1] School of Mechanical Engineering, Shenyang University of Technology, Shenliao West Road, Shenyang 110870, China
[2] Department of Mechanical Engineering, Nippon Institute of Technology, 4-1 Gakuendae Miyashiro-cho Minami-Saitama-gun, Saitama 345-8501, Japan
[3] School of Mechanical and Electrical Engineering, Shenyang University of Aeronautics and Astronautics, No. 37 Doyi South Street Shenbei New District, Shenyang 110135, China
[4] School of Systems Engineering, Kochi University of Technology, Kochi 780-8515, Japan
* Correspondence: zhang.xiaoyou@nit.ac.jp

Abstract: Aiming at the efficiency and precision in micro electrical discharge machining (micro-EDM) is affected because the interpole voltage is unstable in conventional micro-EDM. This paper describes a five-degrees-of-freedom (5-DOF) controlled, wide-bandwidth, and high-precision magnetic levitation actuator. The conventional micro-EDM can install the actuator to maintain a stable interpole voltage between the electrode and workpiece to realize the high-speed micro-EDM. In this paper, the structure of the magnetic levitation actuator is designed, and the magnetic field characteristics are analyzed. On this basis, an integrator and regulator are used along with a controller with local current feedback to eliminate steady-state errors, stabilize the control system, and improve the bandwidth and positioning accuracy of the magnetic levitation actuator, and the dynamic performance of the actuator is evaluated. The experimental results show that the developed actuator has excellent positioning performance with micron-level positioning accuracy to meet the demand for the real-time, rapid, and accurate adjustment of the interpole gap during micro-EDM.

Keywords: micro-EDM; magnetic levitation actuator; magnetic field characteristics; dynamic characteristics

Citation: Luan, B.; Zhang, X.; Xu, F.; Yang, G.; Jin, J.; Xu, C.; Sun, F.; Oka, K. High Precision Magnetic Levitation Actuator for Micro-EDM. *Actuators* **2022**, *11*, 361. https://doi.org/10.3390/act11120361

Academic Editor: Kirill Poletkin

Received: 11 November 2022
Accepted: 1 December 2022
Published: 2 December 2022

1. Introduction

A micro-EDM uses the high temperature generated by pulse discharge to melt and vaporize the material; there is no direct contact between the electrode and the workpiece during the processing, and there is no macro-cutting force in the process [1–4]. Compared with conventional machining methods, micro-EDM technology has the advantage of not being limited to the requirements of workpiece material hardness and strength and can machine any conductive material [5–7]. It is widely used in processing difficult materials such as those with high hardness, high toughness, high brittleness, and any conductive materials, as well as those with micro-holes and complex shapes. Overall, it has become a required processing method at this stage and is widely used in mold manufacturing [8–10]. However, conventional micro-EDM has low productivity [11]. When processing occurs in the case of broken or short circuits, the inter-pole gap of the micro-EDM needs to be adjusted quickly, and the conventional EDM machine tool adopts the motor and ball screw drive mode and–due to the large inertia of the mechanical transmission system–cannot ensure the ideal interpole gap in time, affecting the speed of micro-EDM and thereby limiting the efficiency and accuracy of conventional micro-EDM [12,13]. Therefore, it is necessary to improve micro-EDM's response speed and positioning accuracy to achieve a stable discharge gap and interpole voltage.

Magnetic levitation technology has the advantages of high precision, fast response, and no pollution, which provides better technical solutions for several industries, such as magnetic levitation actuators, magnetic levitation bearings, precision positioning, magnetic levitation trains, and so on [14–16]. Serguei Maximov et al. [17] applied Lagrange's equation to the analysis of a magnetic levitation system and experimentally determined the required equilibrium height of the levitated object; this method can be effectively implemented in a magnetic levitation system. Mundher H. A. [18] Yaseen utilized a microcontroller for the real-time control of Simulink functions to achieve stable control of magnetic levitation; the stability of the magnetic levitation plane was examined using phase lead compensation and linear quadratic regulator controllers. The results show that the control system with the linear quadratic regulator controller provides the same stability performance as the phase lead compensation controller. Chuan Zhao et al. [19] designed a magnetic levitation system that uses a zero-power control method, and the results show that the system levitates well and uses constant air gap control to achieve higher safety when changing the levitation mass. Kirill Poletkin [20] investigated the static pull-in of a tilt drive in a hybrid levitation micro-actuator and nonlinear modeling of the calculation of the mutual inductance and the action force between two circular filaments and experimentally verified the accuracy of the developed model, predicting the pull-in parameters of the hybrid levitation actuator with the developed analytical tool. Dongjue HE et al. [21] designed a magnetic levitation lens driving actuator, which is applied in laser processing; the actuator drives the lens to achieve real-time positioning of the laser beam focus point with good control performance and positioning accuracy. Tong Zheng et al. [22] designed a sizing optimization method to determine the magnetic levitation actuator for a rotary table and validated the effectiveness and accuracy of the method using simulation followed by experiments. Hirohisa Kato et al. [23] designed and fabricated a centrifugal pump using bearingless electrodes and permanent magnetic bearings to levitate and rotate the rotor in extremely low-temperature environments. Iwanori Murakami et al. [24] designed a superconducting magnetic bearing that combines a superconducting magnetic levitation system with a repulsive magnetic levitation system, which has a strong axial levitation force and can overcome resonance. The driving flywheel can reach 10,000 rpm by experimental verification. António J. Arsénio et al. [25] studied the optimization of the three-dimensional shape of the horizontal axis radial levitation bearing to find the optimal spacing between the Permanent magnet rings in the rotor and also between the high-temperature superconductor blocks in the stator. The guiding force is maximized while maintaining the minimum suspension force for a given size. Timo Hopf et al. [26] combined a control strategy with a flywheel system by studying an external rotor flywheel suspended by an active magnetic bearing, which improved the operating range and produced higher efficiency and a more stable power supply. Haiyue Zhu et al. [27] designed a magnetic levitation parallel actuated dual-stage motion system for the precision positioning of six axes, and the design improved the system's dynamic characteristics in terms of reduced stage size and weight. Duan Ji An et al. [28] and Zhou Haibo et al. [29] proposed a magnetically levitated platform and combined it with a decoupled controller capable of precision motion with the ability of a comprehensive stroke. Ahn Dahoon et al. [30] designed a dual-servo mechanism, first applied to a magnetic levitation stage system, that can simultaneously achieve high precision and a long stroke with an accuracy of up to 10 nm. The platform is suitable for high-precision manufacturing processes. Feng Sun et al. [31] designed a 3-DOF permanent magnetic levitation platform with a dual closed-loop controller for stable levitation and motion control, and experimental results showed that the system achieved stable levitation and motion control of the platform with high positioning accuracy.

The magnetic levitation motion system–with the advantage of non-contact motion, where the motion target is provided by the magnetic force and both levitation and thrust forces can be generated by the same actuator in the magnetic levitation motion system–uses active control to achieve high positioning accuracy with good dynamic performance [32]. Magnetic levitation actuators can eliminate the mechanical connection between the actuator and the moving parts by remote forces, effectively overcoming frictional forces, leading to improved operability and reduced energy consumption [33]. To improve the electrode positioning response speed and efficiency during micro-EDM and to meet the requirements of high-speed, high-precision, and fine EDM, a magnetic levitation actuator with five degrees of freedom, a wide bandwidth, high accuracy, and large stroke electrode positioning has been developed by combining the magnetic levitation actuator with a micro-EDM machine tool. The actuator is compact and can be connected to the conventional micro-EDM machine tool for micro-EDM. During the machining process, when there is a broken or short circuit between the workpiece and the electrode, the magnetic levitation actuator replaces the spindle of the micro-EDM machine tool and drives the electrode to quickly adjust the interpole gap with radial and axial strokes of several millimeters so that the interpole voltage is always kept within the smooth machining voltage, realizing the rapid adjustment of abnormal discharge.

In this paper, a 5-DOF, wide-bandwidth, high-precision, large-stroke magnetic levitation actuator is introduced, and the magnetic field characteristics of this actuator are analyzed. Based on this, a local current feedback loop is used to improve the bandwidth and positioning accuracy of the magnetic levitation actuator. Finally, the characterization experiments show that the actuator has good dynamic performance and can meet the needs of micro-EDM.

2. 5-DOF Controlled Magnetic Levitation Actuator

2.1. The Structure of Magnetic Levitation Actuator

Figure 1 shows the structure diagram of the proposed 5-DOF controlled magnetic levitation actuator. It is commonly used in magnetic levitation electromagnets of electromagnets consisting of iron-core and noncore electromagnets (air-core coil). Compared with the air-core coil, the iron-core electromagnet has less magnetic leakage so that it can produce a larger electromagnetic force. However, the electromagnetic force is proportional to the square of the coil current and inversely proportional to the square of the gap, so it is highly nonlinear and difficult to control. Compared with the iron-core electromagnet, the air-core coil has more magnetic leakage and weaker electromagnetic force, but it is proportional to the coil current, easy to control, and can achieve a large stroke. Therefore, the actuator is mainly composed of two permanent magnet rings on the spindle and eight sets of air-core coils on the stator that are symmetric with the center of gravity of the spindle and with two adjacent air-core coils connected in tandem as a group. Soft iron rings are caught between two permanent magnetic rings placed opposite each other–a structure that concentrates magnetic flux. Four displacement sensors are installed in each of the X and Y directions, and one sensor in the Z direction; these sensors are used to measure the displacement of the spindle.

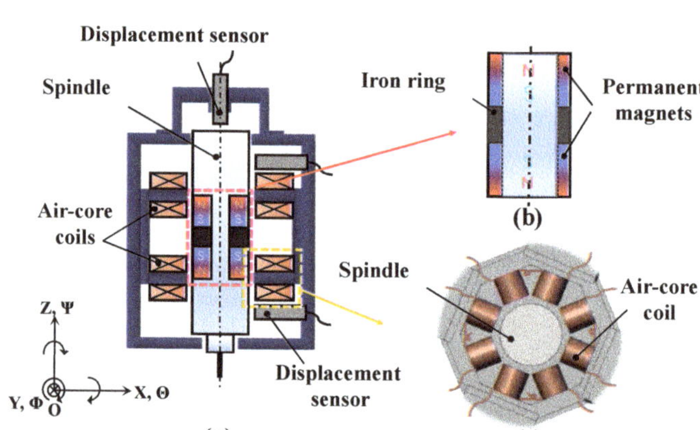

Figure 1. The structure diagram of the 5-DOF controlled magnetic levitation actuator. (**a**) Actuator; (**b**) Permanent magnets; (**c**) Air-core coils.

2.2. The Principle of Magnetic Levitation Actuator

The 5-DOF motion of the spindle is controlled by the attractive or repulsive force between the energized coil and the permanent magnet. For example, in Figure 2a, the combined force generated by coil 1 and coil 2 is in the positive direction of the X-axis, and the combined force generated by coil 3 and coil 4 is in the same direction as the combined force of coils 1 and 2, so that the spindle is driven to move in the direction shown in Figure 2a. In Figure 2b, the combined force generated by coil 1 and coil 2 is in the positive direction of the X-axis, and the combined force generated by coil 3 and coil 4 is in the opposite direction to the combined force of coils 1 and 2, so the spindle is driven to rotate in the direction shown in Figure 2b. In Figure 2c, the combined force generated by coils 1 and 2 and coils 5 and 6 is in the negative Z-axis direction, and the combined force generated by coils 3 and 7 and coils 4 and 8 is in the same direction as the combined force of coils 1, 2, 3, and 4, so the driving spindle moves in the direction shown in Figure 2c.

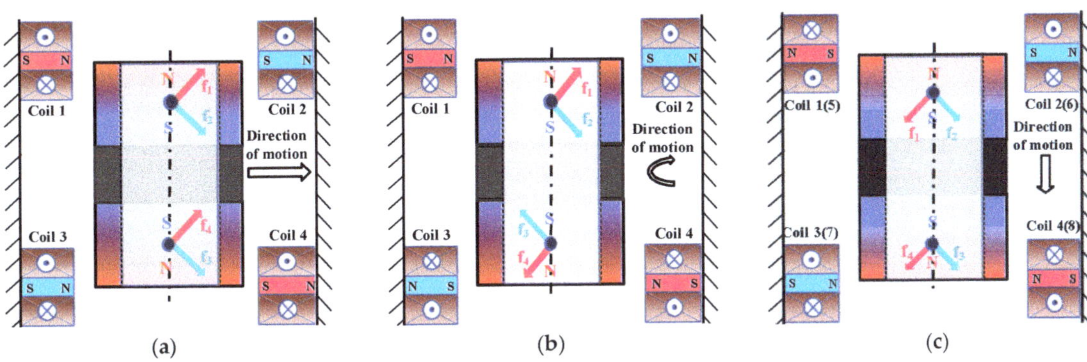

Figure 2. Principle of spindle motion control. (**a**) X direction; (**b**) Φ direction; (**c**) Z direction.

2.3. Mathematical Model of 5-DOF Magnetic Levitation Actuator

The equations of motion in the Z, X, and θ directions of the spindle are shown in Equations (1)–(3). The equations of motion in the Y and Φ directions are the same as those in the X and θ directions, so they are omitted.

$$m\ddot{z} + c\dot{z} + k_z z = F_Z \quad (1)$$

$$m\ddot{x} + c\dot{x} + k_x x = F_x \quad (2)$$

$$J_\theta\ddot{\theta} + c\dot{\theta} + k_\theta\theta = F_\theta \quad (3)$$

where m is the mass of the spindle; J is the rotational inertia in the θ direction; c is the damping coefficient; z is the distance traveled in the Z direction; x is the distance traveled in the X direction; θ is the angle of rotation back in the θ direction; k_z is the air gap stiffness in the Z direction; k_x is the air gap stiffness in the X direction; k_θ is the air gap stiffness in the θ direction; F_z is the electromagnetic force in the Z direction; F_x is the electromagnetic force in the X direction; M_θ is the torque generated in the θ direction; l is the distance from the center of the kinematic to the force point. Meanwhile, the forces generated between the coil and the permanent magnet in the Z and X directions are F_z and F_x, and the torque generated in the θ direction is also M_θ.

$$F_Z = F_1 + F_2 + F_5 + F_6 - F_3 - F_4 - F_7 - F_8 \quad (4)$$

$$F_x = F_1 - F_2 + F_3 - F_4 \quad (5)$$

$$M_\theta = (F_1 - F_2 - F_3 + F_4)\cdot l \quad (6)$$

In addition, the electromagnetic force between the coil and the permanent magnet is

$$f_i = k_i i_k \tag{7}$$

where k_i is the current stiffness factor of the coil and i_k is the drive current supplied to each coil set. In addition, in this actuator, the control currents of the eight groups of coils for controlling vertical motion in the Z direction are the same and set to i_z, the control currents of the four sets of coils for controlling horizontal motion in the X direction are the same and set to i_x, and the control currents of the four sets of coils for controlling rotational motion in the θ direction are the same and set to i_θ. The spindle's equations of motion in the Z, X, and θ directions are

$$m\ddot{z} + c\dot{z} + k_z z = k_i i_z \tag{8}$$

$$m\ddot{x} + c\dot{x} + k_x x = k_i i_x \tag{9}$$

$$J_\theta \ddot{\theta} + c\dot{\theta} + k_\theta \theta = l k_i i_\theta \tag{10}$$

The design principle of the magnetic levitation actuator is that the coil cuts the magnetic induction line, so a counter-electromotive force is generated when the coil is energized. According to Kirchhoff's second law, the dynamic voltage balance of this magnetic levitation actuator can be obtained.

$$U = U_L + U_R + E = L\frac{di}{dt} + Ri + k_v \dot{z} \tag{11}$$

where U is the total electric potential; U_L is the inductor voltage drop; k_v is the reverse electric potential coefficient; U_R is the resistance voltage drop; E is the reverse electric potential; L is the coil inductance; R is the coil resistance. The electrical and mechanical models are coupled, and the Rasch transform can sort out the transfer function of the magnetic levitation actuator.

$$U_z(s) = Ls\frac{ms^2\Delta z + cs\Delta z + k_z\Delta z}{k_i} + R\frac{ms^2\Delta z + cs\Delta z + k_z\Delta z}{k_i} + k_v s\Delta z \tag{12}$$

$$U_x(s) = Ls\frac{ms^2\Delta x + cs\Delta x + k_x\Delta x}{k_i} + R\frac{ms^2\Delta x + cs\Delta x + k_x\Delta x}{k_i} + k_v s\Delta x \tag{13}$$

$$U_\theta(s) = Ls\frac{ms^2\Delta\theta + cs\Delta\theta + k_\theta\Delta\theta}{lk_i} + R\frac{ms^2\Delta\theta + cs\Delta\theta + k_\theta\Delta\theta}{lk_i} + k_v s\Delta\theta \tag{14}$$

2.4. Experimental Magnetic Levitation Actuator

Figure 3 shows a physical view of the magnetic levitation actuator and its spindle. The actuator has a mass of 8 kg, a height of 190 mm, and a width of 134 mm. The hollow coil is made of copper with a wire diameter of 0.7 mm, and the number of turns per coil is 670. The mass of the spindle is 0.8 kg, with a height of 148 mm and a diameter of 45 mm. Through the experiment, the current stiffness of the magnetic levitation actuator is obtained, and the recommended value of the safe ampacity of the copper wire is 5~8 A/mm^2, the cross-sectional area of the copper wire with a diameter of 0.7 mm is 0.38 mm^2, and the safe ampacity is within 3.08 A. From Equation (7) in Section 2.3, it can be obtained that the load of the magnetic levitation actuator is 12.93 N. To improve the accuracy of the displacement sensor displacement detection, the detection material of both sides and the upper end of the spindle is made of stainless steel (SUS304). Considering the remanent magnetism, coercivity, maximum magnetic energy product, and economy, the permanent magnet ring is made of NdFeB–the third-generation permanent magnet material. The spindle displacement in the direction of five degrees of freedom is measured by five eddy current displacement sensors (PU-09, AEC Corp., Dallas, TX, USA), and the actuator is measured by a digital signal processor (DSP; DS1103 PPC Controller Board, dSPACE Corp., Paderborn, Germany) with a sampling rate of 10 kHz.

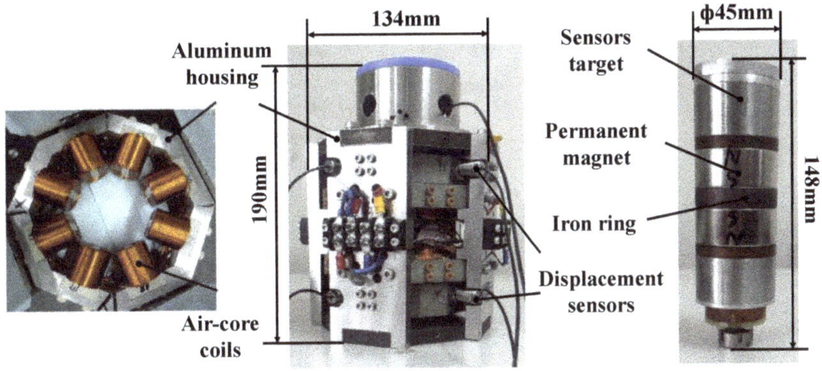

Figure 3. Experimental magnetic levitation actuator and its spindle.

3. Magnetic Field Characteristic Analysis

The most important factor affecting the electromagnetic force in the magnetic levitation actuator is the air-gap magnetic induction intensity. In the actuator designed in this paper, the permanent magnet on the spindle is a circular permanent magnet. The surface current analysis method in calculating the air gap magnetic field is mainly applied to the calculation of the air gap magnetic field of the permanent magnet with a regular shape, so the surface current method is selected in this paper to calculate the air gap magnetic field. Figure 4 shows an axially magnetized cylindrical permanent magnet; the equivalent current model of the permanent magnet can be regarded as the external space magnetic field of the permanent magnet generated by the current of bound surface current on the side surface of the permanent magnet.

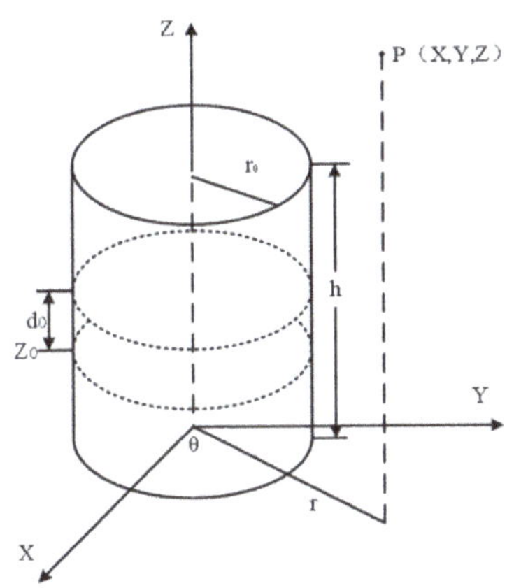

Figure 4. Cylindrical permanent magnet.

First, the current density of the bound surface is obtained. Secondly, after obtaining the magnetic induction intensity at any point P outside the cylinder based on a cylindrical permanent magnet that has been uniformly magnetized to saturation along the same direction, the magnetic induction intensity generated by the electric current loop can be obtained. Finally, the space magnetic induction intensity of the cylindrical permanent magnet can be obtained.

The magnetic induction intensity on the X component is expressed as:

$$B_x = \frac{B_r}{4\pi}\int_0^h \int_0^{2\pi} \frac{r_0(z-z_0)\cos\theta}{K} d\theta dz_0 \tag{15}$$

The magnetic induction intensity on the Y component is expressed as:

$$B_y = \frac{B_r}{4\pi}\int_0^h \int_0^{2\pi} \frac{r_0(z-z_0)\sin\theta}{K} d\theta dz_0 \quad (16)$$

The magnetic induction intensity on the Z component is expressed as:

$$B_z = \frac{B_r}{4\pi}\int_0^h \int_0^{2\pi} \frac{r_0^2 - r_0(x\cos\theta + y\sin\theta)}{K} d\theta dz_0 \quad (17)$$

Among them:

$$K = \left[(x - r_0\cos\theta)^2 + (y - r_0\sin\theta)^2 + (z - z_0)^2\right]^{\frac{3}{2}} \quad (18)$$

where B_r is the residual magnetization intensity of the permanent magnet.

According to the superposition principle, an axially magnetized magnetic ring can be used as an axially magnetized cylinder inside placed a reverse magnetized cylinder, as shown in Figure 5.

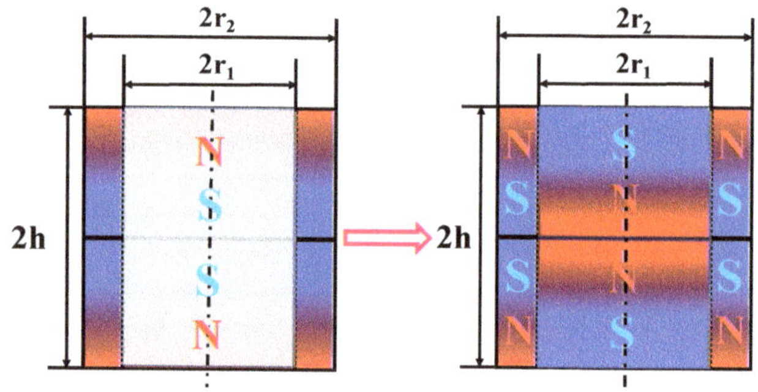

Figure 5. Magnetic ring equivalent diagram.

Based on Equation (15) and the magnetic field superposition principle, the magnetic induction intensity on the X component of the space P point around the axially magnetized magnetic ring can be expressed as:

$$B_x = \frac{B_r}{4\pi}\int_0^{2h} \int_0^{2\pi} \left(\frac{r_2(z-z_0)\cos\theta}{K_2} - \frac{r_1(z-z_0)\cos\theta}{K_1}\right) d\theta dz_0 \quad (19)$$

The magnetic induction intensity on the Y component is expressed as:

$$B_y = \frac{B_r}{4\pi}\int_0^{2h} \int_0^{2\pi} \left(\frac{r_2(z-z_0)\sin\theta}{K_2} - \frac{r_1(z-z_0)\sin\theta}{K_1}\right) d\theta dz_0 \quad (20)$$

The magnetic induction intensity on the Z component is expressed as:

$$B_z = \frac{B_r}{4\pi}\int_0^{2h} \int_0^{2\pi} \left(\frac{r_2^2 - r_2(x\cos\theta + y\sin\theta)}{K_2} - \frac{r_1^2 - r_1(x\cos\theta + y\sin\theta)}{K_1}\right) d\theta dz_0 \quad (21)$$

Among them:

$$K_1 = \left[(x - r_1\cos\theta)^2 + (y - r_1\sin\theta)^2 + (z - z_0)^2\right]^{\frac{3}{2}} \quad (22)$$

$$K_2 = \left[(x - r_2\cos\theta)^2 + (y - r_2\sin\theta)^2 + (z - z_0)^2\right]^{\frac{3}{2}} \quad (23)$$

Two magnetic rings placed relative to each other are combined to form the spindle of the magnetic levitation actuator designed in this paper, as shown in Figure 6.

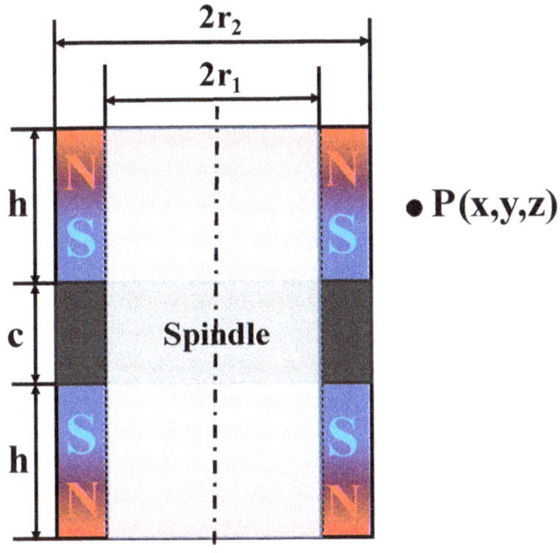

Figure 6. Magnetic array magnetic field model.

Similarly, according to the superposition principle of the magnetic field, the magnetic induction intensity on the X component of $P\ (x, y, z)$ can be expressed as:

$$B_x = \frac{B_r}{2\pi}\left\{\begin{array}{l} \int_{-\frac{c}{2}-h}^{-\frac{c}{2}} \int_0^{2\pi} \left[\frac{r_2(z-z_0)\cos\theta}{K_2} - \frac{r_1(z-z_0)\cos\theta}{K_1}\right] d\theta dz_0 \\ + \int_{\frac{c}{2}}^{\frac{c}{2}+h} \int_0^{2\pi} \left[\frac{r_2(z-z_0)\cos\theta}{K_2} - \frac{r_1(z-z_0)\cos\theta}{K_1}\right] d\theta dz_0 \end{array}\right\} \quad (24)$$

The magnetic induction intensity on the Y component is expressed as:

$$B_y = \frac{B_r}{2\pi}\left\{\begin{array}{l} \int_{-\frac{c}{2}-h}^{-\frac{c}{2}} \int_0^{2\pi} \left[\frac{r_2(z-z_0)\sin\theta}{K_2} - \frac{r_1(z-z_0)\sin\theta}{K_1}\right] d\theta dz_0 \\ + \int_{\frac{c}{2}}^{\frac{c}{2}+h} \int_0^{2\pi} \left[\frac{r_2(z-z_0)\sin\theta}{K_2} - \frac{r_1(z-z_0)\sin\theta}{K_1}\right] d\theta dz_0 \end{array}\right\} \quad (25)$$

The magnetic induction intensity on the Z component is expressed as:

$$B_z = \frac{B_r}{2\pi}\left\{\begin{array}{l} \int_{-\frac{c}{2}-h}^{-\frac{c}{2}} \int_0^{2\pi} \left[\frac{r_2^2-r_2(x\cos\theta+y\sin\theta)}{K_2} - \frac{r_{1_1}^2-r_1(x\cos\theta+y\sin\theta)}{K_1}\right] d\theta dz_0 \\ + \int_{\frac{c}{2}}^{\frac{c}{2}+h} \int_0^{2\pi} \left[\frac{r_2^2-r_2(x\cos\theta+y\sin\theta)}{K_2} - \frac{r_{1_1}^2-r_1(x\cos\theta+y\sin\theta)}{K_1}\right] d\theta dz_0 \end{array}\right\} \quad (26)$$

The magnetic induction intensity in the radial direction of the spindle of the magnetic levitation actuator was simulated and analyzed using the finite element simulation software Ansoft. The magnetic induction intensity of the middle part of the spindle of the magnetic levitation actuator was measured, and the simulation model was established as shown in Figure 7. The distance range of magnetic induction intensity was within 22.5 mm~44.5 mm—that is, the maximum horizontal distance from the outer diameter of the permanent magnet to the coil. Data were collected every 0.5 mm distance, and a total of 45 groups of data were collected.

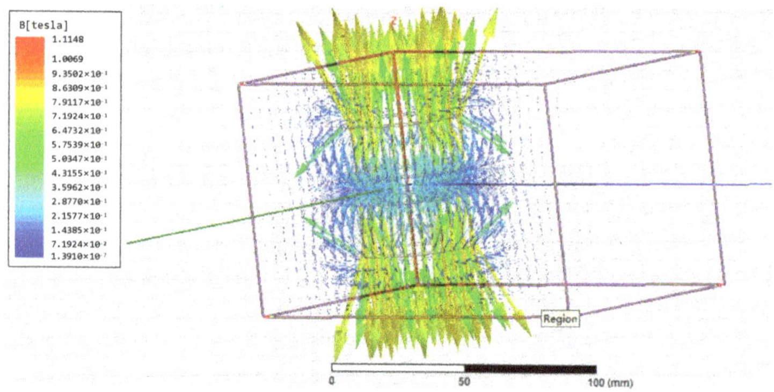

Figure 7. Actuator simulation model diagram.

To understand the magnetic field distribution of the spindle of the magnetic levitation actuator more accurately and to verify the simulation results' reliability and whether the spindle's magnetic induction intensity meets this paper's experimental requirements, a relevant experimental analysis was carried out. An experimental device for magnetic induction intensity was designed, as shown in Figure 8. The device is mainly composed of a Gauss meter (TD8600-E, measuring range 0–2 T, measuring error 0.5%), height gauge, micro platform, and magnetic suspension actuator spindle. The measurement range of magnetic induction intensity is consistent with the simulation analysis.

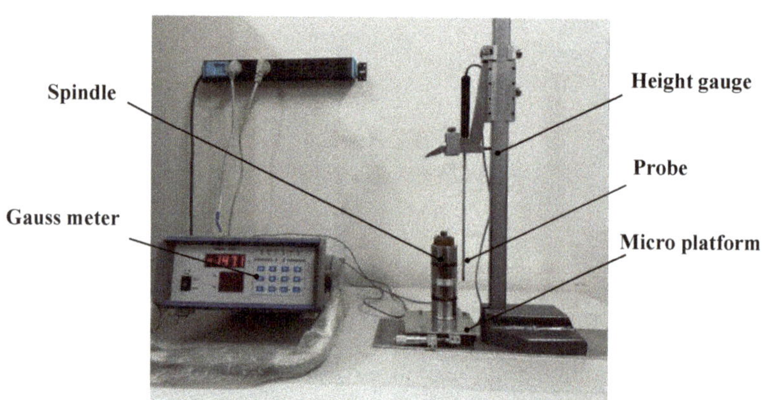

Figure 8. Test platform of magnetic induction strength.

The theoretical calculation, simulation, and experiment are analyzed simultaneously. Figure 9 shows the comparison results between the theoretical calculation, simulation, and experiment of the magnetic induction intensity of the actuator. It can be seen that the curves of theory, simulation, and experiment decrease with the increase in displacement. In addition, the simulation result is larger than the experimental result because the permanent magnet ring has low magnetization and certain magnetic leakage phenomena in the experiment. The results show that the theoretical model of the magnetic induction intensity of the actuator spindle is feasible, and the magnetization intensity of the permanent magnet ring can also meet the experimental requirements. The magnetic flux and induction intensity will be increased by adding a magnetic ring to the spindle.

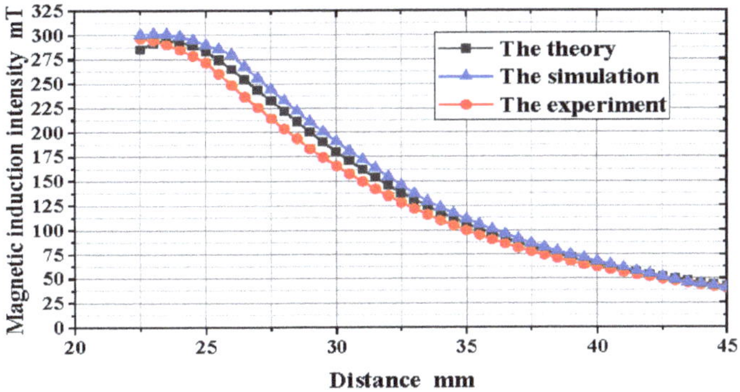

Figure 9. Comparison of simulation and experimental results.

4. 5-DOF Magnetic Levitation Actuator Controller Design and Positioning Performance

4.1. Actuator Motion Control System

Figure 10 shows the designed 5-DOF controllable magnetic levitation actuator controller. An integral compensator is used to eliminate the steady-state error of the control system, and a regulator is used to stabilize the control system. The gain of the integrator is set to δ, a_1 and a_0 are the denominator parameters of the regulator, and b_2, b_1, and b_0 are the numerator parameters of the regulator.

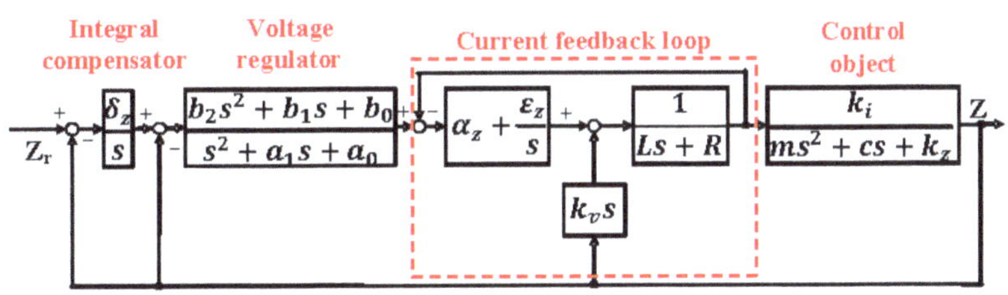

Figure 10. Controller Designed.

In addition, a current feedback loop containing a PI controller was used to improve the response of the current in the coil, as shown in Figure 11. To simplify the controller, the transfer function from the target value of the coil current to the actual current flowing into the coil is approximated as a first-order delay system. The approximate time constant for the first-order delay system is denoted as T_{dz}, and the controller design for the other directions is the same as for the Z-direction. Table 1 shows the model parameters, and Table 2 shows the control parameters of the actuator, which were determined by experimental results and numerical simulations.

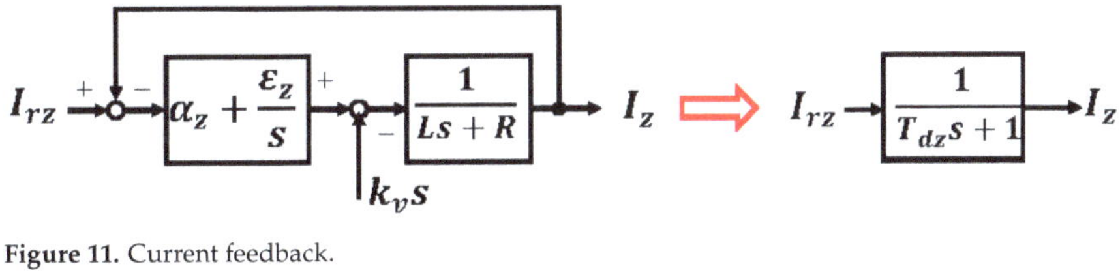

Figure 11. Current feedback.

Table 1. Model parameters.

Variable Name	X (Y) Z θ (Φ) Direction		Unit
mass of the spindle	m	0.80	kg
coil inductance	L	35.4	mH
coil resistance	R	2.6	Ω
torque	l	25	mm
rotational inertia	J_θ	2	kg·m^2
first-order delay system time constants	T_d	3.9×10^{-3}	/
current stiffness	k_i	4.2	N·A^{-1}
air gap stiffness in the X direction	k_x	367.57	N·m^{-1}
air gap stiffness in the Z direction	k_z	170.7	N·m^{-1}
air gap stiffness in the θ direction	k_θ	9.19	N·rad^{-1}
damping coefficient	c	1	N·s·m^{-1}

Table 2. Control parameters.

X (Y) Z θ (Φ) Direction Controller	
δ_x	256.45
δ_z	256.50
δ_θ	3.07×10^4
a_{0x}	3.07×10^5
a_{0z}	1.49×10^5
$a_{0\theta}$	4.53×10^5
a_{1x}	873.31
a_{1z}	1.93×10^3
$a_{1\theta}$	819.73
b_{0x}	1.30×10^8
b_{0z}	6.78×10^9
$b_{0\theta}$	1.79×10^6
b_{1x}	3.52×10^6
b_{1z}	8.50×10^7
$b_{1\theta}$	3.09×10^4
b_{2x}	4.01×10^4
b_{2z}	2.16×10^5
$b_{2\theta}$	93.14
α_x	35
α_z	35
α_θ	35
ε_x	2565
ε_z	2565
ε_θ	2565

4.2. Composition of the Experimental System

Overall, 16 coils are used in groups of two, so eight linear amplifiers are used to amplify the coil currents, and eight current sensors are used to measure the coil currents. These amplifiers and current sensors were integrated into the amplifier box. Figure 12 shows the structure of the experimental system. The magnetic levitation is controlled by a DSP board (ds1103, dSPACE Corp.) with a sampling frequency of 10 kHz. The analog signal obtained by the displacement sensor is collected into the DSP board through the A/D converter (16 bit, ±10 V) for operational processing. The control signal after the operation is amplified by a linear amplifier through the D/A converter (16 bit, ±10 V) and applied to the magnetic levitation actuator. The designed controller was applied to the developed actuator, and the experimental results and numerical simulation parameters in Tables 1 and 2 were used to perform the dynamic characterization experiments of the magnetic levitation actuator.

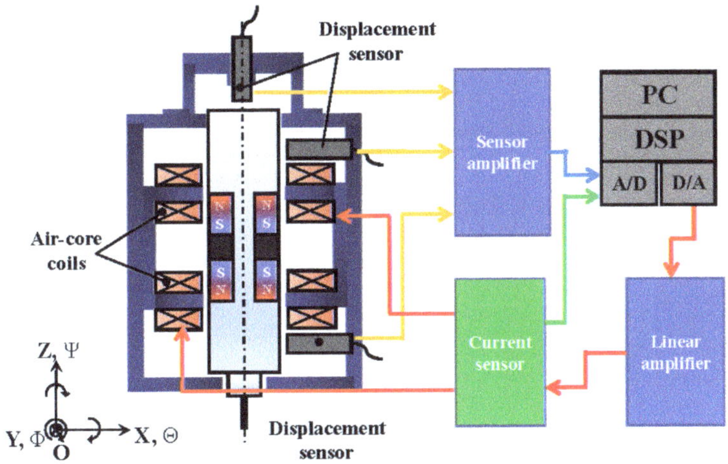

Figure 12. Structure of the experimental system.

4.3. Performance Test

First, a 10 μm step signal was applied to the input reference points in the X, Y, and Z directions, respectively, to measure the response. Then, 1 μrad step signal was applied to the Φ direction and theta direction, respectively, measuring the response, as shown in Figure 13. To avoid the collision between the electrode and workpiece in the process of micro-EDM, the overshoot of the electrode should be set to zero. Therefore, we will adjust the controller parameters in the future to improve the step response of the actuator.

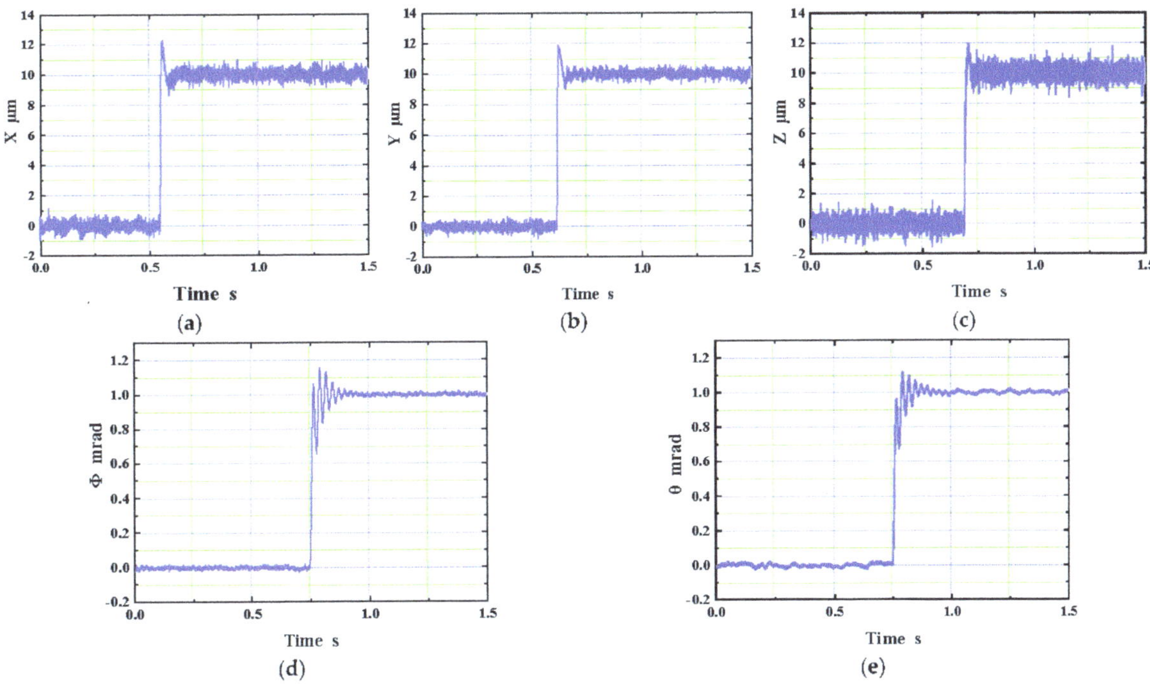

Figure 13. Structure of the experimental system. (**a**) Step response in the X direction; (**b**) Step response in the Y direction; (**c**) Step response in the Z direction; (**d**) Step response in the Φ direction; (**e**) Step response in the θ direction.

Second, the positioning resolution is measured in the X, Y, and Z directions, as well as in the Φ and θ directions. As shown in Figure 14, the experimental results show that the positioning resolution of the X, Y, and Z axes was 1 μm, and the positioning resolution of the Φ, θ direction is 25 μrad and 20 μrad, respectively. The results show that the actuator has good stability, and that the actuator spindle can be maintained at the required position.

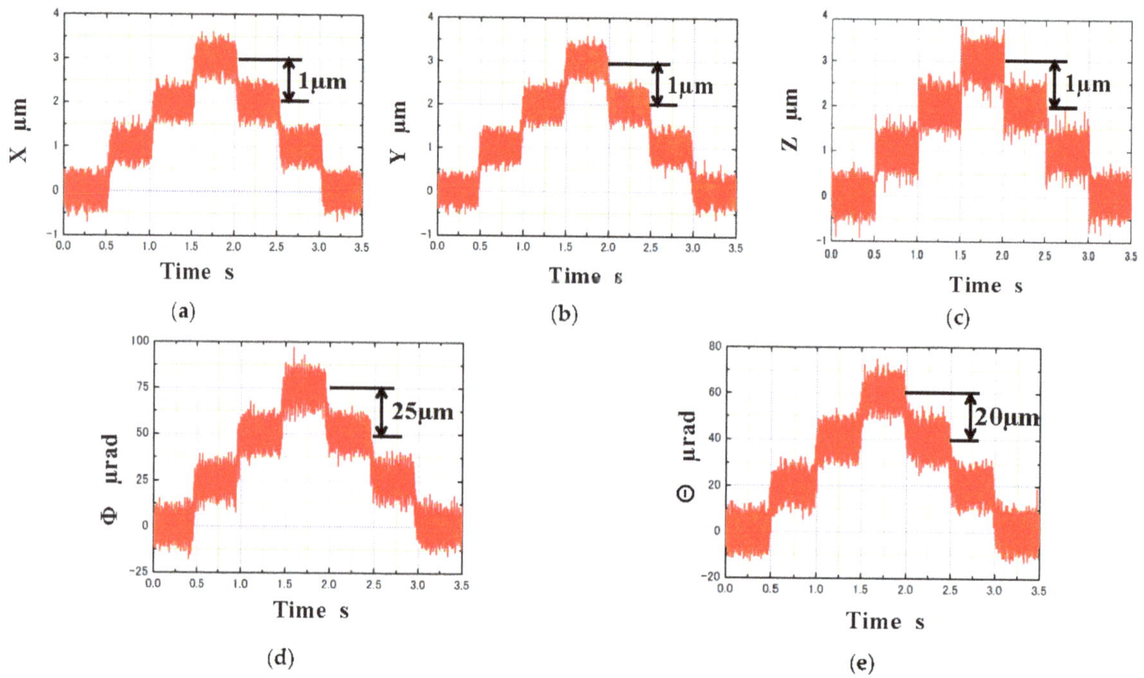

Figure 14. Positioning resolutions. (**a**) Positioning resolution in the X direction; (**b**) Positioning resolution in the Y direction; (**c**) Positioning resolution in the Z direction; (**d**) Positioning resolution in the Φ direction; (**e**) Positioning resolution in the θ direction.

Third, a sine wave with an amplitude of 2 mm and a frequency of 1 Hz was applied to the X, Y, and Z axes, respectively, to measure the travel. Then, 70 mrad amplitude and a frequency of 1 Hz sine wave was applied to the Φ, θ direction measure, respectively. The experimental results are shown in Figure 15. It can be seen from the results that 4 mm of travel was achieved in the X, Y, and Z axes, and a 70 mrad stroke was achieved in the Φ, θ direction. Using the actuator's function, the electrode can swing, the size of the machining hole can be adjusted, and a complex shape-such as the ellipse-can be finely processed.

Fourth, the X, Y, and Z axis directions and to evaluate the frequency response of Φ, θ directions. The frequency response was measured by a frequency response analyzer (FRA 5095, NF Corp., Yokohama, Japan), and the experimental results are shown in Figure 16. Actuator X, Y, and Z direction of the bandwidth is 101 Hz, 101 Hz, and 51 Hz, respectively. The bandwidth of Φ, θ direction was, respectively, 42 Hz, 45 Hz. Based on the above experimental results, the performance evaluation results of the brake are shown in Table 3.

Finally, a multi-axis linkage control was performed. The target values in the X and Y directions were set as sine and cosine waves with frequencies of 0.5 Hz and amplitudes of 0.5, 1.0, and 1.5 mm, respectively. This linkage control caused the spindle to move in a circular manner. The motion trajectory is shown in Figure 17.

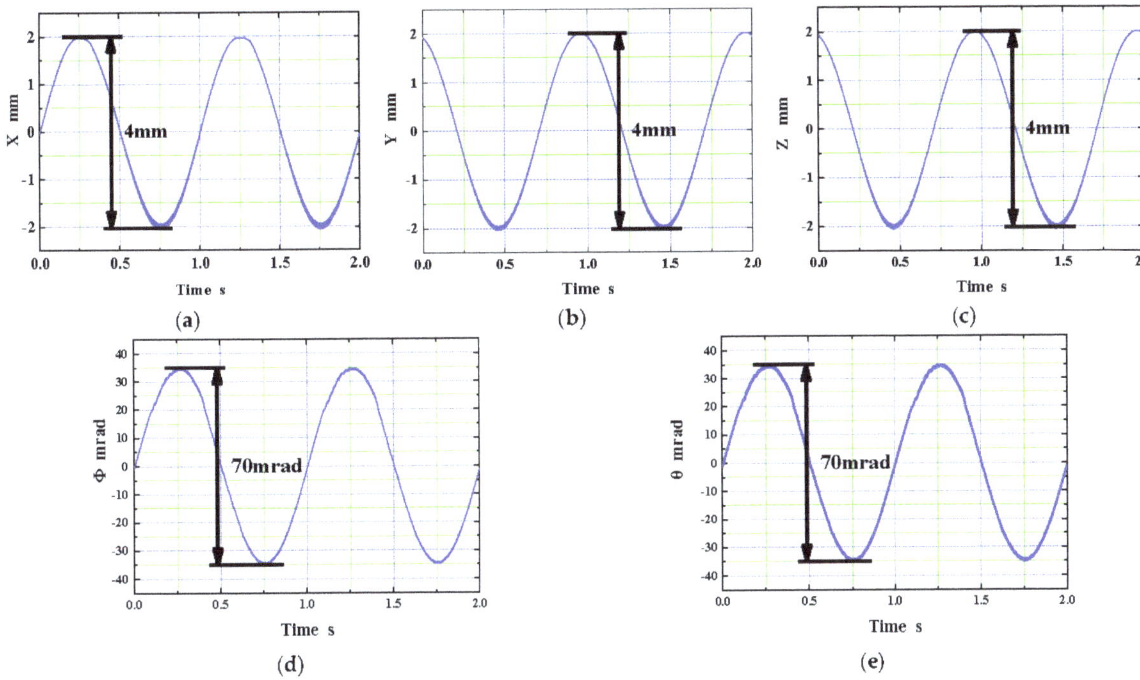

Figure 15. Positioning resolutions. (**a**) The stroke in the X direction; (**b**) The stroke in the Y direction; (**c**) The stroke in the Z direction; (**d**) The stroke in the Φ direction; (**e**) The stroke in the θ direction.

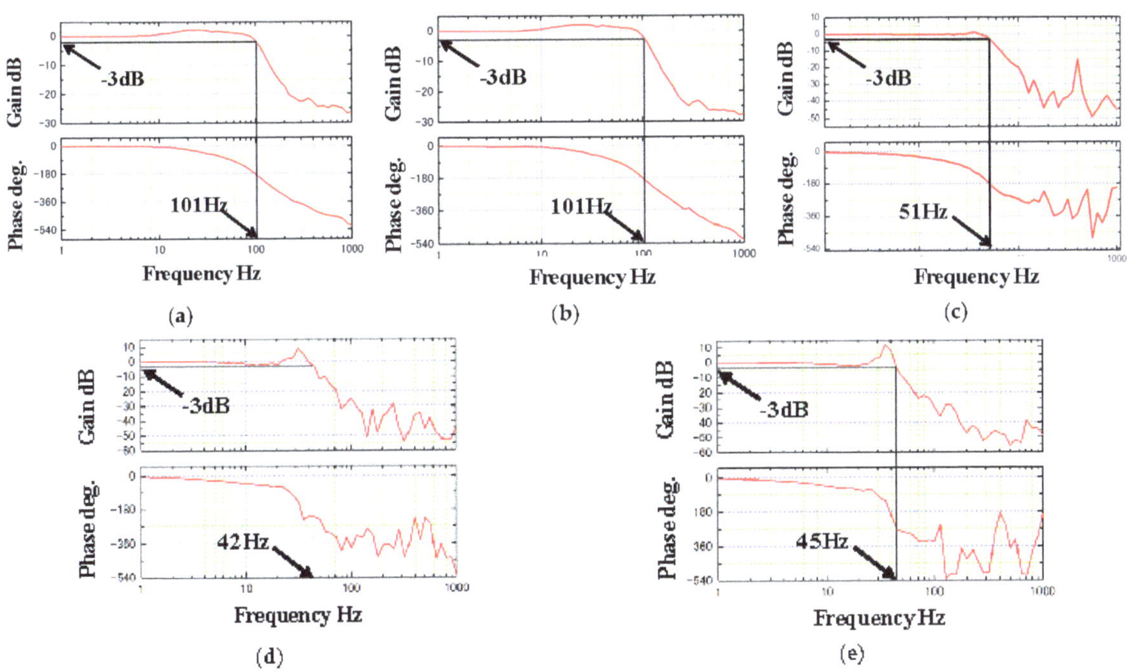

Figure 16. Frequency responses. (**a**) The bandwidth in the X direction; (**b**) The bandwidth in the Y direction; (**c**) The bandwidth in the Z direction; (**d**) The bandwidth in the Φ direction; (**e**) The bandwidth in the θ direction.

Table 3. Performance evaluation results table.

	Response Time	Stroke	Positioning Resolution	Bandwidth
X direction	6.7 ms	4 mm	1 μm	101 Hz
Y direction	6.8 ms	4 mm	1 μm	101 Hz
Z direction	26.3 ms	4 mm	1 μm	51 Hz
Φ direction	39.9 ms	70 mrad	25 μrad	42 Hz
θ direction	16.2 ms	70 mrad	20 μrad	45 Hz

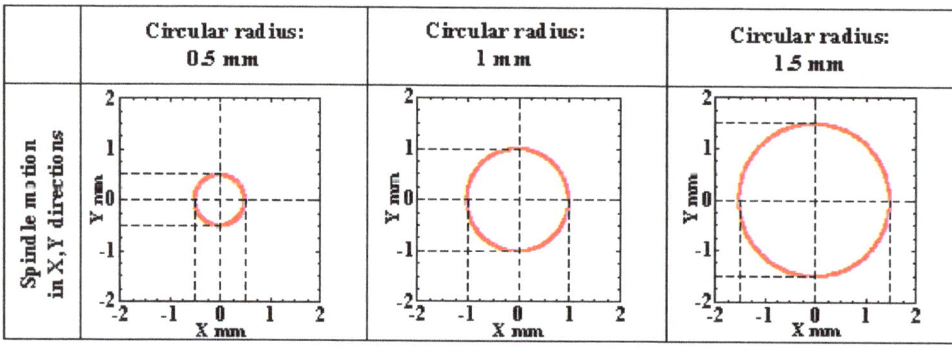

Figure 17. Spindle circular motion.

5. Conclusions

In this paper, a wide-band, high-precision, and large-stroke 5-DOF magnetic levitation actuator is proposed and manufactured, which can be installed in a conventional micro-EDM machine tool. The magnetic field model of the magnetic levitation drive was established, the magnetic field simulation analysis of the magnetic levitation drive was carried out, the related experiments were carried out by building an experimental device, and the magnetic field characteristics of the magnetic levitation drive were analyzed. The controller of the actuator was also designed and evaluated for its response speed, positioning accuracy, motion range, frequency band width, and multi-axis linkage control effect. The developed magnetic levitation actuator has millisecond response speed in all five degrees of freedom; the motion range is up to 4 mm in the translational direction and 70 mrad in the rotational direction; the positioning resolution is up to 1 μm in the translational direction and 25 μrad in the rotational direction; the frequency bandwidth of the actuator is 101 Hz in the radial direction, 51 Hz in the axial direction, and about 40 Hz in the rotational direction. The experimental results show that the developed actuator can achieve a fast response and stable suspension, which can meet the demand for real-time, fast, and accurate adjustment of the interpole gap during micro-EDM processing, thus maintaining the stability of the interpole voltage. It improves the discharge probability of microelectronic processing, thus increasing the processing efficiency.

Author Contributions: B.L.: Conceptualization, Methodology, Software, Validation, Investigation, Data curation, Formal analysis, Writing—original draft, Visualization, Funding acquisition; X.Z.: Supervision, Funding acquisition, Resources, Project administration, Writing—review & editing; F.X.: writing—review & editing; G.Y.: writing-review & editing; J.J.: writing—review & editing; C.X.: writing—review & editing; F.S.: writing—review & editing; K.O.: writing-review & editing. All authors have read and agreed to the published version of the manuscript.

Funding: This research is supported by National Natural Science Fund of China (No. 52005345, No. 52005344), National Key Research and Development Project (No. 2020YFC2006701), China Scholarship Council (No. 202208210073), Scientific Research Fund Project of Liaoning Provincial Department of Education (No. LFGD2020002), Major Project of the Ministry of Science and Technology of Liaoning Province (No. 2022JH1/10400027).

Institutional Review Board Statement: Not applicable.

Informed Consent Statement: Not applicable.

Data Availability Statement: Not applicable.

Acknowledgments: We thank Chuan Zhao, Wenzhe Pei, Fucheng Zhang and Jian Zhang for their suggestions.

Conflicts of Interest: The authors declare no competing interest.

References

1. Prakash, V.; Kumare, P.; Singh, P.K.; Hussain, M.; Das, A.K.; Chattopadhyaya, S. Micro-electrical discharge machining of difficult-to-machine materials: A review. *Proc. Inst. Mech. Eng. Part B J. Eng. Manuf.* **2019**, *233*, 339–370. Available online: https://journals.sagepub.com/doi/10.1177/0954405417718591 (accessed on 5 June 2022). [CrossRef]
2. Urso, D.G.; Giardini, C.; Ravasio, C. Effects of Electrode and Workpiece Materials on the Sustainability of Micro-EDM Drilling Process. *Int. J. Precis. Eng. Manuf.* **2018**, *19*, 1727–1734. Available online: https://link.springer.com/article/10.1007/s12541-018-0200-2 (accessed on 5 June 2022).
3. Suganthi, X.H.; Natarajan, U.; Sathiyamurthy, S.; Chidambaram, K. Prediction of quality responses in micro-EDM process using an adaptive neuro-fuzzy inference system (ANFIS) model. *Int. J. Adv. Manuf. Technol.* **2013**, *68*, 339–347. Available online: https://link.springer.com/article/10.1007/s00170-013-4731-5 (accessed on 5 June 2022). [CrossRef]
4. Yang, F.; Yang, J.; Yao, K.; Hua, H. Adaptive Voltage Position Control for Pulse Power Supply in Electrical Discharge Machining. *IEEE Trans. Ind. Electron.* **2019**, *66*, 5895–5906. Available online: https://ieeexplore.ieee.org/document/8480869 (accessed on 5 June 2022). [CrossRef]
5. Kumar, D.; Singh, N.K.; Bajpai, V. Recent trends, opportunities and other aspects of micro-EDM for advanced manufacturing: A comprehensive review. *J. Braz. Soc. Mech. Sci. Eng.* **2020**, *42*, 2172–2191. Available online: https://link.springer.com/article/10.1007/s40430-020-02296-4 (accessed on 10 July 2022). [CrossRef]
6. Shabgard, M.R.; Gholipoor, A.; Baseri, H. A review on recent developments in machining methods based on electrical discharge phenomena. *Int. J. Adv. Manuf. Technol.* **2016**, *87*, 2081–2097. Available online: https://link.springer.com/article/10.1007/s00170-016-8554-z (accessed on 10 July 2022). [CrossRef]
7. Gostimirovic, M.; Pucovsky, V.; Sekulic, M.; Radovanovic, M.; Madic, M. Evolutionary multi-objective optimization of energy efficiency in electrical discharge machining. *J. Mech. Sci. Technol.* **2018**, *32*, 4775–4785. Available online: https://link.springer.com/article/10.1007/s12206-018-0925-y (accessed on 10 July 2022). [CrossRef]
8. Macedo, F.T.B.; Wiessner, M.; Hollenstein, C.; Kustera, F.; Wegenera, K. Dependence of Crater Formation in Dry EDM on Electrical Breakdown Mechanism. *Procedia CIRP* **2016**, *42*, 161–166. Available online: https://linkinghub.elsevier.com/retrieve/pii/S2212827116004960 (accessed on 12 June 2022). [CrossRef]
9. Nahak, B.; Gupta, A. A review on optimization of machining performances and recent developments in electro discharge machining. *Manuf. Rev.* **2019**, *6*, 2. Available online: https://mfr.edp-open.org/articles/mfreview/full_html/2019/01/mfreview180009/mfreview180009.html (accessed on 12 June 2022). [CrossRef]
10. Gohil, V.; Puri, Y.M. Turning by electrical discharge machining: A review. *Proc. Inst. Mech. Eng. Part B J. Eng. Manuf.* **2017**, *231*, 195–208. Available online: https://journals.sagepub.com/doi/10.1177/0954405415590560 (accessed on 12 June 2022). [CrossRef]
11. Lee, C.H.; Lai, T.S. An Intelligent System for Improving Electric Discharge Machining Efficiency Using Artificial Neural Network and Adaptive Control of Debris Removal Operations. *IEEE Access* **2021**, *9*, 75302–75312. Available online: https://ieeexplore.ieee.org/document/9431212 (accessed on 25 June 2022). [CrossRef]
12. Zhang, Z.; Zhang, Y.; Ming, W.Y.; Zhang, Y.M.; Cao, C.; Zhang, G.J. A review on magnetic field assisted electrical discharge machining. *J. Manuf. Process.* **2021**, *64*, 694–722. Available online: https://www.sciencedirect.com/science/article/abs/pii/S1526612521000955?via%3Dihub (accessed on 18 June 2022). [CrossRef]
13. Guo, Y.F.; Ling, Z.B.; Zhang, X.Y.; Feng, Y.R. A magnetic suspension spindle system for small and micro holes EDM. *Int. J. Adv. Manuf. Technol.* **2018**, *94*, 1911–1923. Available online: https://link.springer.com/article/10.1007/s00170-017-0990-x (accessed on 18 June 2022). [CrossRef]
14. Chen, C.H.; Hu, Y.F.; Wu, H.C.; Song, C.S. Parametric design and experiment of maglev actuators for microgravity vibration isolation system. *Int. J. Appl. Electromagn. Mech.* **2018**, *58*, 319–335. Available online: https://content.iospress.com/articles/international-journal-of-applied-electromagnetics-and-mechanics/jae180037 (accessed on 12 July 2022). [CrossRef]
15. Zhang, T.; Le, Q.Y.; Zhu, W.G. Structure and Suspension Force Analysis of Six-Pole Five Degrees of Freedom AC Hybrid Magnetic Bearing. *IEEE Trans. Magn.* **2021**, *57*, 6. Available online: https://ieeexplore.ieee.org/document/9386118 (accessed on 12 July 2022). [CrossRef]
16. Lee, H.R.; Kim, K.C.; Lee, J. Review of Maglev Train Technologies. *IEEE Trans. Magn.* **2006**, *42*, 1917–1925. Available online: https://ieeexplore.ieee.org/document/1644911 (accessed on 12 July 2022).
17. Maximov, S.; Montañez, F.G.; Perez, R.E.; Galvan, J.C.O.; Mestiza, H.A. Analytical Analysis of Magnetic Levitation Systems with Harmonic Voltage Input. *Actuators* **2020**, *9*, 82. [CrossRef]

18. Yaseen, M.H.A. Investigation on planar electromagnetic levitation system using lead compensation and LQR controllers. *Electr. Eng.* **2020**, *102*, 725–736. Available online: https://link.springer.com/article/10.1007/s00202-019-00905-7 (accessed on 12 July 2022). [CrossRef]
19. Zhao, C.; Sun, F.; Jin, J.J.; Tang, J.H.; Xu, F.C.; Li, Q. Analysis of Quasi-Zero Power Characteristic for a Permanent Magnetic Levitation System with a Variable Flux Path Control Mechanism. *IEEE/ASME Trans. Mechatron.* **2021**, *26*, 437–447. Available online: https://ieeexplore.ieee.org/document/9204846 (accessed on 14 July 2022). [CrossRef]
20. Poletkin, K. On the Static Pull-In of Tilting Actuation in Electromagnetically Levitating Hybrid Micro-Actuator: Theory and Experiment. *Actuators* **2021**, *10*, 256. [CrossRef]
21. He, D.J.; Shinshi, T.; Nakai, T. Development of a Lens Driving Maglev Actuator for ep Piercing. *Key Eng. Mater.* **2012**, *523–524*, 774–779. Available online: https://www.scientific.net/KEM.523-524.774 (accessed on 14 July 2022).
22. Zheng, T.; Lu, X.; Xu, F.Q.; Xu, X.Z. Optimisation method of magnetic levitation actuator for rotary table. *IET Electr. Power Appl.* **2020**, *14*, 893–900. Available online: https://ietresearch.onlinelibrary.wiley.com/doi/10.1049/iet-epa.2019.0788 (accessed on 14 July 2022). [CrossRef]
23. Kato, H.; Komori, M.; Asami, K.; Sakai, N. Development of one-axis controlled bearingless motor and its application to a centrifugal pump for extremely low temperature. *Int. J. Appl. Electromagn. Mech.* **2020**, *64*, 1287–1294. Available online: https://content.iospress.com/articles/international-journal-of-applied-electromagnetics-and-mechanics/jae209447 (accessed on 18 July 2022). [CrossRef]
24. Murakami, I.; Zhao, Y.M.; Tashiro, T. Stabilization of a Magnetic Repulsive Levitation Flywheel System Using a High-Efficiency Superconducting Magnetic Bearing. *Actuators* **2022**, *11*, 180. [CrossRef]
25. Arsénio, A.J.; Silva, F.F.; Fernands, J.F.P.; Branco, P.J.C. Optimization of the Guiding Stability of a Horizontal Axis HTS ZFC Radial Levitation Bearing. *Actuators* **2021**, *10*, 311. [CrossRef]
26. Hopf, T.; Richter, M.; Schüßler, B.; Rinderknecht, S. Control Strategies for Highly Gyroscopic Outer Rotors with Diametral Enlargement in Active Magnetic Bearings. *Actuators* **2022**, *11*, 91. [CrossRef]
27. Zhu, H.Y.; Teo, T.J.; Pang, C.K. Magnetically Levitated Parallel Actuated Dual-Stage (Maglev-PAD) System for Six-Axis Precision Positioning. *IEEE/ASME Trans. Mechatron.* **2019**, *24*, 1829–1838. Available online: https://ieeexplore.ieee.org/document/8764447 (accessed on 14 July 2022). [CrossRef]
28. Duan, J.A.; Zhou, H.B.; Guo, N.P. Electromagnetic Design of a Novel Linear Maglev Transportation Platform with Finite-Element Analysis. *IEEE Trans. Magn.* **2011**, *47*, 260–263. Available online: https://ieeexplore.ieee.org/document/5601781 (accessed on 14 July 2022). [CrossRef]
29. Zhou, H.B.; Deng, H.; Duan, J. Hybrid Fuzzy Decoupling Control for a Precision Maglev Motion System. *IEEE/ASME Trans. Mechatron.* **2018**, *23*, 389–401. Available online: https://ieeexplore.ieee.org/document/8100985 (accessed on 20 July 2022). [CrossRef]
30. Ahn, D.; Jin, J.W.; Yun, H.; Jeong, J. Development of a Novel Dual Servo Magnetic Levitation Stage. *Actuators* **2022**, *11*, 147. [CrossRef]
31. Sun, F.; Pei, W.Z.; Zhao, C.; Jin, J.J.; Xu, F.C.; Zhang, X.Y. Permanent Maglev Platform Using a Variable Flux Path Mechanism: Stable Levitation and Motion Control. *IEEE Trans. Magn.* **2022**, *58*, 8300410. Available online: https://ieeexplore.ieee.org/document/9772652 (accessed on 20 July 2022). [CrossRef]
32. Zhou, L.; Wu, J.J. Magnetic Levitation Technology for Precision Motion Systems: A Review and Future Perspectives. *Int. J. Autom. Technol.* **2022**, *16*, 386–402. Available online: https://www.fujipress.jp/ijat/au/ijate001600040386/ (accessed on 22 July 2022). [CrossRef]
33. Poletkin, K.V.; Asadollahbaik, A.; Kampmann, R.; Korvink, J.G. Levitating Micro-Actuators: A Review. *Actuators* **2018**, *7*, 17. [CrossRef]

actuators

MDPI

Article

Energy-Harvesting Characteristics of a Dual-Mode Magnetic Suspension for Vehicles: Analysis and Experimental Verification

Weikang Jiang [1], Yuanyuan Song [2], Yongming Xu [3], Ran Zhou [2,*], Feng Sun [2] and Xiaoyou Zhang [4]

1 Mechanical Systems Engineering Major, Nippon Institute of Technology, Saitama 345-8501, Japan
2 School of Mechanical Engineering, Shenyang University of Technology, Shenyang 110870, China
3 Shenyang Machine Tool Group Company of Limited Liability, Shenyang 110041, China
4 Department of Mechanical Engineering, Nippon Institute of Technology, Saitama 345-8501, Japan
* Correspondence: zhouran@sut.edu.cn

Abstract: The advantages of the proposed dual-mode magnetic suspension are it having a high level of safety and a compact structure compared with the previous studies. However, the structure parameters can affect the energy harvesting of the suspension system. Therefore, this paper aims to analyze the energy-harvesting characteristics of the proposed dual-mode magnetic suspension. Firstly, the structure and working principle of the proposed suspension are introduced. Then, the influences of the various parameters for the actuator on the energy regeneration characteristics are analyzed by the finite element method, such as the magnetic ring, the fixed plug thickness, the heat dissipation, and the air gap, and the actuator parameters are defined by the orthogonal analysis method. Furthermore, the numerical results of the energy harvesting are calculated. Finally, the vibration experimental setup is manufactured, and the simulation analysis is verified by the experiment. The results demonstrate that the excitation amplitude is 3.1 mm, the frequency is 18 Hz, and the maximum induced EMF is 8.8 V. Additionally, compared with passive suspension, the energy harvesting of the proposed suspension has been verified in the laboratory, which lays the foundation for the design and analysis of the dual-mode magnetic suspension in a real vehicle.

Keywords: vehicle suspension; energy-harvesting characteristics; orthogonal analysis; experimental verification

Citation: Jiang, W.; Song, Y.; Xu, Y.; Zhou, R.; Sun, F.; Zhang, X. Energy-Harvesting Characteristics of a Dual-Mode Magnetic Suspension for Vehicles: Analysis and Experimental Verification. *Actuators* **2022**, *11*, 363. https://doi.org/10.3390/act11120363

Academic Editor: Kirill Poletkin

Received: 14 November 2022
Accepted: 1 December 2022
Published: 3 December 2022

1. Introduction

In energy-harvesting suspension, a set of energy regeneration devices is used to replace the traditional damper to alleviate the impact of the road surface and recover the vibration energy of the suspension caused by the road surface excitation during the running of the vehicle. In order to recover and reuse this part of the energy, various forms of energy-harvesting suspensions have been proposed. Since the end of the 1970s, scholars began to theoretically calculate and analyze the feasibility of vibration energy regeneration for vehicle suspension. The current energy-harvesting suspension types mainly include mechanical energy-harvesting suspension [1,2], electro-hydraulic energy harvester [3,4] and electromagnetic energy-harvesting suspension [5,6]. In the energy-harvesting system, the harvested energy from the suspension vibration is transformed into electrical energy to power the electronic equipment, which is a self-powered technology [7,8], and the generated reaction force from the harvester can reduce the vibration of the system.

The energy-harvesting and active control technologies from the vehicle suspension system have been widely investigated, and different energy-harvesting suspension structures have been proposed. Zhang et al. [9] proposed a ball screw-type energy regenerative active suspension, and they established the mathematical model of the energy regenerative actuator. Florean-Aquino et al. [10] investigated semi-active modern control for a quarter-vehicle suspension with a magnetorheological damper. Yang et al. [11] designed a hybrid vehicle suspension system that can regenerate energy from vibrations, and they verified its energy-harvesting performance. The structure of regenerative suspension combined with the linear generator was considered by Kim et al. [12], and the maximum power and the average power were 586.43 W, and 214.98 W, respectively. Arroyo et al. [13] designed and tested a new electromagnetic generator, and the vibration energy harvesting process with a nonlinear energy extraction circuit was optimized. Sapiński et al. [14,15] designed an energy-harvesting linear MR damper that can harvest energy from the excitations, and the performance of the energy-harvesting and damping processes were investigated and analyzed. A compact stiffness controllable magnetorheological damper was proposed and prototyped by Zhu et al. [16], which can achieve self-powered capacity. Firoozy et al. [17] investigated quasiperiodic energy harvesting in a nonlinear vibration-based harvester. A novel hydraulic energy-regenerative shock absorber was presented by Zou et al. [18], which was applied to the vehicle suspension to generate electrical power. An electromagnetic actuator that concurrently realizes two working functions of vibration suppression and energy regeneration was proposed by Wei et al. [19]. Beltran-Carbajal et al. [20] introduced an active vibration control approach from an online estimation perspective of unavailable feedback signals for a quarter-vehicle suspension system. According to the relevant technologies of the maglev rotor [21,22], a magnetic energy-harvesting suspension was presented by Zhou et al. [23], and the energy-harvesting and self-powered characteristics of the suspension were investigated [24]. However, based on the above research, most of the studied energy-harvesting devices have complex conversion mechanisms, which convert the linear motion into the rotary motion, which can make the structure complex and easy to wear down. Moreover, these energy-harvesting devices replace the damper of the traditional passive suspension, which can decrease the driving safety of vehicles when the energy-harvesting device fails, and the devices have the issues of a complex structure and difficulty in maintaining them.

Based on these, this paper proposes a new dual-mode magnetic suspension with energy harvesting and active control; it has the characteristics of a simple structure, it is easy to control, a quick response time, and a high level of safety. The original passive suspension structure is retained, and a new type of direct-driving actuator is designed based on the principle of Lorentz force. The actuator can obtain linear motion without the intermediate conversion mechanism, and it has a fast response speed, uniform thrust, and it is easy to maintain. The content of this paper is organized as follows: Section 2 introduces the structure and working principle of the dual-mode magnetic suspension. Section 3 analyzes the energy-harvesting characteristics for the dual-mode magnetic suspension. In Section 4, the theoretical model of the induced electromotive force (EMF) is established, and the simulation results of the induced EMF are analyzed. In Section 5, the experiments are carried out to validate the energy-harvesting characteristics of the suspension system. The conclusion is presented in Section 6.

2. Structure and Working Principle of the Dual-Mode Magnetic Suspension

2.1. Structure

The structure of the dual-mode magnetic suspension is shown in Figure 1. The dual-mode magnetic suspension mainly consists of an actuator and a passive suspension assembly. The actuator structure is designed based on the voice coil motor structure, and the actuator includes two parts: a stator and a mover; the stator is mainly composed of the fixed plug, the permanent magnet ring, the heat dissipation ring, the soft iron core, and the outer casing. The mover part is mainly a regeneration energy coil. The actuator is installed in the air gap between the spring and the damper rod. The stator is connected to the wheel through the lower cover, and the mover is connected to the vehicle body through the upper cover. The actuator, the damper, and the spring are ensured to have concentricity during installation. The permanent magnet ring is formed by tiling-type permanent magnets, and the magnetization mode of the magnet is radial magnetization. Therefore, the permanent magnet ring, the air gap, the outer casing, and the soft iron core form the closed loop of the magnetic induction lines, as shown in Figure 2. A uniform parallel magnetic field perpendicular to the axial direction of the damper is formed in the air gap between the soft iron core and the outer casing. The permanent magnet ring and the heat dissipation ring are alternately arranged, and the plug is used to fix the permanent magnet rings and the heat dissipation rings.

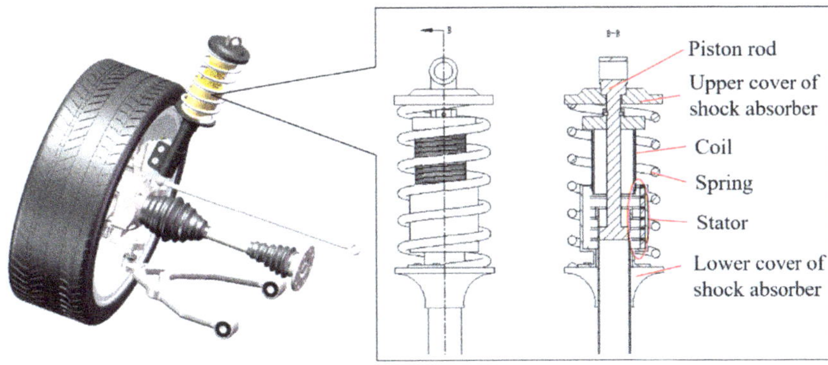

Figure 1. Structure of the dual-mode magnetic suspension.

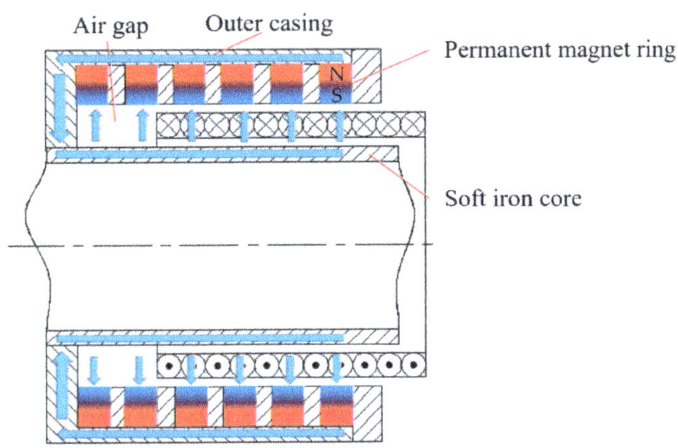

Figure 2. Magnetic circuit in the actuator stator.

2.2. Working Principle

As shown in Figure 3, when the body and the wheel move relative to each other, the relative motion of the regeneration coil and the actuator stator is generated. According to Faraday's law of electromagnetic flux density, the coil can generate positive and negative alternating induced EMFs when it is performing the cutting magnetic induction lines motion. The rectifier circuit rectifies the induced EMF in a single direction. A filter filters out the high-frequency components in the unidirectional EMF and outputs a smoother DC EMF. After the filter process, the voltage stabilizing circuit is added, and the stable EMF is input into the rechargeable battery to realize energy regeneration. In this process, the induced current will generate the damping force that constantly blocks the relative motion of the regeneration coil and the stator of the actuator, and the actuator is equivalent to a damper.

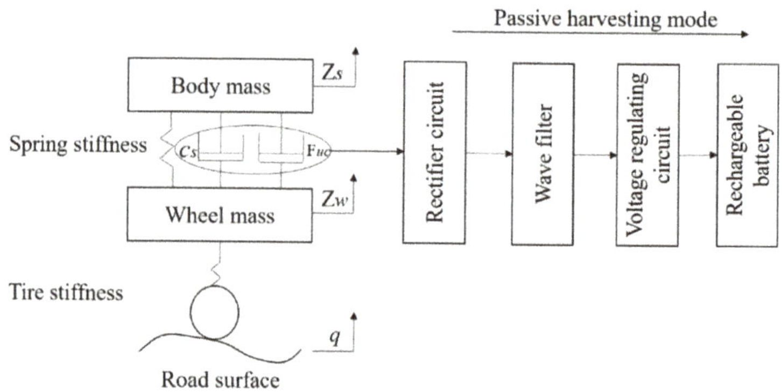

Figure 3. Working principle.

3. Analysis of the Effects of the Stator on the Energy-Harvesting Characteristics

3.1. Effects of the Stator Parameters on the Energy-Harvesting Characteristics

When the structure parameters of the mover are constant, the effects of the stator parameters on the magnetic flux density B in the air gap of the actuator are analyzed by the simulation software.

According to the magnetic flux density model of the stator, the main parameters that affect the magnetic flux density B are the permanent magnet ring's axial height h, the permanent magnet ring's radial thickness l, the fixed plug's thickness d, the heat dissipation ring's thickness s, and the air gap g, as shown in Figure 4. Because many factors affect the magnetic flux density, to turn the problem of multiple factors into multiple single-factor problems, this paper uses the control variable method to analyze the effects of the stator parameters on the magnetic flux density.

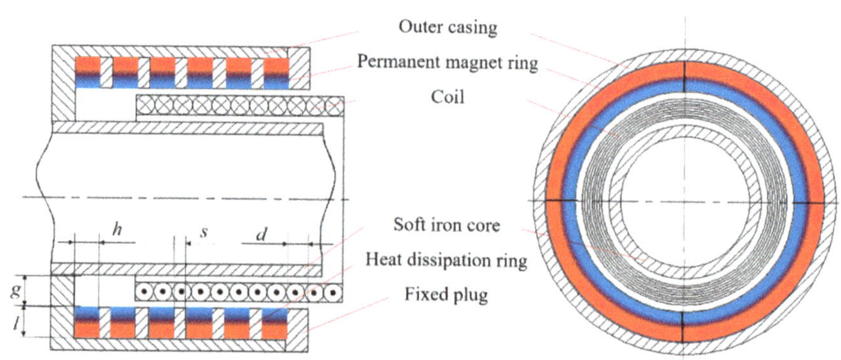

Figure 4. The inner structure of the actuator.

3.1.1. Effect of the Permanent Magnet Ring on the Energy Harvesting Characteristic

There are six permanent magnet rings in the actuator stator, and the magnetic flux density is strongest near the permanent magnet ring, thus, they will produce six peaks. Additionally, the material of the magnetic ring is neodymium iron boron. The investigated range of the magnetic ring radial thickness is from 6 mm to 14 mm, and the increment is 2 mm. Meanwhile, the other conditions are unchanged. The effect of the magnetic ring radial thickness on the magnetic flux density is shown in Figure 5.

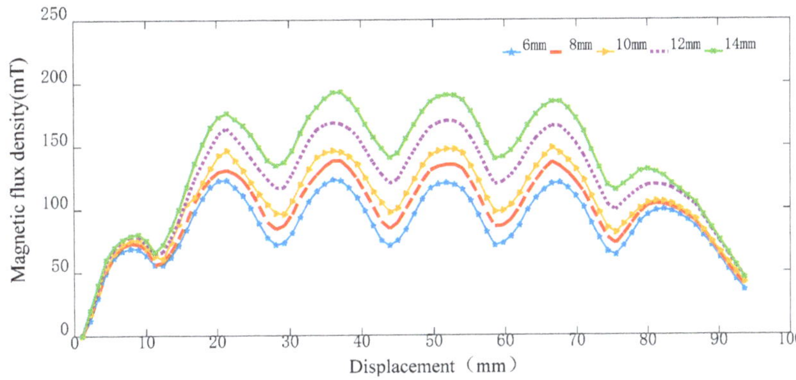

Figure 5. Effect of the magnetic ring radial thickness on the magnetic flux density.

According to the analysis above, the magnetic flux density trend of the five groups of magnetic rings is the same. The permanent magnet ring radial thickness is proportional to the magnetic flux density, and the magnetic flux density of the 14-mm-thick magnetic ring is the largest. The simulation range of the magnetic ring axial height is 8–14 mm, as shown in Figure 6.

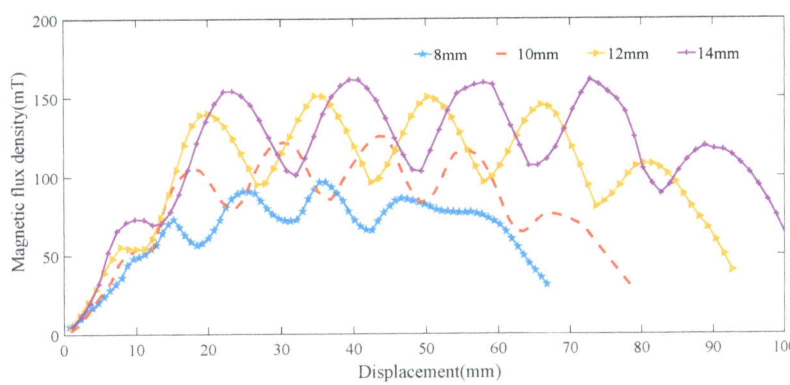

Figure 6. Effect of the magnetic ring axial height on the magnetic flux density.

It can be seen from the figure that as the magnetic ring axial height increases, the axial dimension of the actuator also changes with it. The thicker the magnetic ring is, the greater the corresponding magnetic flux density is. The 14-mm-thick magnetic ring has the maximum magnetic flux density in the air gap, while the 8-mm-thick magnetic ring has the minimum magnetic flux density in the air gap; the magnetic flux density difference between the two is 56 mT. Moreover, the volume is also the largest, and the axial height reaches 100 mm. According to the analysis of the effect of the permanent magnet ring's axial height on the magnetic flux density, the larger the permanent magnet ring volume is, the stronger the magnetic flux density in the air gap is. Furthermore, the position of the magnetic ring in the stator is changed due to the change of the magnetic ring's axial height, thus, the phase of the curve is changed.

3.1.2. Effect of the Fixed Plug's Thickness on the Energy-Harvesting Characteristic

The fixed plug's thickness simulation range is 6–14 mm, and the value interval is 2 mm. The effect of the fixed plug's thickness on the magnetic flux density is shown in Figure 7.

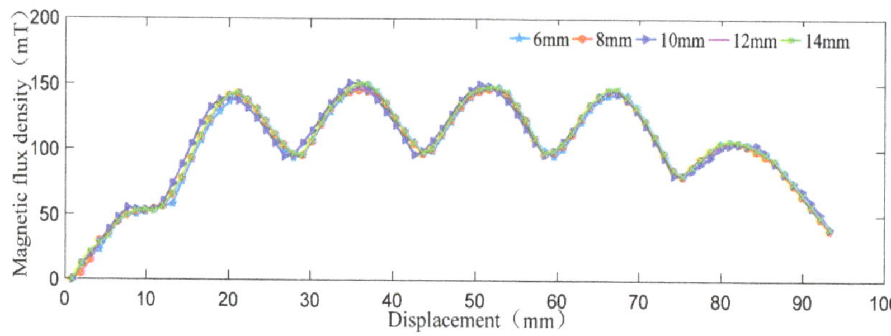

Figure 7. Effect of the fixed plug's thickness on the magnetic flux density.

The fixed plug's thickness is the thickness of the outer casing of the actuator stator, and the fixed plug is an essential part of the design of the actuator. The fixed plug is mainly responsible for the fixation of the permanent magnet ring and heat dissipation ring. It can be seen that when the wall thickness increases, there is no significant change in the magnetic flux density in the air gap, the magnetic flux density of 14 mm is only 0.03% higher than that of the 6 mm one. The outer magnetic fixed plug has more magnetic leakage when it is thinner, which leads to the corresponding weakening of the air gap magnetic field.

3.1.3. Effect of the Heat Dissipation Ring's Thickness on the Energy Harvesting Characteristic

During the movement of the vehicle, the suspension shock absorber generates a lot of heat, and the heat dissipation ring is added between the magnetic rings to slow down the demagnetization phenomenon generated due to the high temperatures. The range of the heat dissipation ring's thickness is 1–4 mm, and the value interval is 1 mm. The simulation analysis is shown in Figure 8.

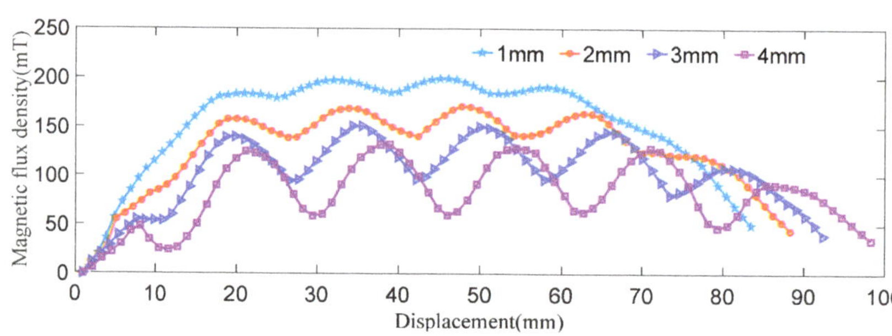

Figure 8. Effect of the heat dissipation ring's thickness on the magnetic flux density.

As can be seen from the analysis of the figure, the heat dissipation ring thickness can produce a more significant effect on the electromagnetic force. The relative magnetic flux density and the magnetic flux density shelf space are increased with the decrease in the height of the heat dissipation ring.

3.1.4. Effect of the Air Gap on the Energy Harvesting Characteristic

The air gap part of the actuator is the moving area of the mover coil, and the range of the investigated air gap size is 8–12 mm, and the interval is 1 mm. The simulation results are shown in Figure 9. It can be seen from the figure that the magnetic flux density shows a decreasing trend with the increase in the air gap.

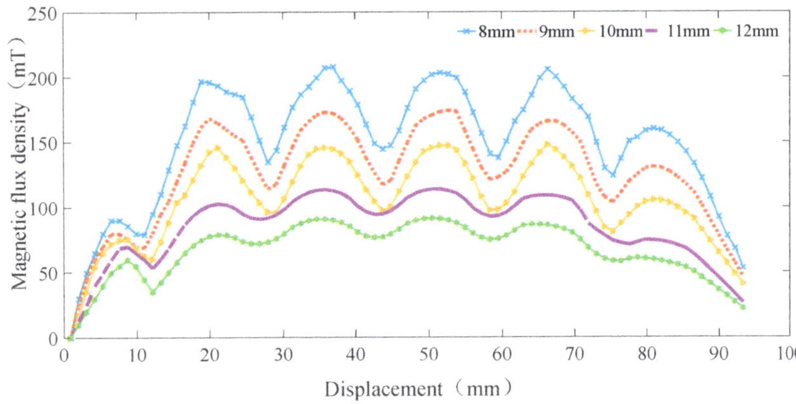

Figure 9. Effect of the air gap on the magnetic flux intensity.

3.2. *Orthogonal Analysis of the Stator Parameters*

3.2.1. Orthogonal Design

The orthogonal analysis has the characteristics of "uniform dispersion, neat and comparable", and it can replace the multi-factor and multi-level test with a large workload with fewer test times, which is highly efficient and economical [25,26]. Based on the previous simulation results, this paper determines the four structural parameters of the permanent magnet ring's radial thickness, the permanent magnet ring's axial height, the heat dissipation ring's thickness, and the air gap for the orthogonal experiment. The parameter ranges are 8–14 mm, 8–14 mm, 1–4 mm, and 9–12 mm, respectively. These parameters are identified in Figure 3. The orthogonal table is designed, as shown in Table 1.

Table 1. Orthogonal design table of level factors.

Level	Magnetic Ring Radial Thickness (mm) A	Magnetic Ring Axial Height (mm) B	Heat Dissipation Ring Thickness (mm) C	Air Gap (mm) D
1	8	8	1	9
2	10	10	2	10
3	12	12	3	11
4	14	14	4	12

3.2.2. Orthogonal Analysis

The orthogonal test scheme and the simulation results of the magnetic flux density are shown in Table 2.

The analysis results of the orthogonal test are shown in Table 3. K_{jk} (k = 1, 2, 3, and 4) is the sum of the simulation results with the same level k in column j. K_{jpk} is the average of the simulation results with the same level k in column j, and R_j is the range. It can be seen that the combination (A4 B3 D2 D1) constitutes the best parameters, and the values of the magnetic ring's thickness, the axial height of the magnetic ring, the height of the heat dissipation ring, the air gap, and the maximum magnetic flux density strength are 14 mm, 12 mm, 3 mm, 10 mm, and 194.4 mT, respectively.

Finally, according to the above analysis, the stator parameters of the actuator are selected, as shown in Table 4.

Table 2. Orthogonal test scheme and simulation results.

Numbering	Magnetic Ring Radial Thickness (mm)	Magnetic Ring Axial Height (mm)	Heat Dissipation Ring Thickness (mm)	Air Gap (N)	Magnetic Flux Density (mT)
1	1	1	1	1	163.625
2	1	2	2	2	165.926
3	1	3	3	3	169.317
4	1	4	4	4	176.305
5	2	1	2	3	178.128
6	2	2	1	4	186.419
7	2	3	3	2	184.526
8	2	4	4	1	188.513
9	3	1	3	4	183.614
10	3	2	4	3	186.325
11	3	1	1	2	188.146
12	3	4	2	1	190.182
13	4	3	4	2	189.379
14	4	2	3	4	192.353
15	4	3	2	1	191.176
16	4	4	1	3	186.479

Table 3. Orthogonal analysis.

Name	Magnetic Ring Radial Thickness (mm) A	Magnetic Ring Axial Height (mm) B	Heat Dissipation Ring Thickness (mm) C	Air Gap (mm) D
K_{j1}	675.173	713.112	724.669	734.373
K_{j2}	747.412	731.023	718.726	737.927
K_{j3}	748.268	737.698	739.760	720.249
K_{j4}	752.736	741.479	740.522	730.864
K_{jp1}	168.775	178.278	181.167	183.668
K_{jp2}	186.853	189.756	179.691	184.482
K_{jp3}	187.067	184.424	184.940	180.062
K_{jp4}	188.184	185.370	185.130	182.716
R_j	19.409	11.478	5.439	3.606
Influencing factors	A > B > C > D			
Best level	A4	B3	C2	D1
Best combination	A4 B3 C2 D1			

Table 4. Stator parameters of the actuator.

Description	Symbol	Numerical Value
Permanent magnet ring length/mm	h	10
Permanent magnet ring height/mm	l	12.5
Cooling ring height/mm	s	3
Outer fixed plug wall thickness/mm	d	10
Air gap length/mm	X_g	90
Air gap width/mm	Y_g	10
NdFeB N42H remanence/T	B_r	1.18
NdFeB N42H coercivity/A/m	H_c	−880,000
Air permeability/H/m	μ_0	$4\pi \times 10^{-7}$

4. Simulation Results of the Dual-Mode Magnetic Suspension

4.1. Theoretical Model of the EMF

During the relative movement between the body and the wheel, the suspension system can generate the relative motion velocity V. Therefore, the induced EMF is expressed as:

$$U_0 = BLV \sin\theta \tag{1}$$

where B is the magnetic flux density produced by the permanent magnet ring, L is the length of wire for cutting magnetic flux density wire, and θ is the angle between the velocity V or L of the damper and the magnetic flux.

$$L = \pi D_c N_1 \tag{2}$$

where N_1 is the number of turns in the coil embedded in the magnetic flux, and D_c is the average diameter of the coil winding.

Combining Formulas (1) and (2) obtains the EMF expression:

$$U_0 = B\pi D_c N_1 V \sin\theta \tag{3}$$

The total turns of the coil are calculated:

$$N_1 = \frac{L_1(d_2 - d_1)}{2d^2} \tag{4}$$

where L_1 is the length of the coil skeleton, d_1 is the inner diameter of the coil, d_2 is the outer diameter of the coil, and d is the diameter of the enameled wire.

In this paper, the sinusoidal signal is used as the road input:

$$X = A\sin\omega t \tag{5}$$

$$\omega = 2\pi f \tag{6}$$

Among these parts, A is the excitation amplitude, ω is the angular velocity, and f is frequency.

Formulas (3) and (7) vary as follows:

$$U_0 = B\pi D_c N_1 V \sin\theta \tag{7}$$

Since the angle θ between the relative motion velocity V of the suspension and the magnetic flux density B is approximately 90°, sin θ = 1, it can be ignored; according to the Formulas (3) and (4), the induced EMF U_0 and the number of turns embedded in the air gap of the coil can be known, and U_0 is proportional to the motion velocity V. Therefore, the regeneration power P is proportional to the square of the number of turns N_1 of the coil embedded in the air gap and the square of the velocity V of motion. When a sinusoidal signal is used as the road surface input, according to the Formula (7), the EMF is related to the three variables of the number of turns, the frequency, and the amplitude.

4.2. Simulation Analysis of the EMF

According to the theoretical model of the dual-mode magnetic suspension, the parameters affecting the energy-harvesting characteristics of the suspension include the vibration frequency, the amplitude, and the turns of the regeneration coil. The EMF is simulated by the control variable method. The simulated coil parameters are shown in Table 5.

Table 5. Parameters of the simulated coil.

Description	Representation Symbol	Numerical Value
Average diameter of the winding	D_c	0.0635 m
Resistivity of copper	ρ	1.75×10^{-8} Ω·m
Coil skeleton length	L	0.09 m
Coil resistance	R	62 Ω
Magnetic flux density	B	0.09 T
Coil turns in the air gap	N_1	1696

4.2.1. Effect of the Number of Coil Turns on the EMF

The turns of the coil entering the stator air gap are positively correlated with the position depth of the coil entering the air gap. The studied depth values of the coil entering the air gap include 50 mm, 60 mm, and 70 mm (the number of coil turns are 188, 1130, and 1319, respectively). The input and output simulation results are shown in Figure 10, and the simulation time is 0.1 s. (From left to right, they are 50 mm, 60 mm, and 70 mm, respectively.)

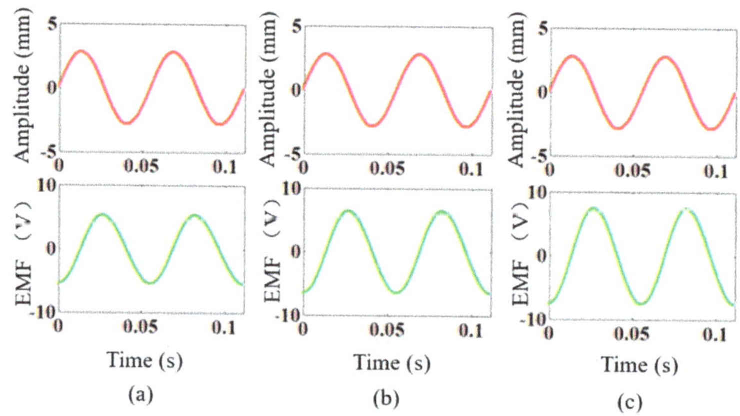

Figure 10. Effect of the number of coil turns on the EMF: (**a**) 188; (**b**) 1130; (**c**) 1319.

The simulation results of the induced EMF are 4.7 V, 6.7 V, and 8.3 V in turn, which are increased by 42.5% and 23.8%, respectively. Thus, the frequency and amplitude remain unchanged, and the coil is embedded deeper into the air gap, which leads to the increase in the induced EMF.

4.2.2. Effect of the Excitation Amplitude on the EMF

The simulation results of the input and output of the coil at 70 mm in the excitation frequency 18 Hz are calculated. The studied excitation amplitudes included 2.5 mm, 2.8 mm, and 3.1 mm, and the simulation time is 0.1 s. (The excitation amplitudes from left to right are 2.5 mm, 2.8 mm, and 3.1 mm, respectively, as shown in Figure 11.)

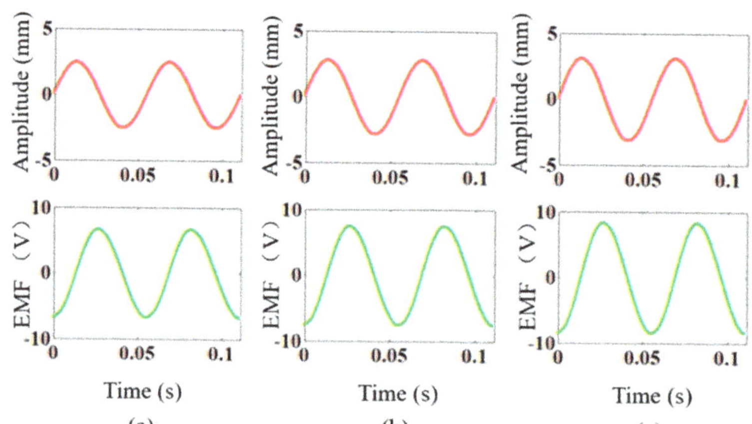

Figure 11. Effect of the excitation amplitude on the EMF: (**a**) 2.5 mm; (**b**) 2.8 mm; (**c**) 3.1 mm.

The simulation results of the induced EMF are 6.4 V, 8.0 V, and 9.2 V, which are increased by 25.0% and 15.1%, respectively. Therefore, the number of turns and frequency remain unchanged, and as the vibration amplitude increases, the induced EMF becomes larger in the sequence.

4.2.3. Effect of the Excitation Frequency on the EMF

When the position of the coil entering the air gap is 70 mm and the excitation amplitude is 2.8 mm, the effect of the excitation frequency on the EMF is analyzed. The input and output simulation results are shown in the excitation frequencies of 15 Hz, 18 Hz, and 21 Hz, and the simulation time is 0.1 s. (The excitation frequencies from left to right are 15 Hz, 18 Hz, and 21 Hz, respectively, as shown in Figure 12.)

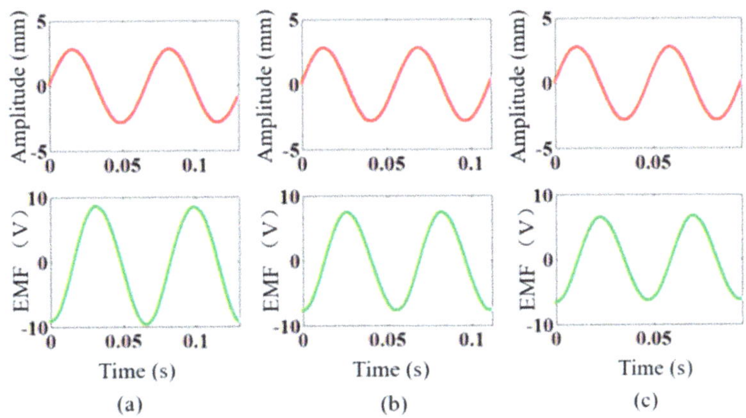

Figure 12. Effect of the excitation frequency on the EMF: (**a**) 15 Hz; (**b**) 18 Hz; (**c**) 21 Hz.

The simulation results of the induced EMF are 7.3 V, 8.4 V, and 7.5 V in sequence, which are first increased by 15.1%, and then, they are decreased by 10.7%. When the turns and amplitude remain unchanged, the EMF is increased more as the frequency is increased.

5. Experiment Verification

5.1. Experimental Setup

In order to verify the correctness of the theoretical model and the simulation analysis for the dual-mode magnetic suspension, a prototype experimental system is built. The actuator is the core component of the dual-mode magnetic suspension, and the sinusoidal signal generated by the exciter is used as the road input to verify the energy-harvesting characteristics of the actuator. The specific signal excitation process is one by which the function signal generator generates a sinusoidal signal through the power amplifier to reach the vibrator. Finally, the vibrator can generate sinusoidal excitation. The Tektronix oscilloscope is used to gather the acceleration and electromotive signals, and the computer receives the synchronization signal through OpenChoice Desktop software (Version: V2.8; Creator: Tektronix; Location: Shenyang, China). The specific experimental setup is shown in Figure 13.

According to the theoretical model, the effects of the turn, the amplitude, and the frequency on the energy-harvesting characteristics are verified. The input is the excitation signal, and the output is the EMF.

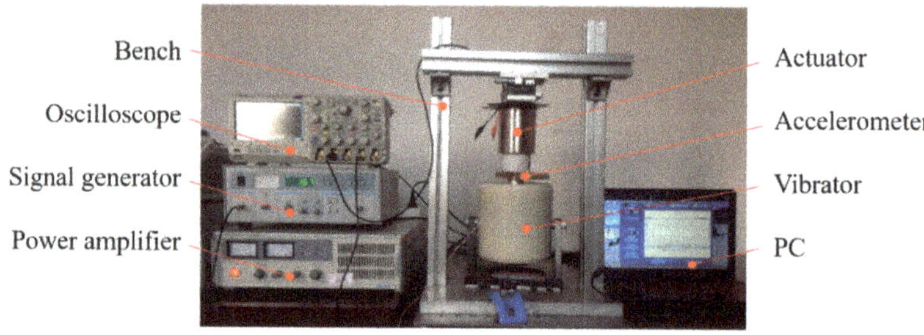

Figure 13. Vibration experimental setup.

5.2. Experimental Results

5.2.1. Experimental Results for the Effect of the Number of Coil Turns on the EMF

Because the turns embedded in the air gap of the actuator can be expressed by the position depth of the coil entering the air gap, the effect of the turns on the energy-harvesting characteristics is verified when the coil is in different positions. When the excitation frequency is 18 Hz, the peak input amplitude is 2.8 mm, and the distance between the coil and the air gap from left to right is 50 mm, 60 mm, and 70 mm (the number of coil turns are 188, 1130, and 1319, respectively); the input and output results are shown in Figure 14.

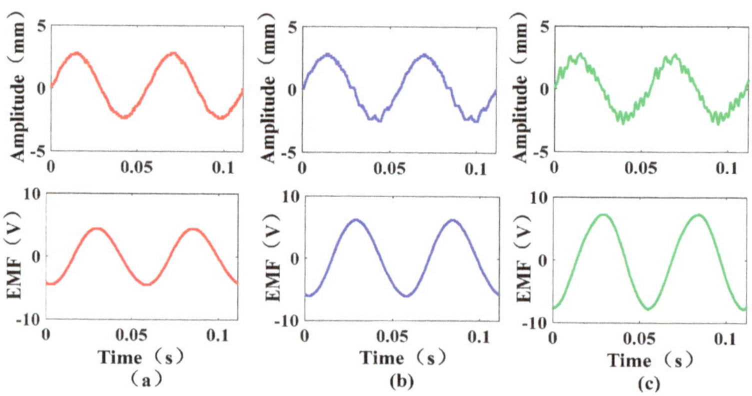

Figure 14. Effect of the number of coil turns on the EMF: (**a**) 188; (**b**) 1130; (**c**) 1319.

According to the experimental results, the peak values of the output EMF are 4.4 V, 6.2 V, and 7.7 V from left to right, which are increased by 40.9% and 24.1%, respectively. Comparing the peaks of the EMF in the three cases, it can be obtained that when the amplitude and frequency are unchanged, the peak value of the induced EMF is greater with the increase in the number of coil turns.

5.2.2. Experimental Results for the Effect of the Excitation Amplitude on the EMF

When the excitation frequency is 18 Hz, the distance between the coil and the air gap is 70 mm, and the input amplitudes from left to right are 2.5 mm, 2.8 mm, and 3.1 mm, respectively; the input and output results are shown in Figure 15.

According to the experimental results, the peak values of the output EMF from left to right are 6.6 V, 7.8 V, and 8.8 V, which are increased by 18.1% and 12.8%, respectively. Comparing the EMF of the different input amplitudes, it can be obtained that when the number of turns and frequency are unchanged, the peak value of the induced EMF is increased as the excitation amplitude is increased.

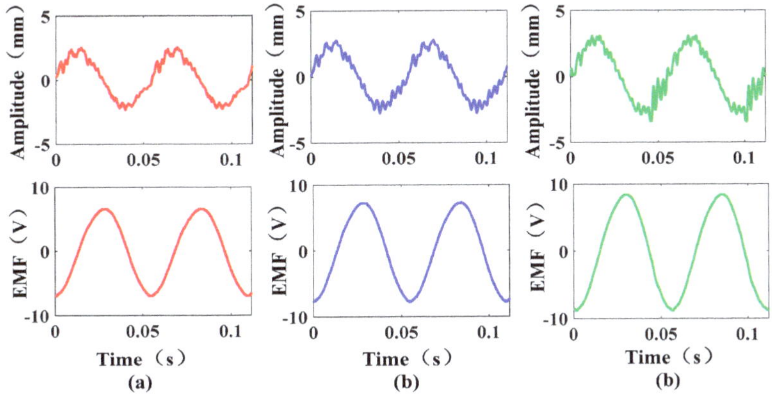

Figure 15. Effect of the excitation amplitude on the EMF: (**a**) 2.5 mm; (**b**) 2.8 mm; (**c**) 3.1 mm.

5.2.3. Experimental Results of the Effect of the Excitation Frequency on the EMF

When the excitation amplitude is 2.8 mm, the distance between the coils and the air gap is 70 mm, and the frequencies from left to right are 15 Hz, 18 Hz, and 21 Hz, respectively; the input and output results are shown in Figure 16.

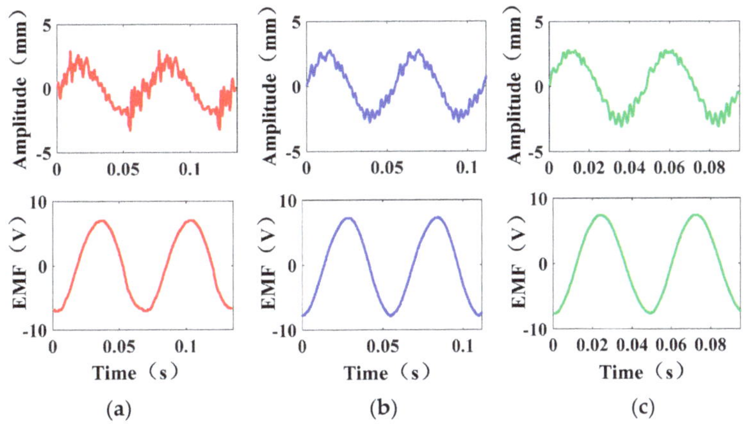

Figure 16. Effect of the excitation frequency on the EMF: (**a**) 15 Hz; (**b**) 18 Hz; (**c**) 21Hz.

According to the experimental results, the peak values of the output EMF are 7 V, 7.8 V, and 7.6 V, respectively. The experimental results show that when the turns and the excitation amplitude remain unchanged with the increase in the frequency, the EMF tends to increase first, and then, they decrease. The main reason is that when the excitation frequency is equal to the resonance frequency of the system, the output EMF is at its maximum.

Comparing the simulation results and the experimental results of the effect of three parameters on the EMF, the simulation results are consistent with the experimental results. In the case where the other variables are consistent, the effect of the number of coil turns and the amplitude on the energy-harvesting characteristics is much greater than the effect of the excitation frequency on the energy-harvesting characteristics. More turns and a greater amplitude can lead to a greater induced EMF.

6. Conclusions

This paper proposes a dual-mode magnetic suspension with energy harvesting and active control. The suspension is stable and safe, and it can complete the conversion of vibration and energy without a complicated conversion mechanism. The optimal structural parameters of the actuator are determined through the finite element simulation analysis, and the experimental prototype is built. Furthermore, the control variable method is used to analyze the effect of the turns of the embedded magnetic field coil and the amplitude and frequency of the excitation signal on the energy-harvesting characteristics of the dual-mode magnetic suspension. Based on the numerical and experimental results, the following conclusions can be obtained:

(1) The number of coil turns have the most significant effect on the energy-harvesting characteristics, with a rate of change up to 40.9%, which is followed by the amplitude of the excitation signal, up to 18.1%, and finally, the frequency of the excitation signal.
(2) With the increase in the number of coil turns and the excitation amplitude, the output-induced EMF of the suspension is increased.
(3) When the excitation frequency is equal to the resonance frequency, the vehicle body resonates with the excitation signal, and the induced EMF is large. Additionally, when the coil enters the air gap of 70 mm (the number of coil turns is 1319), the excitation amplitude is 3.1 mm, and the frequency is 18 Hz, the maximum induced EMF is 8.8 V.

This paper shows the potential good energy regeneration value of this regeneration suspension. In future work, the energy harvesting and dynamic performance of the real vehicle with the dual-mode magnetic suspension will be investigated.

Author Contributions: Conceptualization, W.J. and X.Z.; methodology, W.J., Y.S., F.S. and X.Z.; software, W.J. and Y.X.; validation, W.J., R.Z., F.S. and X.Z.; investigation, W.J. and Y.S.; data curation, X.Z. and Y.X.; writing—original draft preparation, W.J.; writing—review and editing, X.Z., Y.S. and R.Z.; supervision, F.S. and X.Z.; funding acquisition, F.S. and X.Z. All authors have read and agreed to the published version of the manuscript.

Funding: This research was funded by the National Natural Science Fund of China (No. 52005345, No. 52005344), the National Key Research and Development Project (No. 2020YFC2006701), the Scientific Research Fund Project of Liaoning Provincial Department of Education (No. LJGD2020002), and the Major Project of the Ministry of Science and Technology of Liaoning Province (No. 2022JH1/10400027).

Institutional Review Board Statement: Not applicable.

Informed Consent Statement: Not applicable.

Data Availability Statement: Not applicable.

Conflicts of Interest: The authors declare no conflict of interest.

References

1. Long, G.M.; Ding, F.; Zhang, N.; Zhang, J.; Qin, A. Regenerative active suspension system with residual energy for in-wheel motor driven electric vehicle. *Appl. Energy* **2020**, *260*, 114180. [CrossRef]
2. Gu, C.; Yin, J.; Luo, J.; Chen, X.B.; Wang, J.M. Performance-oriented controls of a novel rocker-pushrod electromagnetic active vehicle suspension. *Mech. Syst. Signal Process.* **2018**, *109*, 1–14. [CrossRef]
3. Zou, J.Y.; Guo, X.X.; Abdelkareem, M.A.A.; Xu, L.; Zhang, J. Modelling and ride analysis of a hydraulic interconnected suspension based on the hydraulic energy regenerative shock absorbers. *Mech. Syst. Signal Process.* **2019**, *127*, 345–369. [CrossRef]
4. Zheng, P.; Gao, J.W. Damping force and energy recovery analysis of regenerative hydraulic electric suspension system under road excitation: Modelling and numerical simulation. *Math. Biosci. Eng.* **2019**, *16*, 6298–6318. [CrossRef]
5. Li, S.Y.; Xu, J.; Pu, X.H.; Tao, T.; Gao, H.N.; Mei, X.S. Energy-harvesting variable/constant damping suspension system with motor based electromagnetic damper. *Energy* **2019**, *189*, 116199. [CrossRef]
6. Wei, W.; Sun, F.; Jin, J.Q.; Zhao, Z.Y.; Miao, L.G.; Li, Q.; Zhang, X.Y. Proposal of energy-recycle type active suspension using magnetic force. *Int. J. Appl. Electromagn. Mech.* **2019**, *59*, 577–585. [CrossRef]
7. Khoshnoud, F.; Zhang, Y.C.; Shimura, R.; Shahba, A.; Jin, G.M.; Pissanidis, G.; Chen, Y.K.; De Silva, C.W. Energy regeneration from suspension dynamic modes and self-powered actuation. *IEEE/ASME Trans. Mechatron.* **2015**, *20*, 2513–2524. [CrossRef]
8. Nakano, K.; Suda, Y.; Nakadai, S. Self-powered active vibration control using a single electric actuator. *J. Sound Vib.* **2003**, *260*, 213–235. [CrossRef]
9. Zhang, J.M.; Liu, J.; Liu, B.L.; Li, M. Fractional order PID control based on ball screw energy regenerative active suspension. *Actuators* **2022**, *11*, 189. [CrossRef]

10. Florean-Aquino, K.H.; Arias-Montiel, M.; Linares-Flores, J.; Mendoza-Larios, J.G.; Cabrera-Amado, A. Modern semi-active control schemes for a suspension with MR actuator for vibration attenuation. *Actuators* **2022**, *10*, 22. [CrossRef]
11. Yang, X.F.; Zhao, W.T.; Liu, Y.L.; Chen, L.; Meng, X.P. Design and experimental study of the energy-regenerative circuit of a hybrid vehicle suspension. *Sci. Prog.* **2020**, *103*, 0036850419874999. [CrossRef]
12. Kim, J.H.; Shin, Y.J.; Chun, Y.D.; Kim, J.H. Design of 100W regenerative vehicle suspension to harvest energy from road surfaces. *Int. J. Precis. Eng. Manuf.* **2018**, *19*, 1089–1096. [CrossRef]
13. Arroyo, E.; Badel, A.; Formosa, F. Energy harvesting from ambient vibrations: Electromagnetic device and synchronous extraction circuit. *J. Intell. Mater. Syst. Struct.* **2013**, *24*, 2023–2035. [CrossRef]
14. Sapinski, B. Energy-harvesting linear MR damper: Prototyping and testing. *Smart Mater. Struct.* **2014**, *23*, 035021. [CrossRef]
15. Sapinski, B.; Rosol, M.; Wegrzynowski, M. Investigation of an energy harvesting MR damper in a vibration control system. *Smart Mater. Struct.* **2016**, *25*, 125017. [CrossRef]
16. Zhu, X.; Deng, L.; Sun, S.; Yan, T.; Yu, J.; Ma, Z.; Li, W. Development of a variable stiffness magnetorheological damper with self-powered generation capability. *J. Intell. Mater. Syst. Struct.* **2020**, *31*, 209–219. [CrossRef]
17. Firoozy, P.; Ebrahimi-Nejad, S. Broadband energy harvesting from time-delayed nonlinear oscillations of magnetic levitation. *J. Intell. Mater. Syst. Struct.* **2020**, *31*, 737–755. [CrossRef]
18. Zou, J.Y.; Guo, X.X.; Xu, L.; Tan, G.F.; Zhang, C.C.; Zhang, J. Design, modeling, and analysis of a novel hydraulic energy-regenerative shock absorber for vehicle suspension. *Shock Vib.* **2017**, *2017*, 3186584. [CrossRef]
19. Wei, W.; Li, Q.; Xu, F.; Zhang, X.; Jin, J.; Jin, J.; Sun, F. Research on an electromagnetic actuator for vibration suppression and energy regeneration. *Actuators* **2020**, *9*, 42. [CrossRef]
20. Beltran-Carbajal, F.; Valderrabano-Gonzalez, A.; Favela-Contreras, A.; Hernandez-Avila, J.L.; Lopez-Garcia, I.; Tapia-Olvera, R. An active vehicle suspension control approach with electromagnetic and hydraulic actuators. *Actuators* **2019**, *8*, 35. [CrossRef]
21. Wu, H.C.; Zhou, J.; Xie, C.H.; Zhang, J.Y.; Huang, Y.M. Two-dimensional time series sample entropy algorithm: Applications to rotor axis orbit feature identification. *Mech. Syst. Signal Process.* **2021**, *147*, 107123.
22. Zhou, J.; Wu, H.C.; Wang, W.Y.; Yang, K.Z.; Hu, Y.F.; Guo, X.H.; Song, C.S. Online unbalance compensation of a maglev rotor with two active magnetic bearings based on the LMS algorithm and the influence coefficient method. *Mech. Syst. Signal Process.* **2022**, *166*, 108460.
23. Zhou, R.; Yan, M.; Sun, F.; Jin, J.; Li, Q.; Xu, F.; Zhang, M.; Zhang, X.; Nakano, K. Experimental validations of a magnetic energy-harvesting suspension and its potential application for self-powered sensing. *Energy* **2022**, *239*, 122205. [CrossRef]
24. Zhou, R.; Sun, F.; Yan, M.; Jin, J.; Li, Q.; Xu, F.; Zhang, X.; Nakano, K. Design, analysis and prototyping of a magnetic energy-harvesting suspension for vehicles. *Smart Mater. Struct.* **2020**, *29*, 105034. [CrossRef]
25. Li, Q.; Peng, Z.; Jiang, W.; Ouyang, L.; Wang, H.; Liu, J.; Zhu, M. Optimization of Ti-Zr-Cr-Fe alloys for 45 MPa metal hydride hydrogen compressors using orthogonal analysis. *J. Alloys Compd.* **2021**, *889*, 161629. [CrossRef]
26. Xie, C.; Zhang, T.; Yuan, Z.; Feng, A.; Wu, L. Optimization design and internal flow analysis of prefabricated barrel in centrifugal prefabricated pumping station with double pumps. *Processes* **2022**, *10*, 1877. [CrossRef]

actuators

MDPI

Article

Research on an Intelligent Control Method of a Magnetic Actuator for Micro Electrical Discharge Machining

Chengcheng Xu [1], Fangchao Xu [1,*], Feng Sun [1], Xiaoyou Zhang [1,2], Junjie Jin [1], Boran Luan [1]

1 School of Mechanical Engineering, Shenyang University of Technology, Shenliao West Road, Shenyang 110870, China
2 Department of Mechanical Engineering, Nippon Institute of Technology, 4-1 Gakuendae Miyashiro-cho Minami-Saitama-gun, Saitama 345-8501, Japan
3 School of Systems Engineering, Kochi University of Technology, Kochi 782-8502, Japan
* Correspondence: xufangchao@sut.edu.cn

Abstract: Electrical discharge machining (EDM) is not limited to the strength and hardness of conductive materials, and is a non-contact special processing technology. In micro-EDM, there are problems such as untimely axial positioning, unstable inter-electrode machining voltage, and difficulty in discharging inter-electrode electric erosion products. This paper considers a magnetic actuator with fast response and high accuracy as the local actuator for a micro-EDM. By introducing a domain adjustment mechanism, a variable domain fuzzy PID controller was designed to control the inter-electrode voltage control system of the magnetic actuator for micro-EDM using an intelligent control strategy. During the micro-EDM machining process, the controlled magnetic actuator drives the tool electrode in the axial direction for rapid micro-positioning, thus maintaining effective inter-electrode machining voltage and achieving a high-speed and high precision EDM. Simulation and experimental results showed that compared with traditional micro-EDM, the machining efficiency of the variable domain fuzzy PID control magnetic actuator, and traditional micro-EDM cooperative control, was increased by 40%, the machining process was more stable, and the quality of the machined surface was better.

Keywords: electrical discharge machining (EDM); magnetic actuator; domain adjustment mechanism; variable domain fuzzy PID

Citation: Xu, C.; Xu, F.; Sun, F.; Zhang, X.; Jin, J.; Luan, B.; Oka, K. Research on an Intelligent Control Method of a Magnetic Actuator for Micro Electrical Discharge Machining. *Actuators* **2022**, *11*, 371. https://doi.org/10.3390/act11120371

Academic Editor: Kirill Poletkin

Received: 11 November 2022
Accepted: 7 December 2022
Published: 9 December 2022

1. Introduction

Electrical discharge machining (EDM) is a special machining method that immerses the workpiece in the working medium and conducts electrical erosion on conductive materials through non-contact pulse spark discharge between the tool electrode and the workpiece [1]. In the process of EDM, the two stages of the pulse power supply are connected with the tool electrode and the metal workpiece to provide the required pulse voltage, and the pulse discharge is used to break down the working medium and form a discharge channel. Since the pulse discharge time is in the order of microseconds, the discharge channel is narrow, and the energy at the discharge position is highly concentrated, resulting in instantaneous high temperature, which leads to the melting of metal materials. Different from traditional cutting methods, EDM has no macro cutting force, which is especially convenient for processing complex cavities, and materials with high brittleness, and hardness. EDM has unique processing advantages that cannot be replaced by traditional cutting methods, and has been widely applied in material processing in aviation equipment, electronic instruments, medical engineering and other fields [2–5].

However, the speed and accuracy of traditional EDM are limited by the discharge probability and efficiency. To improve the efficiency of EDM, scholars at home and abroad have conducted a lot of research based on EDM mechanisms. The research to effectively improve the efficiency of EDM includes unconventional electrode designs, electrode rotary or vibratory machining, changing the working medium, improving the response speed of the servo system, and periodic tool electrode lifting. In [6], copper was used as the electrode material and the traditional design was modified by providing relief angles and a land at the electrode face. Primarily, three design types were introduced with eighteen sub-designs by varying the relief angles and the length of the land. Each design type offered a different sparking area. Experimentation was performed in three stages to identify the most appropriate electrode design for the EDM of D2-steel. The results of taper angle, MRR, and TWR were statistically analyzed, and the proposed electrode resulted in an approximately 70% improvement in the MRR, 45–50% reduction in the TWR, and about a 10% decrease in the hole taper angle. The authors of [7–9] used electrode rotation or vibration for EDM machining and evaluated the machining performance of EDM. The experimental results showed that electrode rotation or vibration can effectively improve the machining efficiency of EDM. The authors of [10,11] improved EDM performance by changing the working medium for EDM. The experimental results showed that the working medium greatly influences the material removal rate, electrode wear rate, and surface topography during EDM. The authors of [12] used electrode jumping motion to remove debris from the machining gap by determining the optimal electrode machining time and the optimal electrode jumping height. The experimental results showed that the electrode jumping motion can significantly improve EDM machining efficiency. To obtain a stable and high-speed EDM process, the electrode should have a fast enough positioning response to maintain the inter-electrode voltage's stability and quickly remove the inter-electrode machining debris products. In the traditional EDM machine, a conventional motor with ball screw drive is usually used. This type of drive mechanism has all kinds of intermediate conversion links. There is huge inertia between the mechanical transmission system, which leads to a slow positioning response. Due to the inability to respond to the signal for real-time adjustment of the discharge gap in time, a series of problems, such as unstable discharge between electrodes and difficult discharge of debris products between electrodes, are caused. This is also the main reason for the low efficiency of traditional EDM [13].

To solve the above problems, high-speed, high-precision local actuators and traditional EDM cooperative control have become necessary. As a typical electromechanical coupling system that converts electrical energy into mechanical energy, electromagnetic drive technology has the advantages of high precision, fast response, low noise, and no friction. In precision machining, domestic and foreign scholars have combined the benefits of electromagnetic drive and apply electromagnetic drive technology as a servo drive method in precision machining and manufacturing. The authors of [14] designed a five-degree-of-freedom-controlled, wide-band, and high-precision electromagnetic actuator for driving precision machining instruments to produce a system with high responsiveness, high accuracy, and submicron and micro-radius-level positioning resolution. Reference [15] describes a new linear actuator using electromagnetic suction, which can achieve sub-millimeter travel. The structure of a displacement amplification mechanism applied to extend the stroke of the piezoelectric actuator is also described. The experimental results show that the system has a high response and high accuracy driving performance. The authors of [16,17] proposed a high response, wide band magnetic levitation spindle system (MSSS), which was applied to EDM for micro-hole machining of alloy Inconel 718, and the experimental results showed that the EDM efficiency and accuracy of the system were higher compared to the traditional EDM. Due to the non-linear and time-varying characteristics of the electromagnetic drive, some defects make it difficult to control and be used in industrial transformation. In terms of electromagnetic servo drive control, scholars at home and abroad have also conducted a lot of research. Typical control strategies have included adaptive control [18], fuzzy control [19], sliding mode control [20,21], neural network control [22], fuzzy neural network [23,24], and decoupling control [25].

To meet the needs of high-speed and high-precision EDM, the positioning response speed and processing efficiency of EDM must be improved. This paper introduces a single-degree-of-freedom controllable magnetic local actuator with a compact structure, fast response and high accuracy. The magnetic local actuator was connected with the traditional EDM device to build an inter-electrode voltage control system of the magnetic actuator for micro-EDM. A variable domain fuzzy PID controller was designed to attain intelligence of the control process and improve the performance of the control system. The direct control of the local actuator effectively avoids the effects of errors and hysteresis due to various types of intermediate transformation links, making the magnetic actuator respond in time to drive the electrode for axial micro-positioning and to keep the inter-electrode discharge gap in real-time within the effective discharge range, thereby improving the EDM processing efficiency.

2. 1-DOF Controlled Magnetic Actuator for Micro-EDM

2.1. Magnetic Actuator Structure

The design is shown in Figure 1. A single-degree-of-freedom controllable magnetic actuator is directly connected to the spindle of the EDM machine. The magnetic actuator mainly consists of a coil, spindle, positioning shrapnel, and housing. The coil consists of six copper wire coils connected in parallel. The spindle consists of an electrode and two oppositely placed permanent magnets with a magnetically conductive ring sandwiched between the two oppositely placed permanent magnets, which controls the direction of magnetic flux and reduces flux loss. The spindle is connected to the housing by two positioning shrapnel at the top and bottom, which restrains the radial (X and Y directions) translation and rotation of the spindle and counteracts the gravity of the spindle. The uniform magnetic field inside the coil is generated by two permanent magnet rings placed opposite each other. In such a magnetic field space, a current is applied to the coil of the magnetic actuator, and the resulting electromagnetic force can drive the electrodes to move up and down.

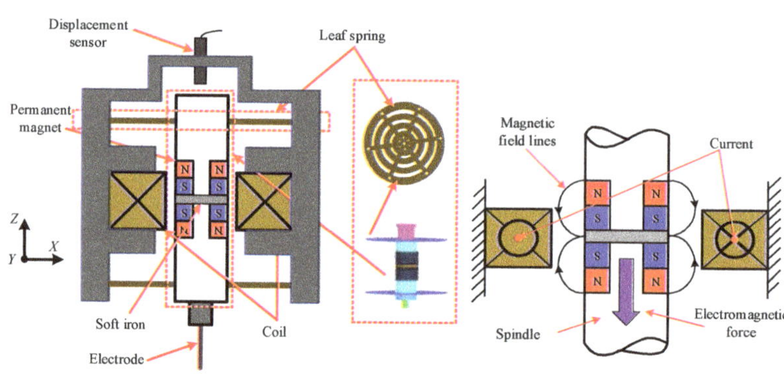

Figure 1. Structure of single-degree-of-freedom controllable magnetic actuator.

In the process of micro-hole EDM, when the inter-electrode voltage is in a broken circuit, the discharge channel between the tool electrode and the workpiece cannot be formed at this time, and EDM cannot be performed. Therefore, the magnetic actuator is required to quickly drive the electrode to move downward slightly, maintain an effective inter-electrode machining voltage, and promote the normal machining of EDM. When the inter-electrode voltage is in a short circuit, the electric erosion products between the tool electrode and the workpiece cannot be discharged in time and the EDM process cannot be carried out normally. Therefore, it is necessary to quickly drive the electrode to move upward slightly by the magnetic actuator to promote the discharge of the electric erosion products and the normal operation of the EDM.

2.2. Dynamic Model of the Magnetic Actuator

Figure 2 shows the dynamics model of the magnetic actuator mover, in which the moving direction of the mover is the state of downward movement.

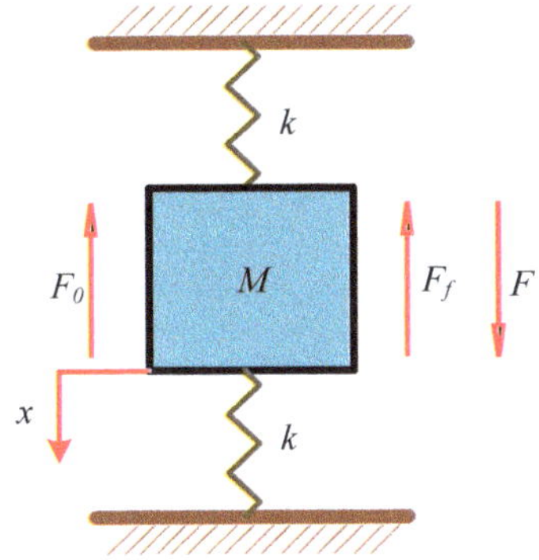

Figure 2. Stress analysis of magnetic actuator mover.

Assume that the downward displacement of the mover of this magnetic actuator in the axial Z direction is x. When ignoring the external disturbance, according to Newton's second law, the axial motion equation of the magnetic actuator mover can be expressed as:

$$Ma = F - F_f - 2kx - cv - F_0 + Mg, \tag{1}$$

$$F - F_0 = F_1, \tag{2}$$

The relationship between the output electromagnetic force of the magnetic actuator and the input current can be expressed as:

$$F_1 = 6K_i i, \tag{3}$$

where F is the combined force, F_1 is the electromagnetic force, F_f is the frictional resistance, F_0 is the discharge force, M is the mass of the mover, a is an acceleration of the motion of the mover, k is the spring constant of the elastic sheet positioning, x is the axial displacement of the mover, c is the friction coefficient, v is the movement speed of the mover, K_i is the stiffness coefficient of the current, and i is the input current of the coil.

Substituting Formula (3) into Formula (1), a mathematical model of the magnetic actuator mover can be obtained:

$$6K_i i = Ma + F_f + 2kx + cv - Mg, \tag{4}$$

The Laplace transform of Formula (4) can be obtained:

$$6K_i I(s) = Ms^2X(s) + 2kX(s) + csX(s), \tag{5}$$

where s is the complex variable.

Then, the transfer function of the mathematical model of the magnetic actuator mover can be obtained as:

$$G(s) = \frac{X(s)}{I(s)} = \frac{6K_i}{Ms^2 + cs + 2k}, \tag{6}$$

Table 1 shows the experimental parameters of the dynamic model of the magnetic actuator mover. Substituting the model parameters into Formula (6), the transfer function between the output displacement $X(s)$ and the input coil current $I(s)$ is:

$$G(s) = \frac{X(s)}{I(s)} = \frac{33.6}{0.27s^2 + 10850}, \tag{7}$$

Table 1. Dynamic model parameters.

System Parameters	Symbol	Value
mover mass	M (kg)	0.27
friction coefficient	c (Ns/m)	0
spring constant	k (N/m)	5.425×10^3
current stiffness coefficient	K_i (N/A)	5.6
frictional resistance	F_f (N)	0

3. Magnetic Actuator Inter-Electrode Voltage Control System

Figure 3 shows the block diagram of the inter-electrode voltage control system of the magnetic actuator for micro-EDM, which is a voltage closed-loop control. The inter-electrode voltage signal is detected by the average voltage detection device and used as the feedback signal of the control system. The feedback voltage between the inter-electrode and the given reference voltage is sent to the controller for processing. The output voltage signal of the controller obtains the control current on the coil of the magnetic actuator through the power amplifier and completes the movement of the electrodes of the magnetic actuator. Through the closed-loop control system, the inter-electrode voltage can be controlled near the ideal machining voltage, which increases the probability of normal discharge, thereby effectively improving the efficiency of micro-EDM.

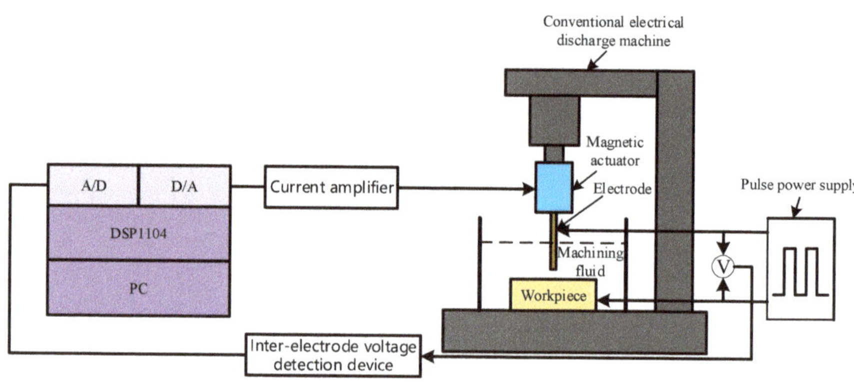

Figure 3. Block diagram of the inter-electrode voltage control system of the magnetic actuator for micro-EDM.

3.1. Conventional PID Controller

Figure 4 shows the structure diagram of the conventional PID controller. In Figure 4, $V_r(t)$ is the target value, which is the desired value of the inter-electrode voltage of the micro-EDM. $V_s(t)$ is the feedback value, which is the current value of the inter-electrode voltage of the micro-EDM. $e(t)$ is the input signal of the PID controller, which is the difference between the feedback value of the inter-electrode voltage of the EDM and the desired value of the inter-electrode voltage. K_p is the proportional adjustment coefficient. K_i is the integral adjustment coefficient. K_d is the differential adjustment coefficient. $u(t)$ is the output signal of the PID controller. $I(t)$ is the value of the coil control current of the magnetic actuator.

The proportional adjustment coefficient K_p can improve the system's control accuracy and response speed. The integration adjustment coefficient K_i can eliminate the steady-state error of the system and improve the control accuracy. The differential adjustment coefficient K_d effectively eliminates oscillations in the regulation process, which affect the steady-state performance and response speed of the control system, and predicts the dynamics of the variation of the input deviation value of the controller. The mathematical model can be described as follows:

$$u(t) = K_p e(t) + K_i \int_0^t e(t)dt + K_d \frac{de(t)}{dt}, \tag{8}$$

The control variables K_p, K_i, and K_d were debugged based on engineering experience. Based on this, the controller of this micro-EDM control system was further designed.

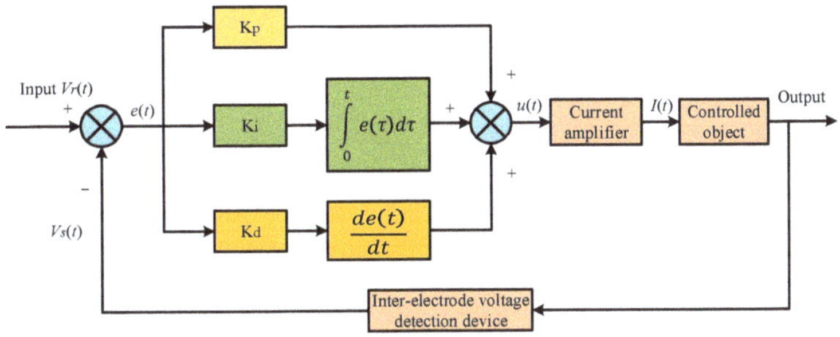

Figure 4. Structure diagram of the conventional PID controller.

3.2. Fuzzy PID Controller

In the control process, the parameters of conventional PID control cannot be modified once they are set. The performance of the controller depends on the three parameters of PID. Generally speaking, the effect of conventional PID control depends on an accurate mathematical model of the controlled object, while the magnetic actuator is a nonlinear system, so an accurate mathematical model of this EDM system is difficult to establish. Fuzzy control does not depend on the precise mathematical model of the controlled object, and its corresponding control rules are established based on the characteristics of the controlled system, expert knowledge, and engineering experience, which can realize the online adjustment of control parameters within a certain range and have better applicability. Therefore, we combine fuzzy control and PID control to design a fuzzy PID controller to realize their complementary advantages.

The fuzzy controller consists of four main modules: the fuzzification module, fuzzy reasoning module, knowledge base module and the defuzzification module [26,27]. Figure 5 shows the schematic diagram of the inter-electrode voltage control system of the fuzzy PID control magnetic actuator for micro-EDM.

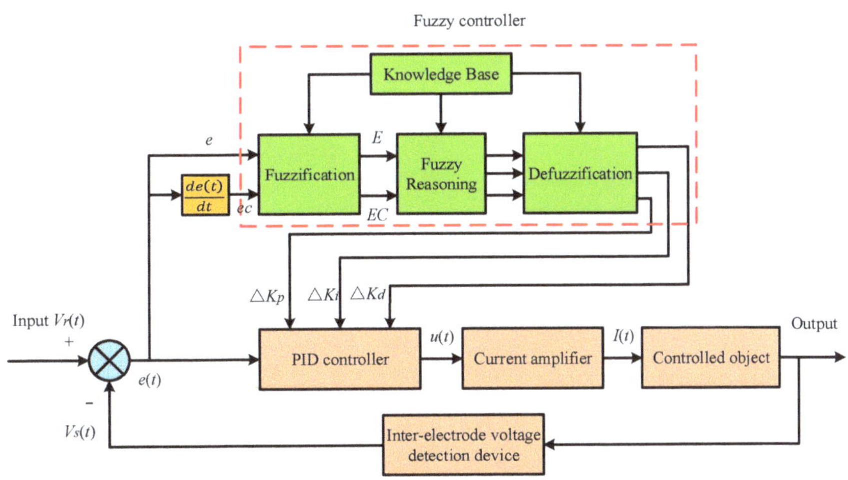

Figure 5. Schematic diagram of the inter-electrode voltage control system of the fuzzy PID control magnetic actuator for micro-EDM.

First, the conventional PID control parameters are adjusted based on engineering experience. Then, based on this, the fuzzy controller corrects the control parameters of the conventional PID online in real-time based on the error value and the rate of change of the error value during the regulation process. Finally, the fuzzy PID controller determines the proportional adjustment coefficient K_p, the integral adjustment coefficient K_i, and the differential adjustment coefficient K_d. The calculation formulas are as follows:

$$\begin{cases} K_p = K_{p0} + \Delta K_p \\ K_i = K_{i0} + \Delta K_i \ , \\ K_d = K_{d0} + \Delta K_d \end{cases} \tag{9}$$

where K_{p0} is the initial proportional coefficient, K_{i0} is the initial integral coefficient, K_{d0} is the initial differential coefficient, ΔK_p is the proportional coefficient correction amount, ΔK_i is the integral coefficient correction amount, ΔK_d is the differential coefficient correction amount.

Selection of the fuzzy domain needs to be combined with the actual control situation, and the key point is the fuzzy partitioning of the input variables and control variables. Since the variables have different ranges of values, the basic domains of each variable are first mapped to a standardized domain with different correspondences, respectively. Usually, the correspondence is taken as a quantization factor. Then, the domain is fuzzily divided and fuzzy subsets are defined. Fuzzy control rules are formulated based on expert knowledge and engineering experience. In this paper, the two input variables of the fuzzy control system are the inter-electrode voltage error *e* and its rate of change *ec*, and the fuzzy domain *E* and *EC* of the input variables are both set to $[-6, 6]$. The three output variables of the fuzzy control system are ΔK_p, ΔK_i, and ΔK_d, and the fuzzy domain of the output variables are set to $(-3, 3)$. The corresponding fuzzy language subsets of the input and output variables of the fuzzy control system are all (negative large (NB), negative medium (NM), negative small (NS), zero (ZO), positive small (PS), positive medium (PM), positive large (PB)). The quantization factors of the input variables of the fuzzy control system are $K_{e0} = 6$, $K_{ec0} = 0.3$. The scaling factors of the output variables of the fuzzy control system are $E_{p0} = 0.002$, $E_{i0} = 0.35$, $E_{d0} = 0.0002$. The affiliation function is trigonometric, the fuzzy reasoning is the Mamdani method, and the defuzzification is the centroid method. The fuzzy rule table of the fuzzy PID control system is shown in Table 2.

Table 2. Fuzzy control rules table of ΔK_p, ΔK_i, ΔK_d.

		EC						
		NB	**NM**	**NS**	**ZO**	**PS**	**PM**	**PB**
E	NB	PB/NB/NS	PB/NB/NS	PM/NM/NB	PM/NM/NB	PS/NS/NB	ZO/ZO/NM	ZO/ZO/PS
	NM	PB/NB/PS	PB/NB/NS	PM/NM/NB	PS/NS/NM	PS/NS/NM	ZO/ZO/NS	NS/PS/ZO
	NS	PM/NB/ZO	PM/NM/NS	PM/NS/NM	PS/NS/NM	ZO/ZO/NS	NS/PM/NS	NS/PM/ZO
	ZO	PM/NM/ZO	PM/NM/NS	PS/NS/NS	ZO/ZO/NS	NS/PS/NS	NM/PM/NS	NM/PM/ZO
	PS	PS/NM/ZO	PS/NS/ZO	ZO/ZO/ZO	NS/PS/ZO	NS/PS/ZO	NM/PM/ZO	NM/PB/ZO
	PM	PS/ZO/PB	ZO/ZO/PM	NS/PS/PM	NM/PS/PM	NM/PM/PS	NM/PB/PS	NB/PB/PB
	PB	ZO/ZO/PB	ZO/ZO/PM	NM/PS/PM	NM/PM/PM	NM/PM/PS	NB/PB/PS	NB/PB/PB

Fuzzy PID control also has limitations. Due to the structure of the fuzzy controller itself, the determination of the domain of fuzzy control in the control process is limited by expert experience and cannot be changed once it is set, so the adaptive ability of the fuzzy controller is limited. If the working conditions of the control system are changed, the control system will not be effectively controlled according to the original parameter settings.

4. Design of the Domain Adjustment Mechanism

The idea of a variable domain overcomes the above limitations. Based on fuzzy PID control, a domain adjustment mechanism is introduced, which expands the fuzzy domain as the error increases and contracts the fuzzy domain as the error decreases while keeping the number of fuzzy subsets and fuzzy rules unchanged. When the error is small, adding appropriate number of fuzzy rules indirectly makes the fuzzy control rules more suitable for the control system and more effective in the whole control process. This solves the problem of low control accuracy in fuzzy control due to fewer control rules near the zero point.

4.1. Design of Variable Domain Contraction-Expansion Factor

The adjustment rule of the variable domain relates to the increase and decrease of the quantization factor being equivalent to the expansion and contraction of the input variable domain, and the increase and decrease of the scaling factor being equivalent to the expansion and contraction of the output variable domain. The domain contraction-expansion factor is the key to the variable domain fuzzy controller design, and plays a crucial role in the control performance. In this paper, the contraction-expansion factor is selected based on fuzzy control. In the domain adjustment mechanism, we use two fuzzy controllers for fuzzy reasoning and obtains the contraction-expansion factor of the adjustment quantization factor and the scale factor, respectively. In this way, the expansion and contraction of the fuzzy domain of input and output of the control system can be realized.

Let the input variables in the fuzzy PID controller be the error e, its rate of change ec, and the output variables are ΔK_p, ΔK_i, and ΔK_d. Then, the domain adjustment mechanism can be divided into the input domain adjustment mechanism and the output domain adjustment mechanism, where the contraction-expansion factors of the input variable fuzzy domain are α_e and α_{ec}, and the contraction-expansion factors of the output variable fuzzy domain are β_p, β_i, and β_d, respectively.

After the contraction-expansion factor adjusts the initial fuzzy domain, the formulas for the two quantization factors after control system regulation are as follows:

$$\begin{cases} K_e = \alpha_e K_{e0} \\ K_{ec} = \alpha_{ec} K_{ec0} \end{cases}, \tag{10}$$

where K_{e0} and K_{ec0} are the initial quantization factors, and K_e and K_{ec} are the adjusted quantization factors.

The formulas for the three scaling factors after control system regulation are as follows:

$$\begin{cases} E_p = \beta_p E_{p0} \\ E_i = \beta_i E_{i0} \\ E_d = \beta_d E_{d0} \end{cases}, \tag{11}$$

where E_{p0}, E_{i0}, and E_{d0} are the initial scaling factors, and E_p, E_i, E_d are the adjusted scaling factors.

The final correction values for the three parameters of the variable domain fuzzy PID control are as follows:

$$\begin{cases} K_p = K_{p0} + \beta_p \Delta K_p \\ K_i = K_{i0} + \beta_i \Delta K_i \\ K_d = K_{d0} + \beta_d \Delta K_d \end{cases}, \tag{12}$$

where K_{p0}, K_{i0}, and K_{d0} are the initial values of proportional, integral, and differential coefficients of conventional PID control, respectively, and K_p, K_i, and K_d are the final corrected values of proportional, integral, and differential coefficients of fuzzy PID control, respectively.

4.2. Rule Design of Variable Domain Regulator

In the domain adjustment mechanism, the two input variables of the fuzzy control system are the inter-electrode voltage error e and its rate of change ec, and the fuzzy domains E and EC of the input variables are both set to (−6, 6), and the fuzzy subset linguistic values of the fuzzy control system input variables are expressed as (negative large (NB), negative medium (NM), negative small (NS), zero (ZO), positive small (PS), positive medium (PM), and positive large (PB)). The contraction-expansion factors to adjust the contraction-expansion change of the input domain are α_e and α_{ec}, respectively. Then, the fuzzy domains α_E and α_{EC} are both (0, 1.5), and their fuzzy subset linguistic values are all expressed as (medium compression (M), light compression (S), keep constant (Z), light expansion (B)). The contraction-expansion factors that adjust the output domain contraction-expansion change are β_p, β_i, and β_d, respectively, then the fuzzy domains β_P, β_I, and β_D are all (0, 2), and their fuzzy subset linguistic values are all expressed as (large compression (CB), medium compression (CM), small compression (CS), keep constant (Z), small expansion (AS), medium expansion (AM), large expansion (AB)). The variable domain fuzzy PID controller is established using MATLAB/Simulink, and the membership function can be selected in the fuzzy control module. Since the triangular membership function has higher sensitivity, it is chosen to use the triangular membership function, as shown in Figures 6 and 7, for the membership functions of the input domain contraction-expansion factors and the output domain contraction-expansion factors, respectively.

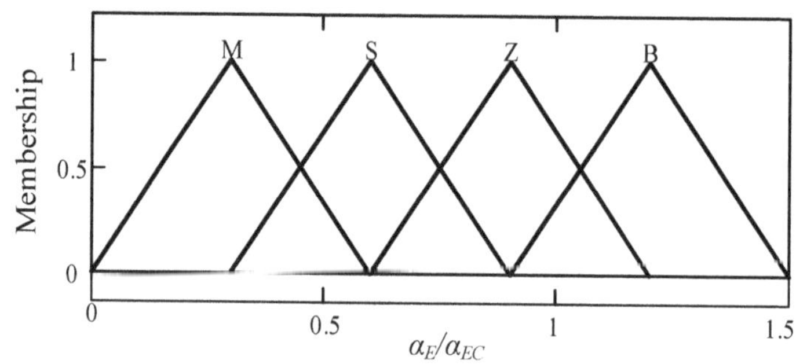

Figure 6. Membership functions of α_E and α_{EC}.

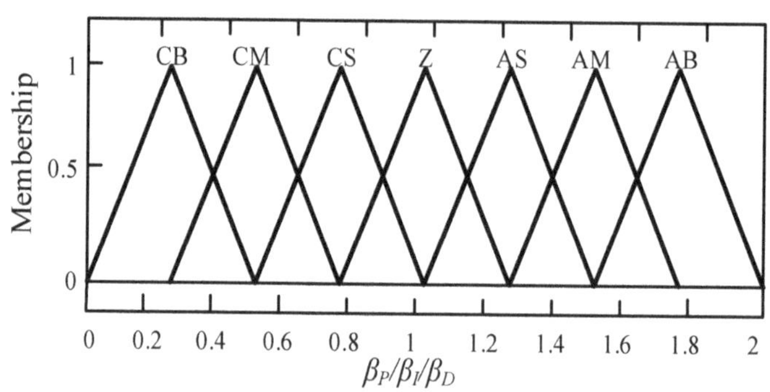

Figure 7. Membership functions of β_P, β_I, and β_D.

Like the fuzzy control in fuzzy PID control, the membership function is a triangular membership function, the fuzzy reasoning is the Mandani method, and the defuzzification is the centroid method. The fuzzy relationship between input and output variables can be determined based on expert knowledge, engineering experience, and the variation relationship between input and output variables. As shown in Tables 3 and 4, the fuzzy rule table of the contraction-expansion factor regulates the contraction-expansion change of the input domain, and the fuzzy rule table of the contraction-expansion factor regulates the contraction-expansion change of the output domain are shown respectively.

Table 3. Fuzzy control rules table of α_e, α_{ec}.

					E_{ec}			
		NB	NM	NS	ZO	PS	PM	PB
	NB	B/B	B/B	M/M	S/S	M/M	B/B	B/B
	NM	B/B	B/M	M/S	S/S	S/M	M/M	B/B
	NS	M/M	M/S	S/S	Z/Z	S/S	M/M	M/M
E_e	ZO	M/M	S/S	Z/Z	Z/Z	Z/Z	S/S	M/M
	PS	M/M	M/M	S/S	Z/Z	S/S	M/M	M/M
	PM	B/B	M/M	M/M	S/S	M/M	M/M	B/B
	PB	B/B	B/M	M/S	S/S	M/M	B/B	B/B

Table 4. Fuzzy control rules table of β_p, β_i, and β_d.

		E_{ec} NB	NM	NS	ZO	PS	PM	PB
E_e	NB	AB/AB/AS	AM/AB/CS	AS/AM/CB	Z/AM/CB	AS/AS/CB	AM/Z/CM	AB/Z/AS
	NM	AM/AB/Z	AS/AB/CS	Z/AM/CB	CS/AS/CM	Z/AS/CM	AS/Z/CS	AM/CS/Z
	NS	AS/AM/Z	Z/AM/CS	CS/AM/CM	CM/AS/CM	CS/Z/CS	Z/CS/CS	AS/CS/Z
	ZO	Z/AM/Z	CS/AS/CS	CM/AS/CS	CM/Z/CS	CM/CS/CS	CS/CM/CS	Z/CM/Z
	PS	AS/AS/Z	Z/AS/Z	CS/Z/Z	CM/CS/Z	CS/CS/Z	Z/CM/Z	AS/CM/Z
	PM	AM/Z/AB	AS/Z/AS	Z/CS/AS	CS/CM/AS	Z/CM/AS	AS/CM/AS	AM/CB/AB
	PB	AB/Z/AB	AM/CS/AM	AS/CS/AM	Z/CM/AM	AS/CM/AS	AM/CB/AS	AB/CB/AB

5. Simulation Analysis of Magnetic Actuator Control System for Micro-EDM

5.1. Variable Domain Fuzzy PID Control System

Figure 8 shows the inter-electrode voltage control system of the variable domain fuzzy PID control magnetic actuator. With the variable domain fuzzy PID controller as the core, the simulation model of the inter-electrode voltage control system of the magnetic actuator for micro-EDM was established in the MATLAB/Simulink module, and the performance and effect of the variable domain fuzzy PID controller were tested.

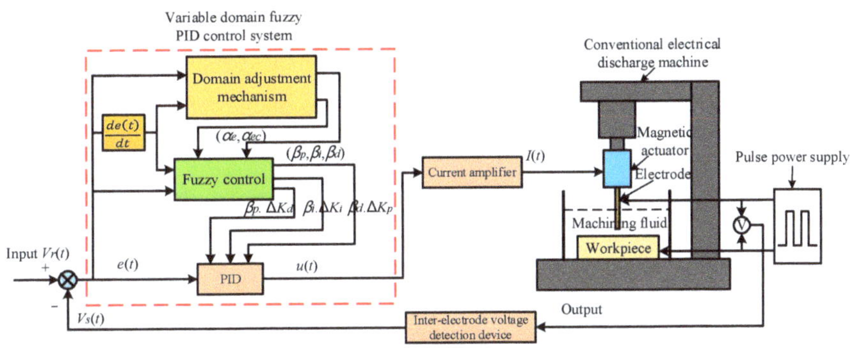

Figure 8. Magnetic actuator inter-electrode voltage control system with variable domain fuzzy PID.

Figure 9 shows the simulation model of the domain regulation mechanism for the variable domain fuzzy PID control, which mainly consists of the input domain contraction-expansion factor (a) and the output domain contraction-expansion factor (b).

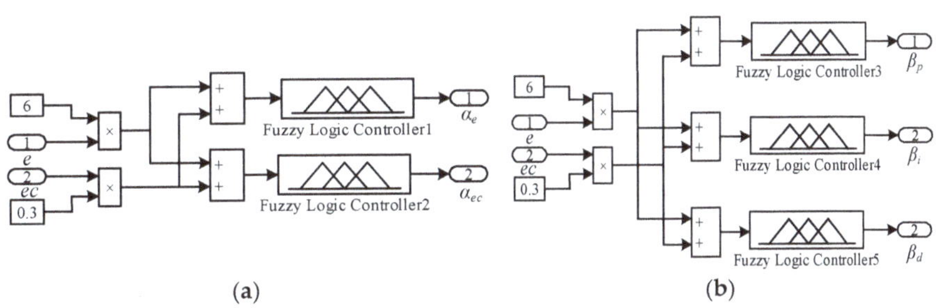

Figure 9. Domain adjustment mechanism. (**a**) Input domain contraction-expansion factor; (**b**) output domain contraction-expansion factor.

5.2. Simulation Results and Analysis

The control system was simulated and analyzed in MATLAB/Simulink module with the following conditions: inter-electrode reference voltage value $V_r(t)$ = 1.50 V, the actual inter-electrode voltage = 46.5 V, the current amplifier scale factor K_w = 0.77, the average inter-electrode voltage detection factor K_u = 20000/31, and the PID parameters adjusted to K_{p0} = 0.15, K_{i0} = 0.0002, K_{d0} = 0.002. Figure 10 shows the simulation curve of inter-electrode voltage control of the magnetic actuator for the micro-EDM.

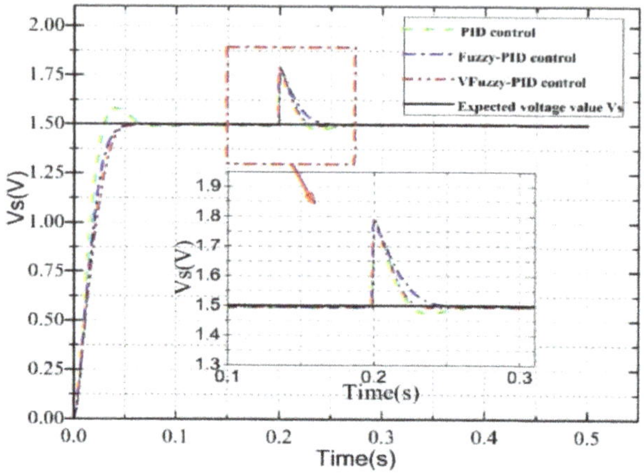

Figure 10. Simulation curve of inter-electrode voltage control of the magnetic actuator for micro-EDM.

Table 5 shows the comparison of various control performance indexes of conventional PID control, fuzzy PID control, and variable domain fuzzy PID control.

Table 5. Comparison of simulation parameters.

Control Method	Adjustment Time (s)	Overshoot (%)
conventional PID	0.07	5.84
fuzzy PID	0.055	0
variable domain fuzzy PID	0.04	0

The simulation comparison analysis showed that increasing the inter-electrode voltage value by 0.3 V at 0.2 s for this control system was equivalent to increasing the inter-electrode voltage value by 9 V in the actual EDM process, which was used to simulate the inter-electrode voltage variation in the non-normal EDM process. Conventional PID control has overshoot, whereas fuzzy PID, and variable domain fuzzy PID control has a strong anti-interference ability and steady-state performance. During abnormal processing, the variable domain fuzzy PID control regulation speed is faster.

6. Experimental Research

To test the actual processing effect of the magnetic actuator control system for micro-EDM, machining experiments were conducted on the micro-EDM machine tool. The experimental study was conducted for four cases: traditional micro-EDM, conventional PID control magnetic actuator and traditional micro-EDM cooperative control, fuzzy PID control magnetic actuator and traditional micro-EDM cooperative control, and variable domain fuzzy PID control magnetic actuator and traditional micro-EDM cooperative control.

6.1. Build an Experimental Platform

Figure 11 shows the micro EDM experimental system, which is mainly composed of a micro EDM machine, magnetic actuator, dSPACE1104, computer, current power amplifier, switching DC power supply, and an inter-electrode voltage detection circuit. This low-pass filter with a cut-off frequency of 300 Hz is added after the inter-electrode voltage detection circuit to filter out the high frequency signal, and the high frequency, high voltage signal is filtered by the detection circuit to obtain a safe voltage signal after bucking, which is input to the controller. The basis for using copper as the electrode material for EDM is that copper has good electrical conductivity, thermal conductivity and high melting point, and copper electrodes are more suitable for micro-EDM.

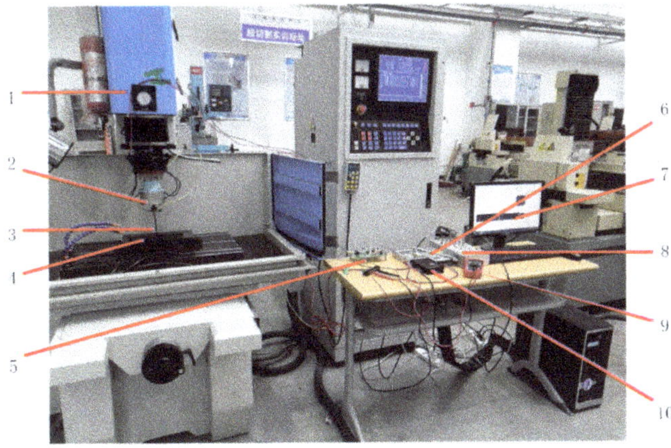

Figure 11. Experimental system of micro-EDM. 1: Micro electrical discharge machining. 2: Magnetic actuator. 3: Tool electrode. 4: Machining workpiece. 5: Discharge gap voltage detection circuit. 6: Switching DC power supply. 7: Computer. 8: dSPACE1104. 9: Multimeter. 10: Current power amplifier.

Table 6 shows the parameter settings of the micro-EDM machine tool.

Table 6. Parameters of micro-EDM machine tool.

Voltage	Current	Pulse Rest	Pulse Discharge	Polarity	Electrode	Workpiece	Working Medium
90 V	3 A	16 μs	90 μs	Positive	copper	45 steel	Coal oil

6.2. Experimental Process

In micro-EDM, the gap voltage signal is detected and processed by the average voltage detection circuit. The detected inter-electrode voltage signal is the feedback signal $V_s(t)$. The reference voltage $V_r(t)$ is set to 1.5 V, which corresponds to the gap voltage between the electrode and the workpiece of 46.5 V in the actual machining. The inter-electrode voltage signal detected by the average voltage detection circuit is used as the feedback signal for the controller. The feedback signal is A/D converted by the dSPACE1104 input port, and the feedback voltage $V_s(t)$ is compared with the reference voltage $V_r(t)$, and the inter-electrode voltage error *e* and its rate of change *ec* are used as the input signal for the controller. The voltage signal $u_r(t)$ output from the controller is D/A converted by the dSPACE1104 output port, and the voltage signal $u_r(t)$ is converted by the power amplifier into a current signal $I_r(t)$ that is fed into the coil of the magnetic actuator, which drives the electrode in the axial direction for fast micro-positioning so that the micro-EDM is always kept within the effective discharge range.

To verify that compared with the traditional micro-EDM, conventional PID control magnetic actuator and traditional micro-EDM cooperative control and fuzzy PID control magnetic actuator and traditional micro-EDM cooperative control, the variable domain fuzzy PID control magnetic actuator and traditional micro-EDM cooperative control has the advantages of good robustness, high adaptive ability, and anti-interference ability, which can significantly improve the dynamic and static performance of the control system.

6.3. Experimental Results and Analysis

6.3.1. Comparison of Discharge Status

To monitor the discharge status in real-time during EDM of micro small holes, the inter-electrode voltage detection circuit detects the feedback voltage $V_s(t)$, which is then used for data acquisition by a digital signal processor (dSPACE1104). The feedback voltage is 0 V when the short-circuit is discharged. When the discharge state is a broken- circuit, the feedback voltage is 3 V, and the actual inter-electrode voltage is 90 V. And the effective discharge range is within the range of 20% (0.6 V) to 80% (2.4 V) of the broken-circuit inter-electrode voltage. Figure 12 shows the inter-electrode feedback voltage diagram for the four machining cases. Comparing the above four machining cases, it can be seen that the variable domain fuzzy PID controlled magnetic actuator can better reduce the number of short-circuit and broken-circuit of the gap voltage and be more sensitive to the short-circuit and broken-circuit, thus effectively improving the EDM efficiency.

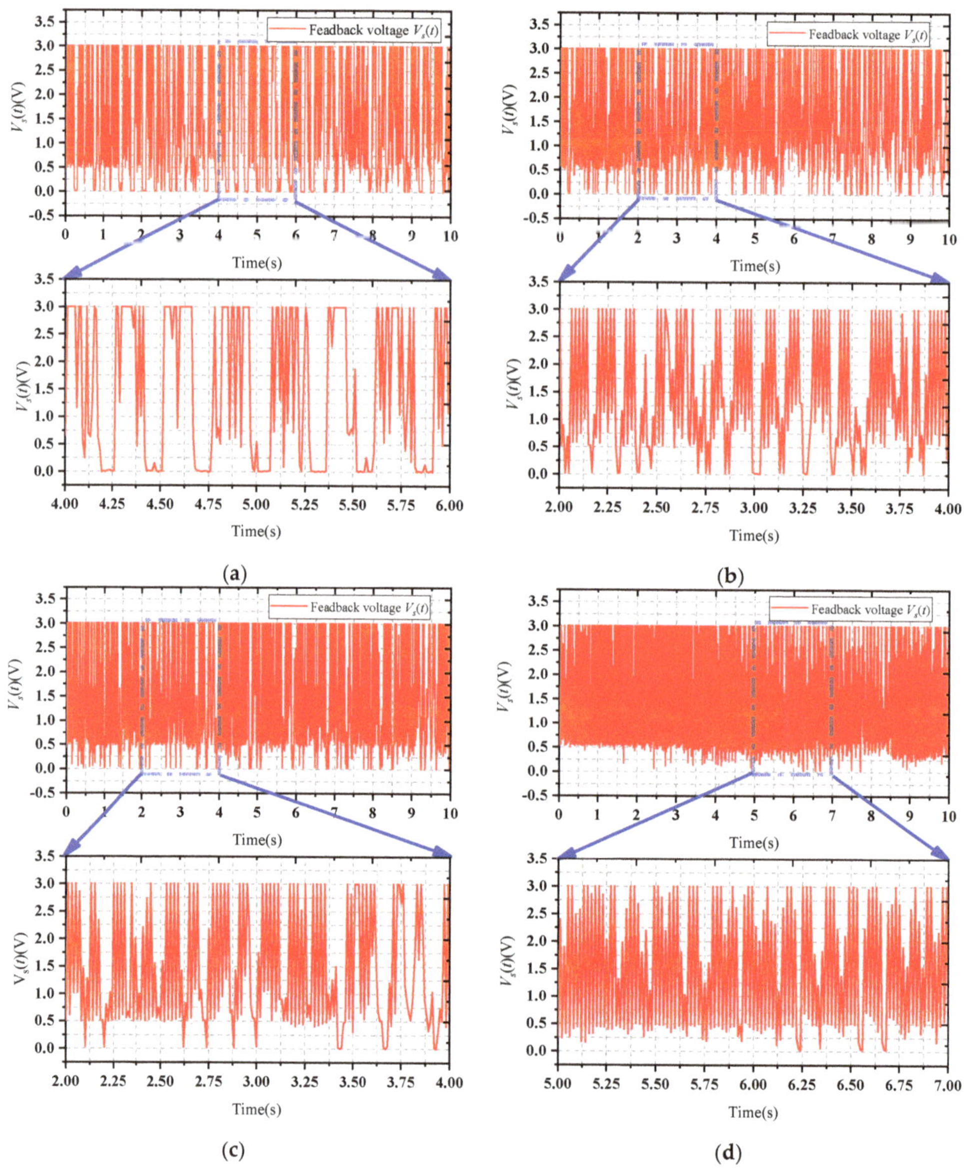

Figure 12. Inter-electrode feedback voltage. (**a**) Traditional micro-EDM. (**b**) Conventional PID control magnetic actuator and traditional micro-EDM cooperative control. (**c**) Fuzzy PID control magnetic actuator and traditional micro-EDM cooperative control. (**d**) Variable domain fuzzy PID control magnetic actuator and traditional micro-EDM cooperative control.

6.3.2. Comparison of Processing Efficiency

To better observe the effect of the magnetic actuator control system on the machining efficiency of micro EDM, four sets of machining experiments were conducted for the above four machining cases. For each group of EDM experiments, under the same processing conditions, the small hole processing experiments were conducted with a small hole depth range of 0~1.5 mm, and the processing time of small holes was recorded every 0.05 mm. To exclude the small holes affected by larger abnormal discharge, a number of processing experiments were conducted, and then the average value of their respective times was taken as the sample data.

Figure 13 shows the relationship between the micro EDM depth and machining time for the above four machining cases. Analysis of the curve in the figure shows that the machining speed slows down with the increase of machining depth. The reason is that when the machining depth is less than 1mm, the machined small hole is very shallow, the machining electric erosion products are easily discharged, and the electric erosion products removal effect has little impact on the machining efficiency. However, when the EDM depth of micro-hole is greater than 1mm, the electric erosion products removal becomes more and more difficult, and the machining electric erosion products cannot be discharged in time, resulting in low EDM machining efficiency. The magnetic actuator can drive the electrodes in the axial direction (Z direction) for rapid micro-positioning, which can effectively change the state of inter-electrode discharge and promote the removal of inter-electrode electric erosion products. In conclusion, the variable domain fuzzy PID control is more effective for this control system, the control system is more stable, and the machining efficiency is higher.

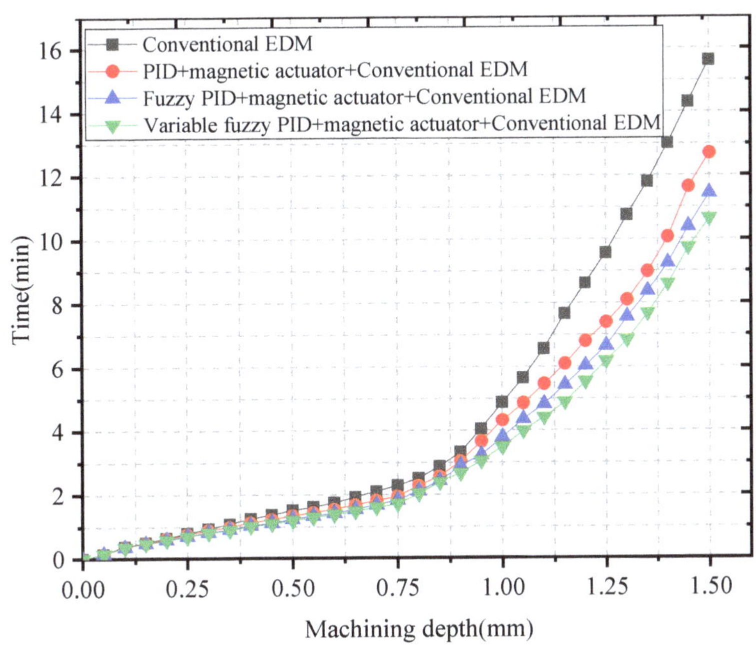

Figure 13. Relationship between micro EDM depth and machining time.

To further observe the effect of the inter-electrode voltage control system of the magnetic actuator for micro-EDM on the machining efficiency, four groups of small holes with a depth of 1 mm were machined in the above four machining cases. The machining speed of eight small holes was selected as sample data for each group of experiments. Figure 14 shows the comparison of machining efficiency.

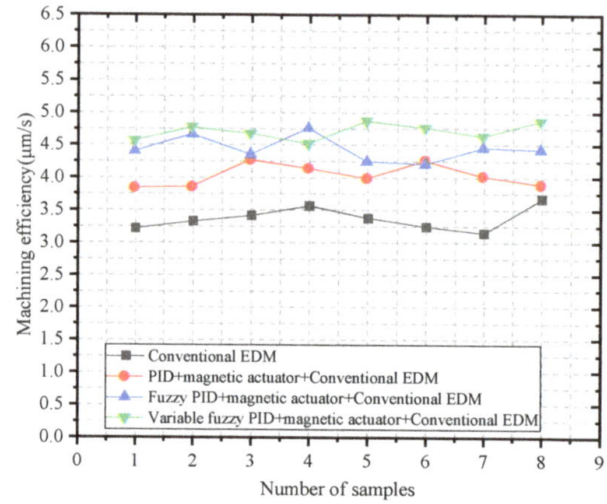

Figure 14. Comparison of micro-EDM efficiency.

Table 7 shows an analysis of experimental results. The experimental results show that, compared with the traditional micro-EDM, the processing efficiency of the conventional PID control magnetic actuator and traditional micro-EDM cooperative control was increased by 20%. The processing efficiency of the fuzzy PID control magnetic actuator and traditional micro-EDM cooperative control was increased by 31.8%. The processing efficiency of the Variable domain fuzzy PID control magnetic actuator and traditional micro-EDM cooperative control was increased by 40%.

Table 7. Experimental result analysis table.

Processing Methods	Average Processing Speed
traditional EDM	3.371 μm/s
PID control magnetic actuator	4.038 μm/s
fuzzy PID control magnetic actuator	4.444 μm/s
variable domain fuzzy PID control magnetic actuator	4.714 μm/s

6.3.3. Comparison of Machining Surface Quality

Figure 15 shows a physical image of the machined workpiece with a machining depth of 1 mm. From left to right are traditional micro-EDM, conventional PID control magnetic actuator and the traditional micro-EDM cooperative control, Fuzzy PID control magnetic actuator and the traditional micro-EDM cooperative control, variable domain fuzzy PID control magnetic actuator and the traditional micro-EDM cooperative control. By comparing the surface quality of the machined workpiece, it can be seen that the variable domain fuzzy PID control magnetic actuator and the traditional micro-EDM cooperative control had better machining results. The surface carbon accumulation of the machined workpiece was significantly reduced, and the surface quality of the machined workpiece was improved.

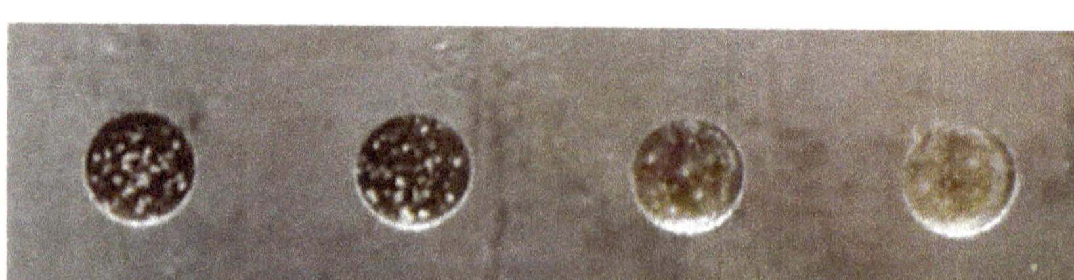

Figure 15. Image of machined workpiece.

7. Conclusions and Future Work

In this paper, a magnetic local actuator with a controllable single degree of freedom, high precision, and fast response for micro EDM is introduced. The magnetic actuator can be directly connected to the spindle of a traditional micro-EDM machine tool. Aiming at the problem of narrow discharge gap servo control, variable domain fuzzy control and PID control were combined, and a variable domain fuzzy PID controller was designed, which brought about intelligence of the control process and improved the performance of the control system. Under the conditions of electrical discharge machining, the system was used to control the magnetic actuator drive electrode to quickly and accurately position in the axial direction, improve the effective discharge probability of micro-EDM, and attain high-speed and high-precision micro-EDM. Simulation and experimental results showed that compared with traditional micro-EDM machine, the processing efficiency of the conventional PID control magnetic actuator and the traditional micro-EDM cooperative control was increased by 20%. The processing efficiency of the fuzzy PID control magnetic actuator and the traditional micro-EDM cooperative control was increased by 31.8%. The processing efficiency of the variable domain fuzzy PID control magnetic actuator and the traditional micro-EDM cooperative control was increased by 40%.

Based on this, in future research, we will consider the use of new type-2 and type-3 fuzzy logic systems to improve this control system, and we will study a multi-degree-of-freedom magnetic actuator and its control method, as well as verifying the control performance of the multi-degree-of-freedom magnetic actuator through EDM experiments.

Author Contributions: C.X.: software, drawing, experiments, data collection and processing, reading and summarization of all literature, writing-original draft, writing-review & editing; F.X.: writing-review & editing; F.S.: supervision, writing-review & editing; X.Z.: supervision, writing-review & editing; J.J.: writing-review & editing; B.L.: writing-review & editing; K.O.: writing-review & editing. All authors have read and agreed to the published version of the manuscript.

Funding: This research is supported by National Natural Science Fund of China (No. 52005345, No. 52005344), National Key Research and Development Project (No. 2020YFC2006701), China Scholarship Council (No. 202208210071), Scientific Research Fund Project of Liaoning Provincial Department of Education (No. LFGD2020002), Major Project of the Ministry of Science and Technology of Liaoning Province (No. 2022JH1/10400027).

Institutional Review Board Statement: Not applicable.

Informed Consent Statement: Not applicable.

Data Availability Statement: Not applicable.

Acknowledgments: We thank Zhou R., Zhao C., Pei W.Z., and Wang D. for their suggestions and recommendations.

Conflicts of Interest: The authors declare no conflict of interest.

References

1. Prakash, V.; Kumar, P.; Singh, P.; Hussain, M. Micro-electrical discharge machining of difficult-to-machine materials: A review. *Proc. Inst. Mech. Eng. Part B J. Eng. Manuf.* **2019**, *233*, 339–370. [CrossRef]
2. Kumar, D.; Singh, N.K.; Bajpai, V. Recent trends, opportunities and other aspects of micro-EDM for advanced manufacturing: A comprehensive review. *J. Braz. Soc. Mech. Sci. Eng.* **2020**, *42*, 2172–2191. [CrossRef]
3. Gostimirovic, M.; Pucovsky, V.; Sekulic, M.; Rodovanovic, M.; Madic, M. Evolutionary multi-objective optimization of energy efficiency in electrical discharge machining. *J. Mech. Sci. Technol.* **2018**, *32*, 4775–4785. [CrossRef]
4. Shabgard, M.R.; Gholipoor, A.; Baseri, H. A review on recent developments in machining methods based on electrical discharge phenomena. *Int. J. Adv. Manuf. Technol.* **2016**, *87*, 2081–2097. [CrossRef]
5. Gohil, V.; Puri, Y.M. Turning by electrical discharge machining: A review. *Proc. Inst. Mech. Eng. Part B J. Eng. Manuf.* **2017**, *231*, 195–208. [CrossRef]
6. Rafaqat, M.; Mufti, N.A.; Ahmed, N.; Rehman, A.U.; AlFaify, A.Y.; Farooq, M.U.; Saleh, M. Hole-Making in D2-Grade Steel Tool by Electric-Discharge Machining through Non-Conventional Electrodes. *Processes* **2022**, *10*, 1553. [CrossRef]
7. Gero, E.; Yan, J.W. Direct Observation of Discharge Phenomena in Vibration-Assisted Micro EDM of Array Structures. *Micromachines* **2022**, *13*, 1286. [CrossRef]

8. Huang, Y.H.; Zhang, Q.H.; Xing, Q.X.; Yao, Z.Y.; Li, J.C. Effects of electrode rotational speed on processing performances of AISI 304 in micro-electrical discharge machining. *Int. J. Adv. Manuf. Technol.* **2019**, *105*, 1665–1674. [CrossRef]
9. Tsai, M.Y.; Fang, C.S.; Yen, M.H. Vibration-assisted electrical discharge machining of grooves in a titanium alloy (Ti-6A-4V). *Int. J. Adv. Manuf. Technol.* **2018**, *97*, 297–304. [CrossRef]
10. Jadam, T.; Sahu, S.K.; Datta, S.; Masanta, M. EDM performance of Inconel 718 superalloy: Application of multi-walled carbon nanotube (MWCNT) added dielectric media. *J. Braz. Soc. Mech. Sci. Eng.* **2019**, *41*, 1–20. [CrossRef]
11. Jiang, Y.; Ping, X.L.; Zhang, Y.; Zhao, W.S. Effects of gas medium on submersed gas-flushing electrical discharge machining of different metal materials. *Int. J. Adv. Manuf. Technol.* **2021**, *115*, 3679–3687. [CrossRef]
12. Wang, J.; Jia, Z.X. Efficiency improvement in electrical discharge machining (EDM) of constant section cavity based on experimental study and numerical calculations. *Prod. Eng.* **2018**, *12*, 567–578. [CrossRef]
13. Guo, Y.F.; Ling, Z.B.; Zhang, X.Y.; Feng, Y.R. A magnetic suspension spindle system for small and micro holes EDM. *Int. J. Adv. Manuf. Technol.* **2018**, *94*, 1911–1923. [CrossRef]
14. Zhang, X.Y.; Shinshi, T.; Shimokohbe, A.; SATO, T.; Miyake, H.; Nakagawa, T. High-speed electrical discharge machining by using a 5-DOF controlled maglev local actuator. *J. Adv. Mech. Des. Syst. Manuf.* **2008**, *2*, 493–503. [CrossRef]
15. Hiroyuki, N.; Toshiro, H. A Novel Electromagnetic Actuator Based on Displacement Amplification Mechanism. *IEEE/ASME Trans. Mechatron.* **2015**, *20*, 1607–1615. [CrossRef]
16. Feng, Y.R.; Guo, Y.F.; Ling, Z.B.; Zhang, X.Y. Investigation on machining performance of micro-holes EDM in ZrB2 -SiC ceramics using a magnetic suspension spindle system. *Int. J. Adv. Manuf. Technol.* **2019**, *101*, 2083–2095. [CrossRef]
17. Feng, Y.R.; Guo, Y.F.; Ling, Z.B.; Zhang, X.Y. Micro-holes EDM of superalloy Inconel 718 based on a magnetic suspension spindle system. *Int. J. Adv. Manuf. Technol.* **2019**, *101*, 2015–2026. [CrossRef]
18. Yang, F.; Yang, J.; Yao, K.; Hua, H. Adaptive Voltage Position Control for Pulse Power Supply in Electrical Discharge Machining. *IEEE Trans. Ind. Electron.* **2019**, *66*, 5895–5906. [CrossRef]
19. Chen, X.L.; Tong, W.; Mao, Y.; Zhao, T. Interval Type-2 Fuzzy Dynamic High Type Control of Permanent Magnet Synchronous Motor with Vector Decoupling Method. *Actuators* **2021**, *10*, 293. [CrossRef]
20. Zheng, T.; Xu, X.Z.; Lu, X.; Xu, F. Learning adaptive sliding mode control for repetitive motion tasks in maglev rotary table. *IEEE Trans. Ind. Electron.* **2021**, *69*, 1836–1846. [CrossRef]
21. Lu, Y.T.; Tan, C.; Ge, W.Q.; Li, B.; Lu, J.Y. Improved Sliding Mode-Active Disturbance Rejection Control of Electromagnetic Linear Actuator for Direct-Drive System. *Actuators* **2021**, *10*, 138. [CrossRef]
22. Chen, D.F.; Chiu, S.P.C.; Cheng, A.B.; Ting, J.C. Electromagnetic Actuator System Using Witty Control System. *Actuators* **2021**, *10*, 65. [CrossRef]
23. Zhu, H.Q.; Gu, Z.W. Active disturbance rejection control of 5-degree-of-freedom bearingless permanent magnet synchronous motor based on fuzzy neural network inverse system. *ISA Trans.* **2020**, *101*, 295–308. [CrossRef] [PubMed]
24. Repinaldo, J.P.; Koroishi, E.H.; Lara-molina, F.A. Neuro-fuzzy control applied on a 2DOF structure using electromagnetic actuators. *IEEE Lat. Am. Trans.* **2021**, *19*, 75–82. [CrossRef]
25. Chen, J.W. Modeling and Decoupling Control of a Linear Permanent Magnet Actuator Considering Fringing Effect for Precision Engineering. *IEEE Trans. Magn.* **2021**, *57*, 1–15. [CrossRef]
26. Wang, N.; Li, Z.; Liang, X.L.; Zhang, J.Q. Design of UAV yaw Angle controller based on variable domain fuzzy PID. *J. Phys. Conf. Ser.* **2021**, *1754*, 012112. [CrossRef]
27. Lin, C.L.; Liu, H.F.; Sun, H.J.; Zheng, Z.Y.; Chen, Z. Implementation of Fuzzy Self-tuning Proportional Integral Derivative Controller on Sample-tube Spin Control System in Nuclear Magnetic Resonance Spectrometer. *Chin. J. Anal. Chem.* **2011**, *39*, 506–510. [CrossRef]

actuators

MDPI

Article

Characteristics Analysis of an Electromagnetic Actuator for Magnetic Levitation Transportation

Junjie Jin, Xin Wang, Chuan Zhao, Fangchao Xu, Wenzhe Pei, Yuhang Liu and Feng Sun *

School of Mechanical Engineering, Shenyang University of Technology, Shenyang 110870, China
* Correspondence: sunfeng@sut.edu.cn; Tel.: +86-189-0404-0100

Abstract: In this article, an electromagnetic actuator is proposed to improve the driving performance of magnetic levitation transportation applied to ultra-clean manufacturing. The electromagnetic actuator mainly includes the stator with the Halbach array and the mover with a symmetrical structure. First, the actuator principle and structure are illustrated. Afterward, in order to select a suitable secondary structure and analyze the characteristics of the actuator, the electromagnetic characteristics of actuators with different secondary structures are analyzed by the finite element method (FEM). Analysis results show that the actuator adopting the secondary structure with a Halbach array will increase the electromagnetic force and working stability. The secondary with the three-section Halbach array is selected for the electromagnetic actuator. Then, the influence of secondary permanent magnet (PM) thickness on the electromagnetic force is analyzed by FEM. The results indicate that the increase in PM thickness will lead to increased electromagnetic force and decreased utilization ratio of PM. Finally, a prototype of an electromagnetic actuator is built, and experiments are implemented. The correctness of the theoretical analysis and the effectiveness of the electromagnetic actuator are verified by experimental results.

Keywords: electromagnetic actuator; Halbach array; characteristic analysis; finite element method

Citation: Jin, J.; Wang, X.; Zhao, C.; Xu, F.; Pei, W.; Liu, Y.; Sun, F. Characteristics Analysis of an Electromagnetic Actuator for Magnetic Levitation Transportation. *Actuators* **2022**, *11*, 377. https://doi.org/10.3390/act11120377

Academic Editor: Kirill Poletkin

Received: 7 November 2022
Accepted: 13 December 2022
Published: 15 December 2022

1. Introduction

Magnetic levitation is a support technology with no contact and no lubrication. It has been widely applied in the industrial fields, such as maglev trains, precision positioning platforms, and magnetic bearings [1–3]. The development of the semiconductor industry needs more stringent requirements in the manufacturing environment, so magnetic levitation technology has broad application prospects in ultra-clean transportation. In 1998, K.H. Park et al. [4] proposed the maglev conveyor system for ultra-clean manufacturing, which combined AGV (Automated Guided Vehicle) and maglev technology.. Without mechanical contact, the conveyor can effectively improve the air cleanliness in cleanrooms. Moreover, the electromagnetic levitation combination of a planar motor supports rapid responses and high-speed motion [5–8]. They are mainly divided into the moving magnet type and the moving coil type. The moving magnet type has a simple mover structure but requires multiple controllers to achieve precise motion. Conversely, the moving coil type does not require a complex controller. However, serious copper losses cause high power consumption and significant heat emission. Therefore, the cooling mechanism is necessary, which leads to the complex structure and large mass of the magnetic levitation platform. Kim et al. designed a hybrid electromagnetic-permanent magnetic levitation transport system proposed in the reference [9,10]. The magnetic levitation system can realize suspension work with low energy consumption. Furthermore, linear motors were used to drive the magnetic levitation platform. It is impossible to change the levitation gap of the magnetic levitation platform. Moreover, it is harmful to ultra-clean transportation due to dependence on a mechanical guide rail.The permanent magnetic levitation transportation system of variable flux path has low steady-state energy consumption, high controlling stiffness, and anti-eccentric load characteristics [11,12]. However, a permanent magnetic levitation transport system is sensitive to external disturbance and has high requirements for the stability of the drive system. Magnetic levitation transportation requires a suitable drive system. This drive system has the advantages of lower disturbance, no contact, high precision, and low mass.

The contact drive device is not suitable for magnetic levitation transportation. Electromagnetic drive technology converts electrical energy into mechanical energy, which is the operation of electromagnetic force. It is a contactless drive technology [13,14]. Electromagnetic drive technology has the advantages of fast response, high controllability, and high precision [15–17]. With the development of rare-earth permanent magnetic materials, the electromagnetic drive technology with PM has the advantages of simple structure and high efficiency. Therefore, it is widely researched and applied, such as traffic, delivery platforms, and machine tools [18–20]. Electromagnetic drive technology with PM is categorized into two broad groups: the core type and the coreless type. The iron core type has greater electromagnetic force, but the cogging effect will produce significant disturbances. In addition, the mover and stator have enormous suction, which is detrimental to magnetic levitation transmission [21]. In contrast, the coreless type has less disturbance but less force [22]. Jansen et al. [23] proposed an electromagnetic actuator: a U-shaped stator structure is adopted, and the mover coil is located in the center of the stator. The structure can increase the thrust of the coreless electromagnetic actuator. Furthermore, much research used Halbach arrays for electromagnetic actuators to improve thrust [24–26]. The above research will increase the thrust of the coreless electromagnetic actuator. Moreover, the normal force will increase, and this will increase disturbance. In addition, the normal force was controlled by a highly complex scheme, which caused detrimental effects on the precision. Many accurate control models have been studied to reduce this damage [27–29]. Generally, the above electromagnetic actuator has a minimal air gap (0.3 mm~1 mm). However, the minimal air gap will cause defective effect to the floating of the magnetic levitation platform. In addition, they rely on the guide rail, which limits the application of magnetic levitation transmission.

To make magnetic levitation transportation applicable to the ultra-clean manufacturing environment, an electromagnetic actuator is proposed. The electromagnetic actuator has the advantages of small mass, big electromagnetic thrust, and low disturbance, and can realize automatic guiding used for magnetic levitation transportation. Therefore, it can be combined with a magnetic levitation platform to allow magnetic levitation transport to be used in an ultra-clean manufacturing environment. This paper is organized as follows. Firstly, the electromagnetic actuator principle and structure are illustrated. The objective is to select the appropriate secondary structure and analyze the characteristics of the actuator, therefore, the actuator is presented with three secondary structures of ordinary radial magnetization, a Halbach three-section array, and a five-section array. The resulting models are analyzed by the FEM. Afterward, considering the actuator's electromagnetic characteristics and cost, it determined the three-section Halbach magnet array is an appropriate secondary structure to the electromagnetic actuator. Subsequently, the influence of magnetic thickness on the electromagnetic force of the electromagnetic actuator is analyzed. Finally, a prototype of the electromagnetic actuator is built, and experiments are implemented. The experiment results show the prototype's effectiveness and the analysis's correctness.

2. Structure and Principle of Electromagnetic Actuator

2.1. Structure and Principle of Halbach Magnet Array

In the Halbach magnet array, the radially magnetized PM plays a dominant role in the magnetic circuit, and the tangentially magnetized PM compensates for the magnetic circuit. The air-gap magnetic field of this PM array tends to sinusoidal distribution, and the harmonic content is less, especially in the continuous magnetization mode, which can produce ideal magnetic field characteristics of the sinusoidal magnetic field waveform.

The Halbach magnet array can enhance the unilateral magnetic field. Figure 1 is the magnetization diagram of the Halbach magnet array, where (a) is a three-section Halbach magnet array, and (b) is a five-section Halbach magnet array. The magnetization angle of two adjacent PMs is θ, and k is the number of PMs contained in a magnetization cycle.

$$\theta = \frac{2\pi}{k} \tag{1}$$

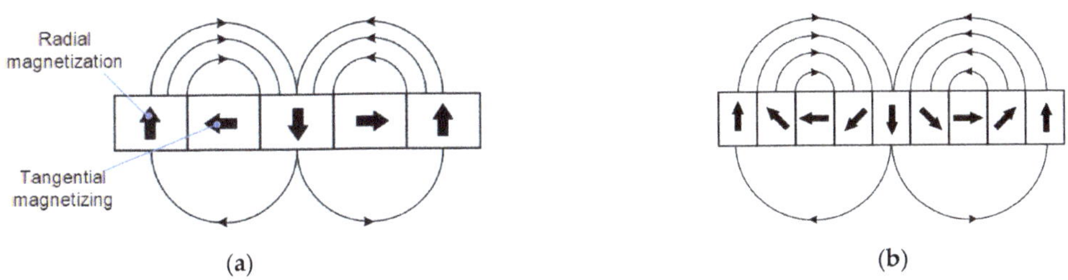

Figure 1. Halbach magnetization diagram: (**a**) Three-section Halbach magnet array; (**b**) Five-section Halbach magnet array.

The Halbach magnet array structure is calculated using the equivalent surface current method based on Ampere's molecular current hypothesis. According to the research results of [30], the magnetic field generated by the PM monomer at any point $p(x, y)$ outside the magnet was expressed as:

$$\begin{cases} B_{1x}(x,y,J) = \frac{\mu_0 J}{4\pi} \ln \frac{(y-h/2)^2+(x-l/2)^2}{(y+h/2)^2+(x-l/2)^2} \\ B_{2x}(x,y,J) = -\frac{\mu_0 J}{4\pi} \ln \frac{(y-h/2)^2+(x+l/2)^2}{(y+h/2)^2+(x+l/2)^2} \\ B_{1y}(x,y,J) = \frac{\mu_0 J}{2\pi} \left[\arctan\frac{(y-h/2)}{(x-l/2)} - \arctan\frac{(y+h/2)}{(x-l/2)}\right] \\ B_{2y}(x,y,J) = -\frac{\mu_0 J}{2\pi} \left[\arctan\frac{(y-h/2)}{(x+l/2)} - \arctan\frac{(y+h/2)}{(x+l/2)}\right] \end{cases} \tag{2}$$

$$\begin{cases} B_x = B_{1x}(x,y,J) + B_{2x}(x,y,J) \\ B_y = B_{1y}(x,y,J) + B_{2y}(x,y,J) \end{cases} \tag{3}$$

where J is the surface current density; μ_0 is the vacuum permeability; h is the PM height; l is the PM width; B_x is the tangential flux density; and B_y is the radial flux density.

According to the coordinate rotation theory, when the PM is tangentially magnetized along the origin, it can be equivalent to the counterclockwise rotation of the coordinate system along the origin by 90°. When the PM is magnetized at any angle θ, the magnetization direction is decomposed into radial and tangential magnetization directions for calculation (where the angle between the magnetization directions and the horizontal direction is θ).

$$\begin{cases} J_{x\theta} = J_\theta \times \cos\theta \\ J_{y\theta} = J_\theta \times \sin\theta \end{cases}, \tag{4}$$

For any group of Halbach magnet arrays, any point $p(x, y)$ magnetic induction intensity distribution can be expressed as

$$\begin{cases} B_x(x,y,J_{x\theta/y\theta}) = \sum\limits_{n=1}^{N} \left[x-(n-1/2)l, y-h/2, J_{x\theta/y\theta}\right] \\ B_y(x,y,J_{x\theta/y\theta}) = \sum\limits_{n=1}^{N} \left[x-(n-1/2)l, y-h/2, J_{x\theta/y\theta}\right] \end{cases}, \tag{5}$$

2.2. Electromagnetic Actuator Structure

Figure 2a is the structure of the electromagnetic actuator. It consists of secondary PM and mover primary winding, in which the actuator bilateral primary symmetrical installation adopts a conjugate structure and the secondary adopts the Halbach magnet array. The primary yoke of the mover adopts non-metallic materials to reduce the heat generated by the actuator due to the eddy current effect during operation. This structure is shown in Figure 2b. In order to compensate for the low density of the electromagnetic thrust generated by the coreless structure, yoke iron is added to the back of the winding. The primary employs a fractional-slot concentrated winding distribution and a 12-slot 10-pole structure, and the winding on both sides are in reverse series. The schematic diagram of its electromagnetic structure is shown in Figure 3. The primary stage adopts coreless armature structure, which eliminates the cogging effect. The bilateral mover can stabilize the bilateral air gap by the normal force of the same size and the opposite direction. Therefore, the structure is applied to the magnetic levitation transport system and can work without relying on the guide rail.

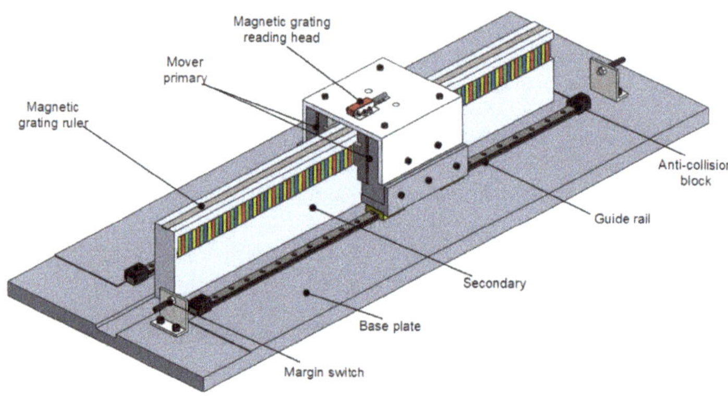

(**a**)

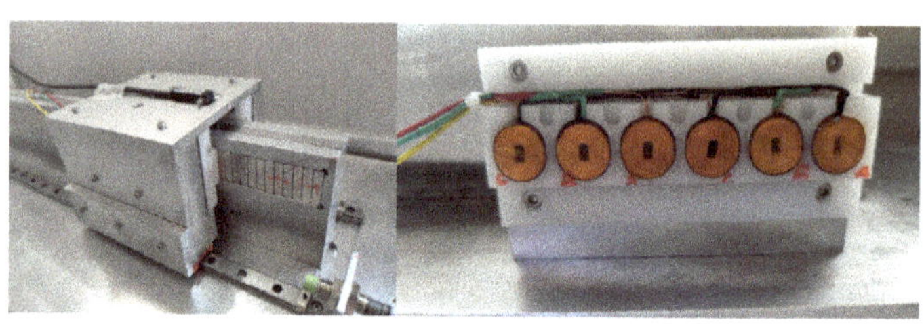

(**b**)

Figure 2. Structure diagram of electromagnetic actuator: (**a**) Mechanical structure; (**b**) Mover primary structural.

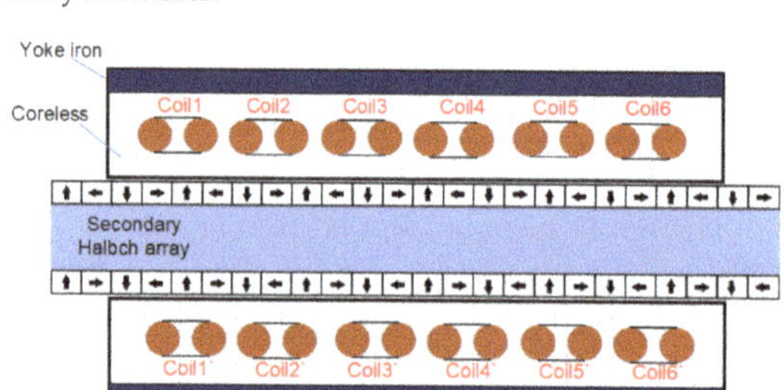

Figure 3. Electromagnetic structure of the electromagnetic actuator.

2.3. Electromagnetic Actuator Heterogeneous Secondary Structure

Figure 4 shows the cross-section diagram of (a) radial magnetization, (b) a Halbach three-section magnet array magnetization, and (c) a Halbach five-section magnet array magnetization as secondary for the electromagnetic actuator. They have the same slot to pole ratio. The main parameters of the actuator are shown in Table 1.

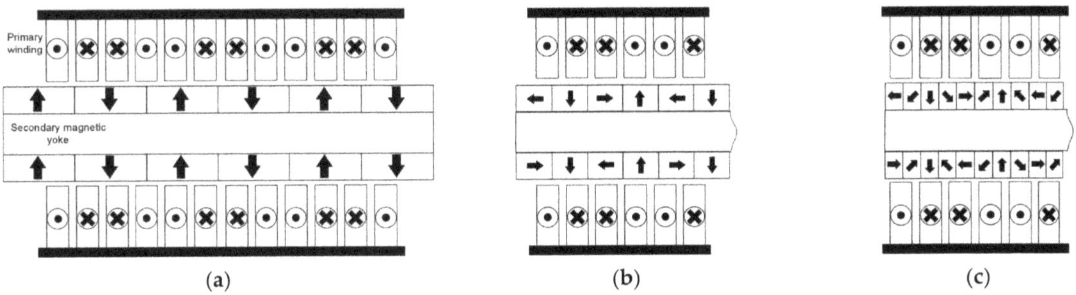

Figure 4. Cross-section diagram of three secondary structures: (**a**) Structure of electromagnetic actuator with radially magnetized secondary; (**b**) Structure of electromagnetic actuator with Halbach three-section magnetized secondary; (**c**) Structure of electromagnetic actuator with Halbach five-section magnetized secondary.

Table 1. The parameters of the electromagnetic actuator.

Parameters	Radial Magnetization	Halbach Three-Section	Halbach Five-Section
Pole pitch (mm)	12	12	12
Magnet height (mm)	5	5	5
Magnet width (mm)	12	6	3
Primary winding height (mm)	19	19	19
Primary winding width (mm)	127	127	127
Winding turns	489	489	489
Current (A)	5	5	5
gap length (mm)	2	2	2
PM	NdFb54	NdFb54	NdFb54

3. Characteristic Analysis of Electromagnetic Actuator

This paper compares the electromagnetic characteristics of the electromagnetic actuator with the same pole to slot ratio which has different secondary structures. The electromagnetic actuator mainly includes the bilateral U-type actuator's primary (mover) and secondary (stator), as shown in Figure 1. The secondary stator is located in the middle of the primary bilateral mover, forming the air gap on both sides, and the air gap on both sides is 2 mm. The tree different secondary structures are a radially magnetized magnet array structure, a three-section Halbach magnet array structure, and a five-section Halbach magnet array structure. Electromagnetic actuator electromagnetic characteristics of three secondary structures are analyzed using the two-dimensional time-step finite element method. In practical application, the adopted joint action of multiple primary units and the cooperation of each unit actuator can be considered as the infinite primary length of the actuator. The end effect caused by the start of each actuator unit has little impact on the electromagnetic characteristics of the drive [31]. Therefore, this paper ignores the influence of the actuator end effect on the actuator.

FEM model sizes for three electromagnetic actuators are shown in Table 1. Silicon steel is used as yoke material with a thickness of 3 mm. The winding material property is copper, and aluminum is used by secondary permanent magnet yoke. In the PM and winding area, the mesh is divided into 0.1 mm. The primary yoke and permanent magnet yoke area divided the mesh into 0.3 mm, and another area mesh is 0.5 mm. Firstly, the electromagnetic actuator finite element model is divided into two regions to solve, respectively. Then, the solution results of the two regions are synthesized and compared with the complete solution results. The calculation is based on the infinite boundary condition.

3.1. FEM of Magnetic Field Analysis

The no-load magnetic field distributions of three different secondary structures of electromagnetic actuators are shown in Figure 5. The (a) shows that the magnetic field distribution generated by the PM is the same on the working and yoke magnet sides. Furthermore, there is more significant magnetic field intensity between adjacent PM s. Therefore, the electromagnetic actuator of the radially magnetized secondary generates significant flux leakage on the contact part of two adjacent PMs and magnet yoke. Conversely, the electromagnetic actuator of the Halbach magnetized secondary has significant magnetic field intensity on the working sides. Because the Halbach magnet can enhance the magnetic field intensity of the working side, the magnetic flux leakage is less. Compared with the magnetic field distributions of two Halbach magnetized secondaries, the magnetic field distribution is more uniform for the five-section Halbach magnet array, and the magnetic flux leakage is the least.

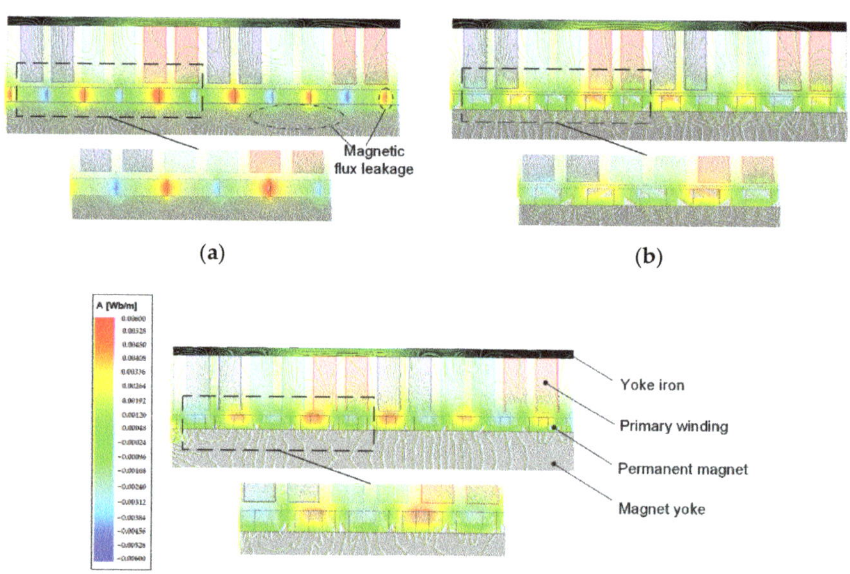

(c)

Figure 5. Magnetic field distribution: (**a**) Electromagnetic actuator with radially magnetized secondary; (**b**) Electromagnetic actuator with Halbach three-section magnetized secondary; (**c**) Electromagnetic actuator with Halbach five-section magnetized secondary.

Figure 6 shows the magnetic flux density distribution of the horizontal midline of the upper air gap. The Halbach magnet array combination can enhance the air gap magnetic field. Additionally, the more magnets at a pole pitch, the better the sinusoidal air gap flux density. According to Formula (5), the flux density peak at the center of the air gap is calculated. The comparison results are shown in Table 2. The simulated calculation does not consider magnetic flux leakage and assumes that the magnetic field is evenly distributed in the air gap. The Electromagnetic actuator of the radially magnetized secondary has severe magnetic flux leakage, so there are significant differences between the simulation and FEM data. Conversely, the electromagnetic actuator of the Halbach-magnetized secondary is similar between simulation and FEM data.

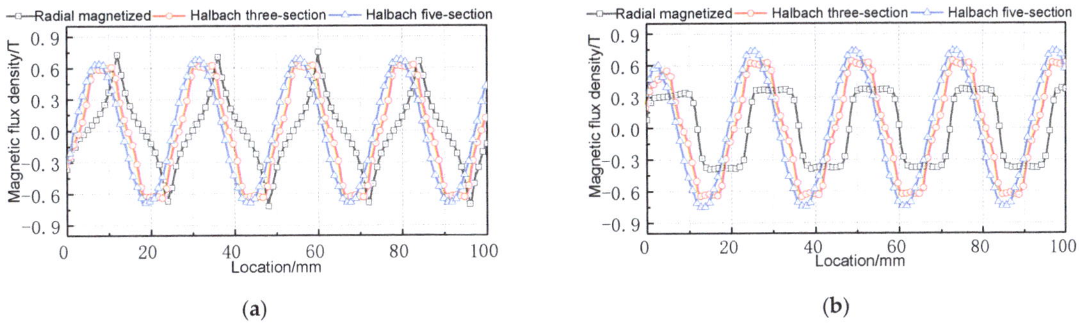

Figure 6. The magnetic flux density of air gap centerline: (**a**) Tangential magnetic flux density; (**b**) Radial magnetic flux density.

Table 2. Results of simulation data and FEM data.

Magnetic Flux Density		Radial Magnetization	Halbach Three-Section	Halbach Five-Section
Tangential magnetic flux density (T)	Simulation	0.5	0.6	0.55
	FEM	0.7	0.61	0.58
	deviation	40%	1.6%	5.4%
Radial magnetic flux density (T)	Simulation	0.29	0.6	0.75
	FEM	0.37	0.62	0.76
	deviation	28%	3.3%	1.3%

3.2. Electromagnetic Actuator Electromagnetic Thrust Analysis

Electromagnetic thrust is one of the essential parameters of the electromagnetic actuator. This section analyzes the electromagnetic actuator when the primary size is the same different secondary structure. The effects of different secondary structures, namely radial magnetization, the three-section Halbach magnet array, and the five-section Halbach magnet array, on the electromagnetic actuator's electromagnetic thrust performance are compared.

Figure 7 shows the electromagnetic thrust generated by three different secondary structure actuators when the three-phase winding current is 5 A, and the frequency is 10 Hz. When a U-type symmetrical structure is used, the thrust waveform and phase of bilateral primary are the same. Figure 7a is the unilateral electromagnetic thrust and (b) is the regional thrust of (a). Because Figure 7b is a regional thrust enlarged diagram, the thrust fluctuation of the three electromagnetic actuators can be clearly shown. However, the peak and valley values of the overall thrust range cannot be shown in Figure 7b. The calculation formula for the electromagnetic thrust fluctuation is [32]:

$$K_F = \frac{F_{max} - F_{min}}{F_{avg}} \times 100\%, \tag{6}$$

where F_{max} is the peak electromagnetic thrust; F_{min} is the valley value of electromagnetic thrust; F_{avg} is the average electromagnetic thrust.

After analysis and calculation, the same total PM volume, the electromagnetic actuator of radially magnetized secondary electromagnetic thrust peak value is 21.62 N, the valley value is 21.1 N, the average thrust is 21.42 N, and the thrust fluctuation is 2.4%. For the electromagnetic actuator of the Halbach three-section magnetized secondary, the electromagnetic thrust peak value is 34.83 N, the valley value is 34.5 N, the average thrust is 34.76 N, and the thrust fluctuation is 0.84%. For the electromagnetic actuator of the Halbach five-section magnetized secondary, the electromagnetic thrust peak value is 30.84 N, the valley value is 30.71 N, the average thrust is 30.8 N, and the thrust fluctuation is 0.41%. The utilization ratio of PM materials is defined as the ratio of the actuator's average thrust to the

PM's volume. This index measures the utilization of PM materials with high cost. According to this, the PM utilization rate of the radial magnetization secondary is 0.67 N/cm^3. The PM utilization ratio of the Halbach three-section magnet array secondary is 1.10 N/cm^3. The PM utilization ratio of the Halbach five-section magnet array secondary is 0.978 N/cm^3.

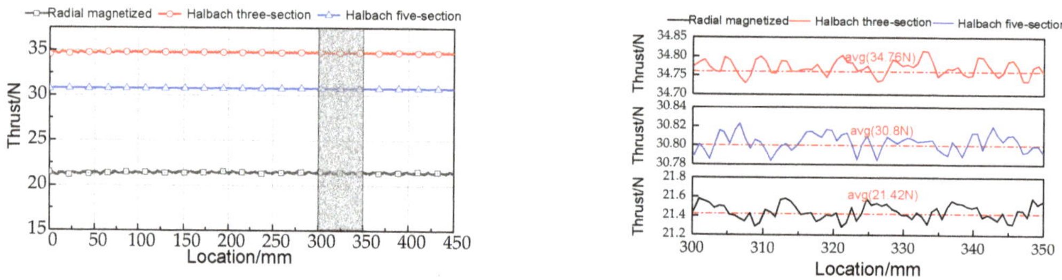

(a) (b)

Figure 7. Electromagnetic thrust: (**a**) Bilateral combined electromagnetic thrust; (**b**) Regional thrust.

The above analysis shows that the PM utilization rate of the electromagnetic actuator with Halbach three-section magnet array secondary structure is the highest. The more PM blocks at a pole pitch, the smaller the thrust fluctuation and the more stationary the actuator works.

3.3. Electromagnetic Actuator Normal Force Analysis

The electromagnetic actuator will generate electromagnetic thrust and normal force. In order to improve the operation accuracy of the actuator, it is necessary to consider the influence of friction disturbance caused by periodic fluctuation of normal force on the horizontal electromagnetic thrust [33].

Figure 8 shows the electromagnetic force results of three different secondary structures for the electromagnetic actuator, when the three-phase winding current is 5 A and the frequency is 10 Hz.

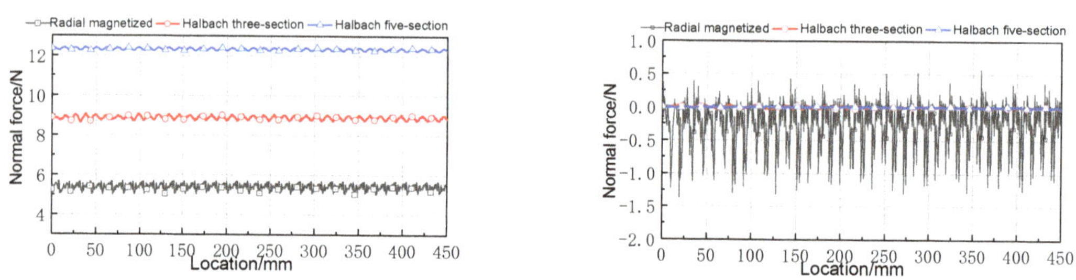

(a) (b)

Figure 8. Normal electromagnetic force: (**a**) Unilateral normal force; (**b**) Bilateral combined normal force.

After analysis and calculation, the unilateral normal force of the electromagnetic actuator with the same total PM volume is shown in Figure 8a. In the electromagnetic actuator of the radially magnetized secondary, the normal force peak value is 5.69 N, the valley value is 5.02 N, the average normal force is 5.41 N, and the normal force fluctuation is 12%. In the electromagnetic actuator of the Halbach three-section magnetized secondary, the normal force peak value is 9.15 N, the valley value is 8.62 N, the average normal force is 8.88 N, and the normal force fluctuation is 6%. In the electromagnetic actuator of the Halbach five-section magnetized secondary, the normal force peak value is 12.41 N, the valley value is 12.25 N, the average normal force is 12.32 N, and the normal force fluctuation is 1%.

Electromagnetic actuator adopts symmetrical bilateral primary, so the primary on both sides generates equal normal force with opposite direction. It is specified that the normal force generated on the upper side is negative and on the lower side is positive. The bilateral combined normal force is shown in Figure 8b. In the electromagnetic actuator of the radially magnetized secondary, the normal force peak value is −1.37 N, the valley value is 0.57 N, and the average normal force is −0.4 N. In the electromagnetic actuator of the Halbach three-section magnetized secondary, the normal force peak value is −37 mN, the valley value is 35 mN, and the average normal force is 1 mN. In the electromagnetic actuator of Halbach five-section magnetized secondary, the normal force peak value is 27.6 mN, the valley value is 26 mN, and the average normal force is 0.8 mN.

The above analysis shows that the Halbach magnet array will enhance the radial air gap flux density according to Section 3.1. The normal force increases with the increase of the normal component of air gap flux density. The normal force generated by the bilateral winding of the U-type structure can be offset by each other. The more PM blocks will have minor normal force fluctuation. The smaller the influence of the normal force of bilateral primary synthesis on the thrust fluctuation, the more stable the electromagnetic actuator will be.

3.4. Electromagnetic Actuator No-Load Back EMF and Self-Inductance Analysis

The no-load back EMF is one of the parameters that needs to be considered in the design of the electromagnetic actuator, which has an important influence on the static and dynamic performance of the electromagnetic actuator. The three-phase no-load EMF of the electromagnetic actuator with different secondary structures at the synchronous speed of 0.24 m/s is shown in Figure 9. Each phase no-load back EMF is 120° mutual difference. For the electromagnetic actuator with the radially magnetized secondary, the no-load back EMF amplitude is 0.39 V. For the electromagnetic actuator with the Halbach three-section magnetized secondary, the no-load back EMF amplitude is 0.57 V. For the secondary of the electromagnetic actuator with the Halbach five-section magnetized, the no-load back EMF amplitude is 0.61 V. The no-load back EMF sinusoidal waveform quality of the Halbach magnet array secondary structure is higher than that of the actuator with the radial magnetization secondary structure. The no-load back EMF of the actuator with the Halbach five-section magnet array magnetized secondary is the largest, and the sinusoidal waveform is the best.

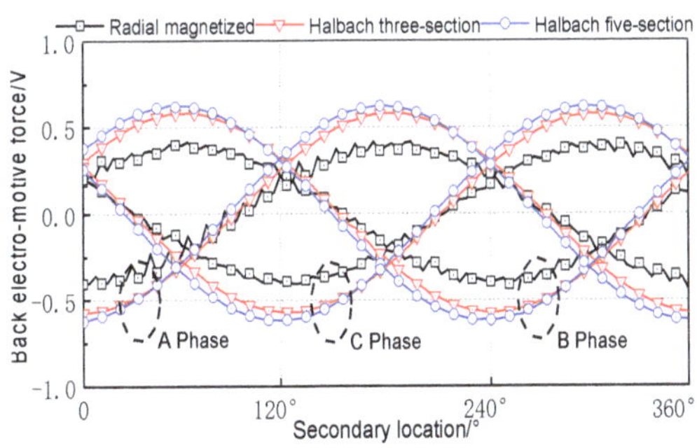

Figure 9. Three-phase no-load back EMF.

Electromagnetic actuator self-inductance is one of the critical parameters in actuator design. The self-inductance fluctuation will produce thrust fluctuation, which will harm the thrust fluctuation of the actuator. It has an important effect on the static and dynamic performance of the actuator. Figure 10 shows the actuator's self-inductance curves of A phase primary winding with three different secondary structures. When the primary winding of the A phase moves to the position with the maximum PM flux interlinkage, the self-inductance is the smallest. When the primary winding of the A phase moves to the minimum position of PM flux interlinkage, the self-inductance is the largest. The self-inductance varies periodically with pole pitch.

Figure 10a is the self-inductance of the electromagnetic actuator with a radially magnetized secondary. The average self-inductance of unilateral A phase winding is 3.663 mH, and the change rate is 0.12%. The Figure 10b is the self-inductance of the electromagnetic actuator with a Halbach three-section magnetized secondary. The average self-inductance of unilateral A phase winding is 3.664 mH, and the change rate is 0.1%. The Figure 10c is the self-inductance of electromagnetic actuator with a Halbach five-section magnetized secondary. The average self-inductance of unilateral A phase winding is 3.663 mH, and the change rate is 0.08%. The self-inductance amplitude and phase of the A phase winding on both sides are the same for the electromagnetic actuator. The self-inductance is superimposed after the series connection. The average self-inductance of the superimposed electromagnetic actuator of the radially magnetized secondary is 7.327 mH, and the change rate is still 0.12% as that of the unilateral. For the electromagnetic actuator of the Halbach three-section and five-section magnetized secondary, the average self-inductance of A phase winding is 7.328 mH and 7.321 mH, respectively, and the change rate is the same as that of a unilateral.

According to the above analysis, the electromagnetic actuator by the Halbach magnet array secondary structure self-inductance waveform is closer to the sinusoidal waveform. The Halbach magnet array secondary structure generation change rate of self-inductance is lower. It is further explained that the Halbach magnet array type secondary can reduce the thrust fluctuation. The pole pitch has more PM blocks, and the actuator thrust fluctuation will be smaller.

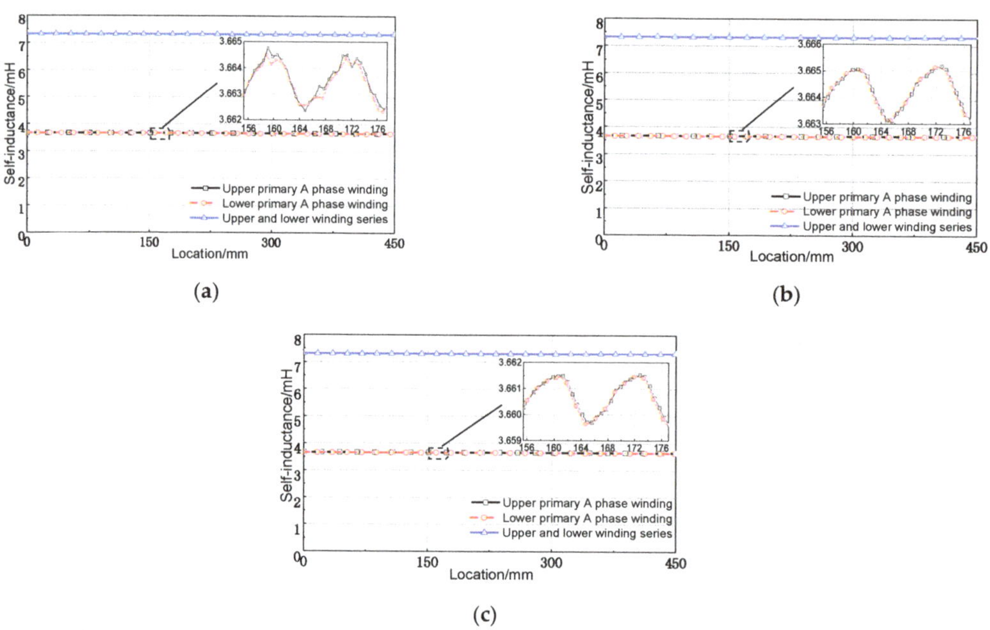

Figure 10. Electromagnetic actuator self-inductance: (**a**) Electromagnetic actuator with radially magnetized secondary; (**b**) Electromagnetic actuator with Halbach three-section magnetized secondary; (**c**) Electromagnetic actuator with Halbach five-section magnetized secondary.

Through the analysis of Sections 3.1–3.4, the performance comparison parameters of three different electromagnetic actuators are obtained, as shown in Table 3.

Table 3. Comparison of different electromagnetic actuators.

Parameters	Radial Magnetization	Halbach Three-Section	Halbach Five-Section
Average thrust (N)	21.42	34.76	30.8
Average normal force (N)	−0.4000	0.0010	0.0008
No-load back EMF (V)	0.39	0.57	0.61
Thrust fluctuation	2.4%	0.84%	0.41%
PM utilization ratio (N/cm^3)	0.67	1.10	0.97
Self-inductance fluctuation	0.12%	0.1%	0.08%

4. Analysis of the Influence of PM Thickness on Electromagnetic Force

Compared to electromagnetic actuators with three-section and five-section, both disturbances can be accepted by magnetic levitation transportation. Furthermore, electromagnetic actuator with Halbach three-section magnetized secondary has significant thrust and the PM utilization ratio. Moreover, it also possesses the low cost.The Halbach three-section magnetized secondary is selected for the electromagnetic actuator. The pole size is shown in Figure 11. Pole pitch τ is defined as the distance between two proximity radially magnetized centers. The τ = 12 mm. When the single magnet width w = 12 mm, the influence of 3~12 mm thickness h on actuator performance is analyzed.

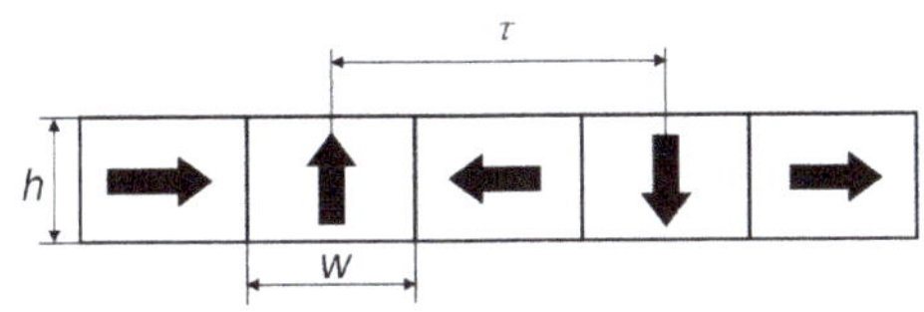

Figure 11. Schematic diagram of pole size.

Figure 12 shows the average electromagnetic thrust of the electromagnetic actuator with PM thickness h. The figure shows that the average electromagnetic thrust increases with the increase of PM thickness h but the curve slope of the average electromagnetic thrust decreases. After analysis, the electromagnetic actuator tangential air gap flux density increases with the magnet thickness. When the thickness is $h > 9$ mm, the growth trend of tangential magnetic density tends to be gradual. The utilization rate of the PM is shown in Table 4. When the secondary pole pitch τ of the Halbach magnet array type is constant, the utilization ratio of the PM decreases with the increase of magnet thickness h.

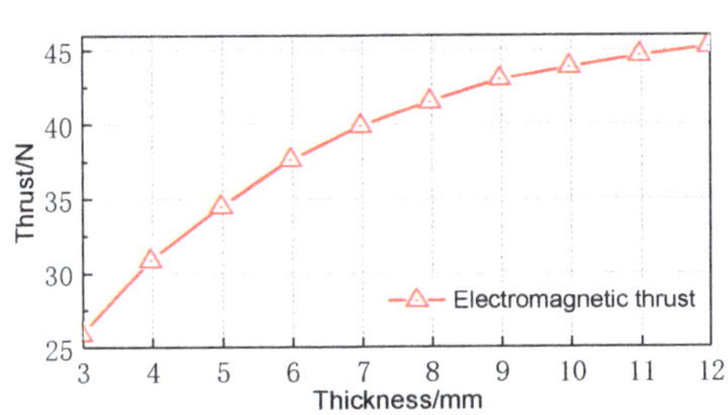

Figure 12. The curve of thrust variation with h.

Figure 13 shows the variation curves of the unilateral average normal force and the bilateral combined average normal force of the electromagnetic actuator with the magnetic pole thickness h. The relationship between radial magnetic density and magnet thickness is the same as tangential magnetic density. So, the average normal force also increases with the magnetic pole thickness h. Like the electromagnetic thrust, the average normal force's curve slope becomes smaller. Since the U-shaped symmetrical structure is adopted, the normal force of the bilateral synthesis tends to zero. Therefore, the normal force is insensitive to the change in magnet thickness.

Table 4. PM utilization ratio of different magnet thicknesses.

Thickness (mm)	PM Utilization Ratio (N/cm³)
3	1.362
4	1.215
5	1.103
6	0.987
8	0.817
9	0.753
10	0.691
11	0.639
12	0.595

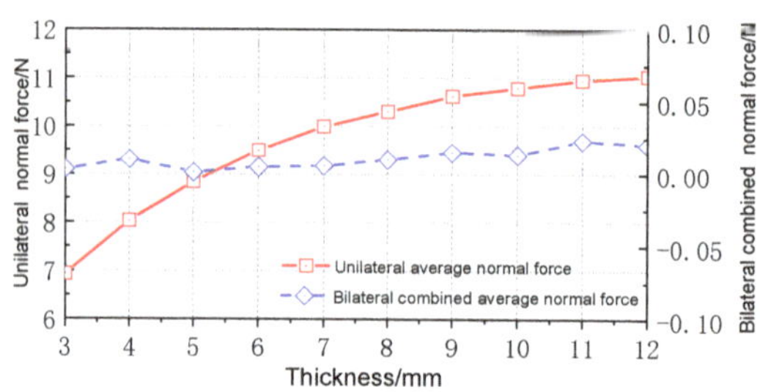

Figure 13. The curve of normal force variation with h.

5. Experimental Validations

To verify the correctness of the above results, design an electromagnetic actuator prototype with a three-section Halbach magnet array secondary structure, as shown in Figure 14. The prototype parameters are shown in Table 5. The prototype's no-load back EMF and the average thrust of different current excitations are tested. The FEM value is consistent with the measured value of the prototype. Still, the FEM value is slightly higher than the measured value because the two-dimensional finite element calculation fails to consider the influence of the transverse end effect and the end magnetic flux leakage.

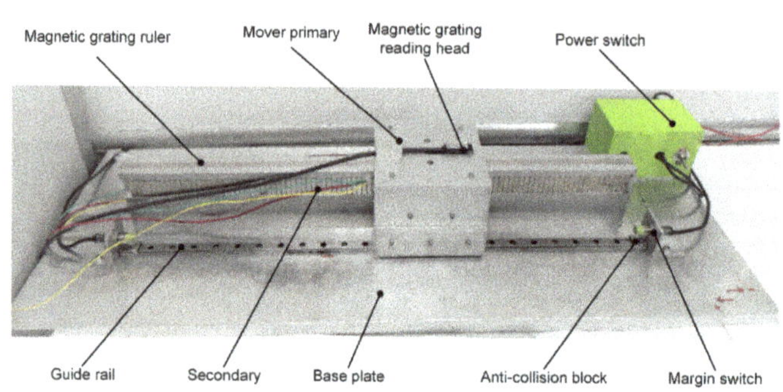

Figure 14. Electromagnetic actuator prototype.

Figure 15 is the no-load back EMF test platform. The linear module is controlled by PLC, which drives the prototype to move synchronously at a speed of 0.24 m/s. The no-load back EMF of the prototype at a speed of 0.24 m/s can be observed from the oscilloscope. The no-load back EMF of the electromagnetic actuator prototype at the synchronous speed of 0.24 m/s is shown in Figure 16. Each phase no-load back EMF is 120° mutual difference, the amplitude is 0.53 V, and they have better sinusoidal waveform. Compared with other windings, the B-phase winding has less amplitude than FEM. This error is due to the large installation gap between the stator and the B-phase winding coil.

Table 5. The parameters of the electromagnetic actuator prototype.

Parameter	Value
Mover mass (Kg)	2.99
Secondary length (mm)	600
Magnet height (mm)	5
Pole pitch (mm)	12
Mover length (mm)	127
Mover width (mm)	109
Distance of move (mm)	460
Resistance (Ω)	7.1
Gap length (mm)	2
Pole pitch (mm)	12

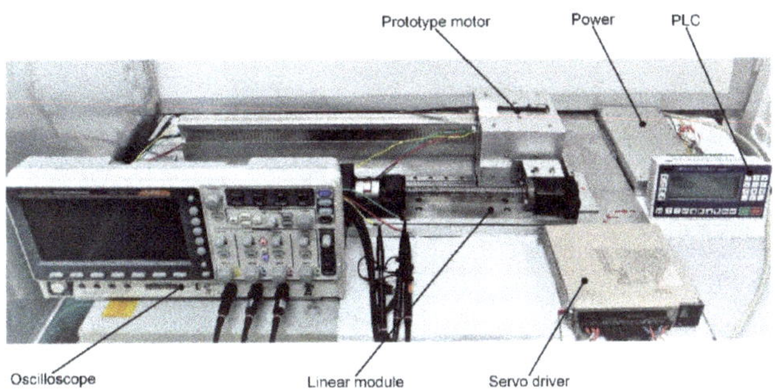

Figure 15. The no-load back EMF test platform of the prototype.

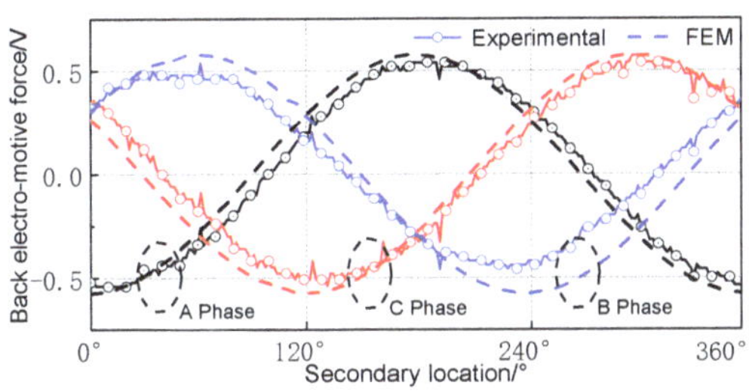

Figure 16. No-load back EMF.

Figure 17 is the electromagnetic thrust test platform of the prototype. Thrust measurement tests under different current excitations were performed Firstly, one side of the force sensor is fixed on the stator, and the other is fixed on the mover. Secondly, the prototype is set to the current control mode by the computer (PC). Thirdly, the PC gives the servo drive different current signals, and the servo drive controls the prototype to work with different currents. Finally, the electromagnetic thrust generated by the prototype at different currents is displayed by the force sensor. The unilateral and bilateral electromagnetic thrusts are measured. The measured values of the prototype compared with FEM are shown in Figure 18: (a) unilateral electromagnetic thrust and (b) bilateral electromagnetic thrust. When the current is less than 1.3 A, the electromagnetic force generated by the prototype is less than the starting thrust, and the measured thrust does not match the FEM value. When the current is more than 1.3 A, the measured thrust of the prototype is consistent with the FEM value.

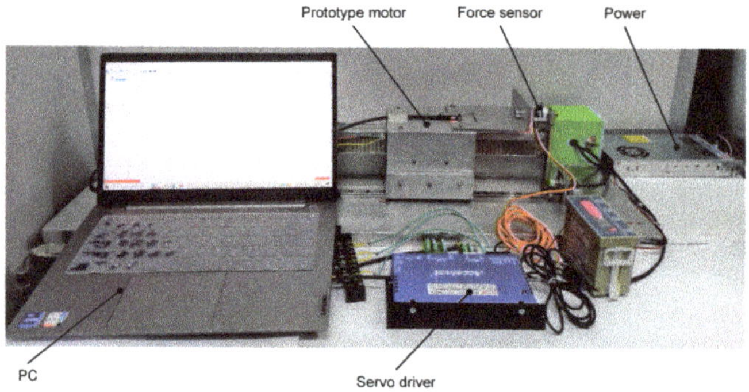

Figure 17. The electromagnetic thrust test platform of the prototype.

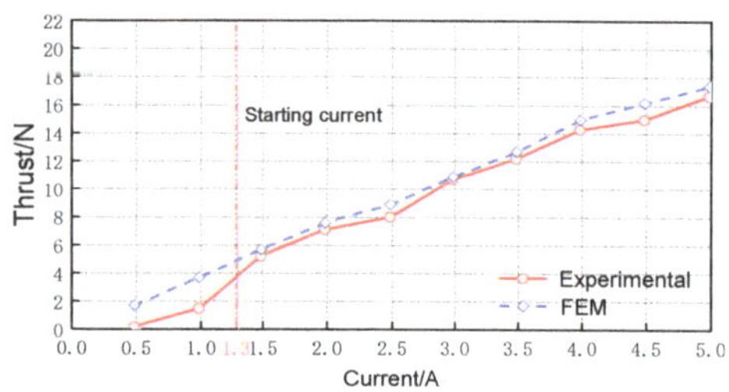

(**a**)

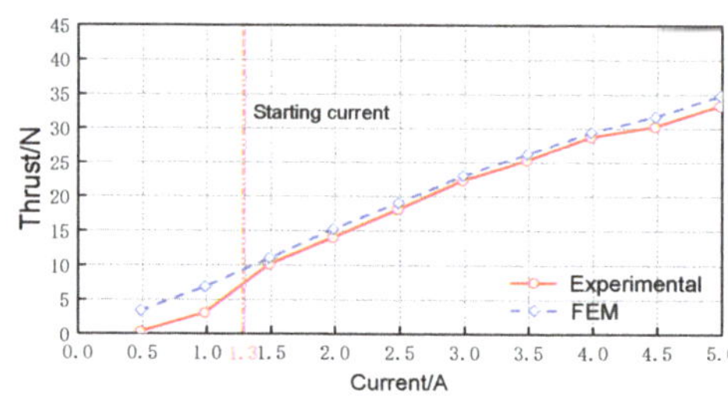

(**b**)

Figure 18. Electromagnetic thrust: (**a**) Unilateral electromagnetic thrust; (**b**) Bilateral electromagnetic thrust.

6. Conclusions

This paper proposes a self-direction electromagnetic actuator applied in magnetic levitation transportation. Electromagnetic actuator characteristics of different secondary structures are analyzed by the FEM. This paper explained the electromagnetic actuator has a big electromagnetic thrust and low disturbance. Afterward, the Halbach three-section magnet array secondary was selected as the secondary structure of the electromagnetic actuator. This structure's secondary thickness's influence on the electromagnetic force was further analyzed. Finally, the experimental prototype was made to verify the correctness of the analysis and effective application to magnetic levitation transport systems. The conclusions are as follows:

(1) For the electromagnetic actuator with the Halbach three-section magnet array secondary compared with the radial magnetization secondary, the electromagnetic thrust improved by 62.1%, the normal force reduced from 0.4 N to 1 mN, and the thrust fluctuation reduced from 2.4% to 0.84%. Moreover, for the secondary structure of the Halbach magnet arrays, the PM utilization ratio improved by 64.2%. This shows the Halbach magnet array can effectively improve the electromagnetic thrust and reduce the disturbance.

(2) Comparing the electromagnetic actuator of the Halbach three-section magnet array secondary and the Halbach five-section magnet array secondary, the Halbach three-section magnet array secondary electromagnetic thrust improved by 12.9%; the normal force improved from 0.8 mN to 1 mN; the thrust fluctuation improved from 0.41% to 0.84%; and the PM utilization ratio improved by 13.4%. The above description electromagnetic actuator by Halbach three-section magnet array secondary has a larger electromagnetic thrust and a higher PM utilization ratio. However, the stability of the motion will be reduced in comparison to the electromagnetic actuator of the Halbach five-section magnetized secondary. In addition, the Halbach five-section magnetized secondary has a high cost. Thus, the Halbach three-section magnet array is more suitable for practical application.

(3) The air gap flux density increases with the thickness of the PM. Hence, the electromagnetic thrust and normal force increases with the thickness. The PM utilization ratio decreases with increasing thickness.

Author Contributions: Conceptualization, J.J., X.W. and C.Z.; methodology, F.S., X.W., C.Z. and J.J.; software, J.J., X.W. and F.X.; validation, F.S., C.Z., W.P., Y.L. and J.J.; formal analysis, F.S., F.X., W.P. and J.J.; investigation, F.S., X.W., F.X. and J.J.; resources, F.S., J.J. and F.X.; data curation, X.W. and W.P.; writing—original draft preparation, J.J., X.W., F.S., Y.L. and C.Z.; writing—review and editing, J.J., X.W., and F.S.; visualization, F.S., X.W., F.X. and J.J.; supervision, F.S. and J.J.; project administration, F.S.; funding acquisition, F.S. All authors have read and agreed to the published version of the manuscript.

Funding: This research is supported by National Natural Science Fund of China (No.52005345, No. 52005344), National Key Research and Development Project (No.2020YFC2006701), Scientific Research Fund Project of Liaoning Provincial Department of Education (No. LFGD2020002), Major Project of the Ministry of Science and Technology of Liaoning Province (No.2022JH1/10400027).

Conflicts of Interest: The authors declare no conflict of interest.

References

1. Lamablawa, F.; Aritonang, S. Application of Magnetic Force to Means of Transportation Magnetic Levitation High-speed Railway. *AJIRSS Asian J. Innov. Res. Soc. Sci.* **2022**, *1*, 68–73. [CrossRef]
2. Vikrant, K.S.; Jayanth, G.R. Diamagnetically levitated nanopositioners with large-range and multiple degrees of freedom. *Nat. Commun.* **2022**, *13*, 3334. [CrossRef]
3. Zhou, J.; Wu, H.; Wang, W.; Yang, K.; Hu, Y.; Guo, X.; Song, C. Online unbalance compensation of a maglev rotor with two active magnetic bearings based on the LMS algorithm and the influence coefficient method. *Mech. Syst. Signal Process.* **2022**, *166*, 108460. [CrossRef]
4. Park, K.; Ahn, K.; Kim, S.; Kwak, Y. Wafer distribution system for a clean room using a novel magnetic suspension technique. *IEEE/ASME Trans. Mechatron.* **1998**, *3*, 73–78. [CrossRef]
5. Ahn, D.; Jin, J.W.; Yun, H.; Jeong, J. Development of a Novel Dual Servo Magnetic Levitation Stage. *Actuators* **2022**, *11*, 147. [CrossRef]
6. Yang, F.; Zhao, Y.; Li, H.; Mu, X.; Zhang, W.; Yue, H.; Liu, R. Design and Analysis of a 2-DOF Electromagnetic Actuator with an Improved Halbach Array for the Magnetic Suspension Platform. *Sensors* **2022**, *22*, 790. [CrossRef]
7. Pandey, A.; Adhyaru, D.M. Control techniques for electromagnetic levitation system: A literature review. *Int. J. Dyn. Control.* **2022**, *10*, 1–11. [CrossRef]
8. Zhang, K.; Xu, F.; Xu, X. Robust iterative learning model predictive control for repetitive motion of maglev planar motor. *IET Electr. Power Appl.* **2022**, *16*, 1189–1200. [CrossRef]
9. Ha, C.-W.; Kim, C.-H.; Lim, J. Experimental Verification of a Magnetic Levitation Transport System for the OLED Display Evaporation Process stipulation Vacuum. *IEEE Robot. Autom. Lett.* **2018**, *4*, 2786–2791. [CrossRef]
10. Kim, J.; Ha, C.-W.; King, G.B.; Kim, C.-H. Experimental development of levitation control for a high-accuracy magnetic levitation transport system. *ISA Trans.* **2020**, *101*, 358–365. [CrossRef]
11. Zhao, C.; Oka, K.; Sun, F.; Harada, A.; Jin, J.; Zhang, M. Design of Zero-Power Control Strategy with Resisting Tilt of Hybrid Magnetic Levitation System. *IEEE Trans. Ind. Electron.* **2022**, *69*, 11394–11402. [CrossRef]
12. Sun, F.; Pei, W.; Zhao, C.; Jin, J.; Xu, F.; Zhang, X. Permanent Maglev Platform Using a Variable Flux Path Mechanism: Stable Levitation and Motion Control. *IEEE Trans. Magn.* **2022**, *58*, 8300410. [CrossRef]
13. Chen, L.Y.; Li, L.C. Development of Linear Motor and Its Key Technologies for Compressor. *Proc. CSEE* **2013**, *33*, 52–68+15. [CrossRef]
14. Zhao, X.; Kou, B.; Huang, C.; Zhang, L. Optimization Design and Performance Analysis of a Reverse-Salient Permanent Magnet Synchronous Motor. *Machines* **2022**, *10*, 204. [CrossRef]
15. Tang, R.Y. *Modern Permanent Magnet Motor: Theory and Design*; Mechanical Industry Press: Beijing, China, 1997; pp. 1–11.
16. Lakhe, R.K.; Chaoui, H.; Alzayed, M.; Liu, S. Universal Control of Permanent Magnet Synchronous Motors with Uncertain Dynamics. *Actuators* **2021**, *10*, 49. [CrossRef]
17. Hammoud, I.; Hentzelt, S.; Xu, K.; Oehlschlagel, T.; Abdelrahem, M.; Hackl, C.; Kennel, R. On Continuous-Set Model Predictive Control of Permanent Magnet Synchronous Machines. *IEEE Trans. Power Electron.* **2022**, *37*, 10360–10371. [CrossRef]
18. Polater, N.; Tricoli, P. Technical Review of Traction Drive Systems for Light Railways. *Energies* **2022**, *15*, 3187. [CrossRef]
19. Jiao, Z.; Cao, Y.; Yan, L.; Li, X.; Zhang, L.; Li, Y. Advancing Motivation Feedforward Control of Permanent Magnetic Linear Oscillating Synchronous Motor for High Tracking Precision. *Actuators* **2021**, *10*, 128. [CrossRef]
20. Xu, J.; Zhang, L.; Meng, D.; Su, H. Simulation, Verification and Optimization Design of Electromagnetic Vibration and Noise of Permanent Magnet Synchronous Motor for Vehicle. *Energies* **2022**, *15*, 5808. [CrossRef]

21. Zhang, T.; Mei, X. Research on Detent Force Characteristics of Winding Segmented Permanent Magnet Linear Synchronous Motor Based on Analytical Model. *Symmetry* **2022**, *14*, 1049. [CrossRef]
22. Habib, A.; Mohd Zainuri, M.A.A.; Che, H.S.; Ibrahim, A.A.; Rahim, N.A.; Alaas, Z.M.; Ahmed, M.M.R. A systematic review on current research and developments on coreless axial-flux permanent-magnet machines. *IET Electr. Power Appl.* **2022**, *16*, 1095–1116. [CrossRef]
23. Jansen, J.W.; Lomonova, E.A.; Compter, J.C. Analysis of the parasitic forces and torques in coreless linear motors. In Proceedings of the 2011 International Conference on Electrical Machines and Systems, Beijing, China, 20–23 August 2011; IEEE: Piscataway, NJ, USA, 2011. [CrossRef]
24. Li, Z.; Wu, Q.; Liu, B.; Gong, Z. Optimal Design of Magneto-Force-Thermal Parameters for Electromagnetic Actuators with Halbach Array. *Actuators* **2021**, *10*, 231. [CrossRef]
25. Ling, Z.; Ji, J.; Zeng, T.; Zhao, W. Design optimization and comparison of linear magnetic actuators under different topologies. *Chin. J. Electr. Eng.* **2020**, *6*, 41–51. [CrossRef]
26. Guo, L.Y.; Wang, H.M.; Gu, X.; Shi, T.N. Modeling and Analyzing of Magnetic Field in Unequal Segmented Halbach Array Permanent Magnet Machine. *Int. J. Appl. Electromagn. Mech.* **2021**, *65*, 333–353. [CrossRef]
27. Krämer, C.; Kugi, A.; Kemmetmüller, W. Optimal force control of a permanent magnet linear synchronous motor based on a magnetic equivalent circuit model. *Control. Eng. Pract.* **2022**, *122*, 10507. [CrossRef]
28. Saarakkala, S.E.; Sokolov, M.; Hosseinzadeh, R.; Hinkkanen, M. Levitation control for a double-sided bearingless linear motor based on feedback linearization. In Proceedings of the 2019 IEEE Energy Conversion Congress and Exposition (ECCE), Baltimore, MD, USA, 29 September–3 October 2019; IEEE: Piscataway, NJ, USA, 2019. [CrossRef]
29. Zhou, H.; Deng, H. Hybrid fuzzy decoupling control for a precision maglev motion system. *IEEE/ASME Trans. Mechatron.* **2017**, *23*, 389–401. [CrossRef]
30. Chen, Y.; Zhang, K. Electromagnetic Force Calculation of Conductor Plate Double Halbach Permanent Magnet Electrodynamic Suspension. *Appl. Comput. Electromagn. Soc. J. (ACES)* **2014**, *29*, 916–922. Available online: https://journals.riverpublishers.com/index.php/ACES/article/view/10855 (accessed on 5 June 2022).
31. Tan, Q. Electromagnetic Force Fluctuation Analysis and Suppression of Precision Permanent Magnet Linear Synchronous Motor. Ph.D. Thesis, Harbin Institute of Technology, Harbin, China, June 2021. [CrossRef]
32. Liu, G.; Chen, M.; Zhao, W.; Chen, Q. Design and Analysis of Five-Phase Fault-Tolerant Interior Permanent-Magnet Vernier Machine. *IEEE Trans. Appl. Supercond.* **2016**, *26*, 0604805. [CrossRef]
33. Jian, L. Research on Permanent Magnet Linear Synchronous Motor with Low Normal Force Fluctuation. Master's Thesis, Shenyang University of Technology, Shenyang, China, May 2020. [CrossRef]

actuators

MDPI

Article

Improvement in Position Response of Laser Focus Controlled Magnetic Actuator Based on Mixed Sensitivity Robust Control

Liping Wu [1], Ling Tong [1,*], Guang Yang [1,2], Qi Zhang [1], Fangchao Xu [1], Junjie Jin [1], Xiaoyou Zhang [1,3] and Feng Sun [1]

1 School of Mechanical Engineering, Shenyang University of Technology, Shenyang 110870, China
2 School of Mechanical Engineering, Shenyang Aerospace University, Shenyang 110136, China
3 Department of Mechanical Engineering, Nippon Institute of Technology, Saitama 345-8501, Japan
* Correspondence: tongling@sut.edu.cn

Abstract: The relative position between the laser beam and the nozzle is controlled by laser-focus-controlled magnetic actuators to achieve non-coaxial laser cutting and improve laser cutting efficiency. In this paper, a 3-DOF (degrees of freedom) magnetic actuator is designed to solve the inconsistency of the laser beam focus and the nozzle focus in off-axis laser cutting. A mixed sensitivity robust controller is designed, and its simulation analysis and experimental research are carried out. First, the kinetic mathematical equations are established according to the structure of the actuator. Then, a mixed sensitivity robust controller is designed and analyzed using MATLAB/Simulink. The control performance is simulated and analyzed under 20% parameter variation and pulse disturbance with an uncertain mathematical model and external disturbance, respectively. Finally, the experimental study of the step response of the actuator is carried out. The experimental results show that the step response of the actuator in the Y, X, and θ directions can quickly reach the steady-state value. Furthermore, the steady-state error in the X is 1.6%; the steady-state error in the Y is 0.39%; the steady-state error in the θ is 0.45%. Their errors are all less than 0.025 mm, so they meet the position performance requirements. It can provide technical support for laser off-axis cutting.

Keywords: electromagnetic drive; mixed sensitivity; robust control; differential control

Citation: Wu, L.; Tong, L.; Yang, G.; Zhang, Q.; Xu, F.; Jin, J.; Zhang, X.; Sun, F. Improvement in Position Response of Laser Focus Controlled Magnetic Actuator Based on Mixed Sensitivity Robust Control. *Actuators* **2023**, *12*, 4. https://doi.org/10.3390/act12010004

Academic Editor: Kirill Poletkin

Received: 5 November 2022
Revised: 6 December 2022
Accepted: 19 December 2022
Published: 21 December 2022

1. Introduction

Laser cutting has become the most widely used cutting technology [1–3] in the current manufacturing industry due to its characteristics such as no contact with the workpiece during processing [4,5], concentrated energy, low contamination of the workpiece, easy guidance, and easy automation control. The manufacturing industry places higher demands on machining accuracy and processing efficiency. Traditional laser cutting technology cannot meet the required cutting quality and efficiency. Therefore, improving laser cutting quality [6,7] and cutting efficiency [8] has become one of the hot spots of research and concern for scholars in related fields [9,10]. Elsheikh A. et al. conducted cutting experiments on CO_2 laser cutting of PMMA sheets, analyzed the factors affecting the kerf geometry, and used a genetic algorithm to select the optimal cutting parameters. The results showed that this method could significantly improve kerf quality [11]. Vora J. et al. analyzed the influence of different laser cutting process parameters on the incision quality. The results showed that the gas pressure had the greatest influence on the cutting quality [12]. Hong L. et al. studied the laser cutting of silicon steel sheets. They verified the operation of rotating airflow laser cutting by adding a cyclone deslagger at the bottom of the cut workpiece [13]. Quintero F. et al. designed a cutting experiment that was performed by changing the angle between the laser lens axis and the auxiliary gas axis. The results showed that the quality and efficiency of the cut could be improved [14,15]. The laser cutting head and the auxiliary gas of variable angle laser cutting are separated, so the cutting quality and efficiency differ in each feeding direction. At the same time, there is no gas pressure at the laser beam, which may cause slag splashing and contamination of the lens, seriously affecting the cutting quality and efficiency. The problem of variable angle cutting can be solved by non-coaxial cutting of the laser beam and the gas flow beam. The principle of this cutting method is to separate the laser cutting head's auxiliary gas from the laser beam's central axis for laser cutting. Riveiro A. et al. conducted an experimental study of cutting aluminum alloys using an off-axis supersonic nozzle, analyzed the effect of machining parameters on cutting speed and cutting quality, and obtained the best results at high-frequency pulses [16]. They also developed an off-axis supersonic rectangular nozzle. They compared it with a coaxial nozzle, and the results showed that the newly designed off-axis supersonic rectangular nozzle gas injection system significantly improved cutting quality and efficiency [17]. Yagi A. et al. investigated the effect of off-axis position on laser cutting quality in nitrogen-assisted gas fiber laser cutting. Laser cutting was studied by varying the relative positions of the laser beam's central axis and the nozzle's central axis. The experimental results showed that the nozzle shifting setup could over obtain a comparable dross height for the coaxial setup and reduce the consumption of auxiliary gas [18]. These studies can effectively improve the quality of laser cutting. Based on the above literature review, it is necessary to design a high-speed, high-precision, interference-resistant, and compact actuator to control the position of the laser lens for automatic non-coaxial laser cutting.

Conventional mechanical actuators are characterized by complex mechanical structures between components, the need for lubrication, slow response speed, and low positioning accuracy [19]. Therefore, they cannot meet the high speed, precision, and compact structure standards required. Electromagnetic drive technology has the characteristics of a simple structure, compact structure, and no need for lubrication [20–26], which can meet the needs of non-coaxial laser cutting actuators. Zhang X. et al. designed a high-speed, high-precision electromagnetic actuator that can be connected with traditional laser cutting machine tools to control the relative position of the laser beam's axis and the auxiliary gas's axis [27]. He D. et al. proposed a 6-DOF (degrees of freedom) magnetically levitated lens-driven actuator for off-axis laser cutting control [28,29]. M Y. et al. investigated a 2-DOF electromagnetic actuator for laser off-axis cutting and achieved a positioning stroke of ± 500 μm [30]. In the magnetic drive system, there are some unmodeled parts and uncertain disturbances, and robust control can stabilize the control under model uncertainty [31,32]. Due to the characteristics of robust control, there have been many studies in the past decades, among which hybrid sensitivity robust control is the commonly used method [33–36]. However, most of the existing research on magnetically driven actuators was carried out on the basis of actuator model determination. The electromagnetic actuator is subject to strong nonlinearity and uncertainty disturbances. Considering the exact model of nonlinearity will increase the difficulty of controller design.

In summary, this paper takes the LCY-YAG laser cutting machine, a Beijing Zhengtian Hengye Numerical Control Technology Co., LTD. product, as the object and designs a 3-DOF actuator with electromagnetic drive and permanent magnet self-reset. The actuator uses three groups of differentially arranged electromagnets to provide driving force, a pair of axially placed and axially magnetized permanent magnets to provide recovery force, and the position of the actuator is detected by the eddy current displacement sensor. There are some parameters of the actuator that cannot be accurately modeled. The mixed sensitivity robust control strategy is selected to control the system stably. By choosing the weighting function, the controller model is obtained [37–40], and the position response simulation and experiment of the actuator are carried out. The actuator structure is simple and reliable. Experimental results show that the actuator can achieve position response and trajectory centering in the xOy plane and has anti-interference characteristics. The steady-state error is 0.016 mm, which meets the design requirement of the laser cutting machine's precision of 0.025 mm. The novelty of this paper is the design of a 3-DOF electromagnetic actuator for off-axis laser cutting applications. The self-resetting module of the actuator adopts a single magnetic spring structure, which solves the deviation of the position of the electromagnetic actuator caused by inaccurate processing or installation caused by the self-resetting module composed of three magnetic springs designed in the literature [41].

This paper consists of six sections. In Section 2, the structure and mathematical model of the actuator are presented. The design of the hybrid sensitivity robust controller is given in Section 3. In Section 4, the simulation analysis is performed. Experimental verification is carried out in Section 5. Conclusions are given in Section 6.

2. Structural and Mathematical Models

2.1. Structure of the Actuator

The purpose of the actuator is to drive the position of the laser lens to achieve off-axis laser cutting. The actuator mainly consists of a drive module, a self-resetting module, a support module, and a detection module, as shown in Figure 1. The driving part consists of three pairs of differential electromagnets (EM), permalloy, frame, and moving plate. The self-resetting part consists of two ring-shaped permanent magnets (PM) of the same size, axially magnetized and axially mounted in parallel. One ring-shaped permanent magnet is fixed to the frame, and the other is fixed to the moving plate. The support component consists of a ceramic thrust ball bearing, the frame, and the moving plate. The detection component consists of an eddy current sensor and a position detection block.

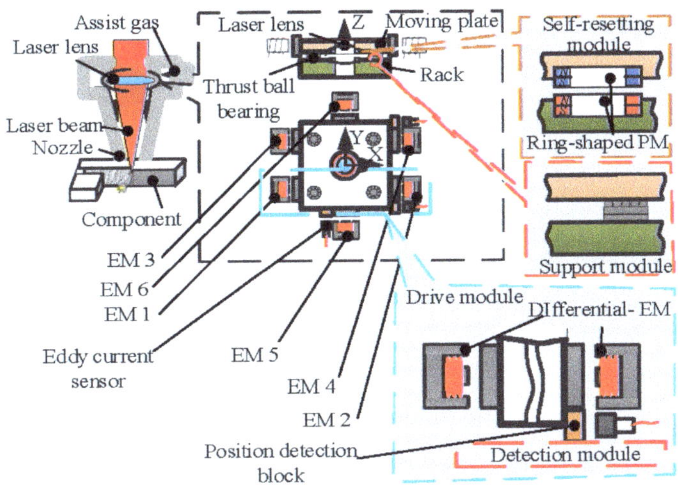

Figure 1. Actuator structure. The mechanical structure of the actuator is shown.

2.2. The Principle of the Actuator

The actuator can translate in the X and Y directions and rotate around the Z axis. The actuator principle relies on a differential electromagnet to provide the driving force, which makes the moving plate move, as shown in Figure 2. The recovery force and initial stiffness are provided by two axially magnetized permanent magnets. The position real-time feedback signal is provided by an eddy current displacement sensor. The motion in the X direction is driven by two pairs of differential electromagnets consisting of electromagnets 1 and 2 and electromagnets 3 and 4. The motion in the Y direction is driven by a pair of differential electromagnets consisting of electromagnets 5 and 6. The rotation motion around the Z axis is acted by three pairs of differential electromagnets.

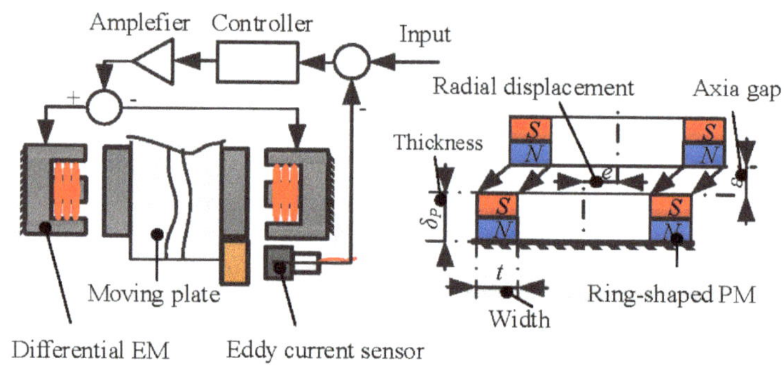

Figure 2. Working principle of the actuator.

2.3. *The Mathematical Model of the Actuator*

The force analysis of the actuator is shown in Figure 3. According to the structure of the actuator, when taking the equilibrium position, the center of mass O of the moving plate is the origin of the coordinates, and its coordinates are represented by (x, y, θ). The X and Y are pointing, and counterclockwise rotation around the Z axis is shown in Figure 3 as the positive direction. When an input signal is given to the actuator, the actuator moves the moving plate, and the signals detected by the sensor are x_1, x_2, and y_1; the driving forces provided by the three pairs of differential electromagnets are F_{x1}, F_{x2}, and F_{y1}; the ring-shaped permanent magnet The provided restoring force is F_e (which can be decomposed into F_{ex} and F_{ey}); frictional resistance f_x, f_y; frictional resistance torque M_θ. Under the action of these resultant forces (or resultant torque), the moving plate moves to a new equilibrium position O' (x_1, x_2, y_1). N_c is the detection point of sensor 1; M_c is the detection point of sensor 2; P_c is the detection point of sensor 3; N is the action point of the differential electromagnet force composed of electromagnet 3 and electromagnet 4; M is the action point of the differential electromagnet force composed of electromagnet 1 and electromagnet 2; P is the action point of the differential electromagnet force composed of electromagnet 5 and electromagnet 6.

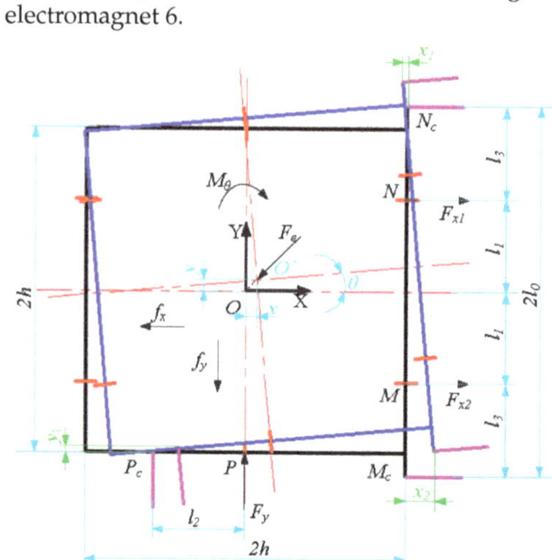

Figure 3. Force analysis diagram.

According to the force analysis, the movement of the center of mass of the moving plate of the actuator is X, Y, and θ, and the signals detected by the sensor are x_1, x_2, and

y_1. To facilitate modeling, the coordinate system is transformed. It can be known from the geometric relationship:

$$\begin{cases} x = [(x_1 + \varepsilon x_1) + (x_2 + \varepsilon x_2)]/2 \\ y = y_1 + \varepsilon y_1 \\ \theta \approx tan\theta = (x_2 - x_1)/2l_0 \end{cases} \tag{1}$$

where ε_{x1}, ε_{x2}, and ε_{y1} are the influence of the rotation angle θ on the detected values x_1, x_2, and y_1.

It can be known from the geometric relationship (as shown in Figure 4).

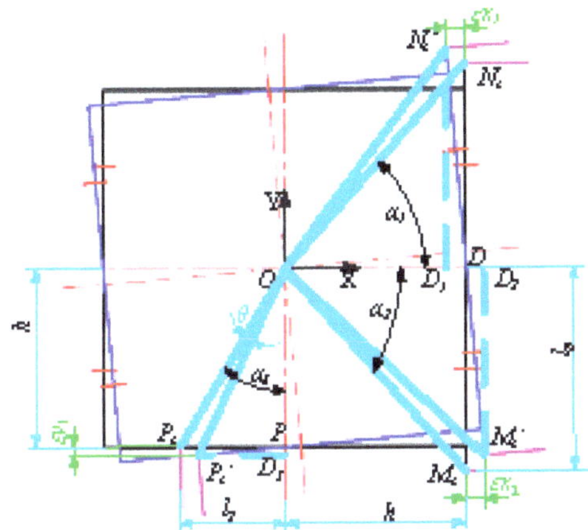

Figure 4. Analysis of the influence of the rotation angle on the detection position.

In RtΔ*ODNC* and RtΔ*OD*1*NC*′,

$$|\varepsilon x_1| = (l_0/\sin\alpha_1) \cdot [\cos\angle(\alpha_1 + \theta) - \cos\angle\alpha_1] \tag{2}$$

In RtΔ*ODMC* and RtΔ*OD*2*MC*′, the geometric relationship:

$$|\varepsilon x_2| = (l_0/\sin\alpha_2) \cdot [\cos\angle(\alpha_2 - \theta) - \cos\angle\alpha_2] \tag{3}$$

In RtΔ*OPPC* and RtΔ*OD*3*PC*′,

$$|\varepsilon y_1| = (l_2/\sin\alpha_3) \cdot [\cos\angle(\alpha_3 - \theta) - \cos\angle\alpha_3] \tag{4}$$

The detection position in the x direction is symmetrical, so $\alpha_1 = \alpha_2$. After arranging Formulas (2)–(4) and bringing them into Formula (1), we obtain:

$$\begin{cases} x = (x_1 + x_2)/2 + h \cdot (\cos\theta - 1) \\ y = y_1 + h \cdot (\cos\theta - 1) + l_2 \sin\theta \\ \theta \approx tan\theta = (x_2 - x_1)/(2l_0) \end{cases} \tag{5}$$

The value of θ is extremely small, so if $\cos\theta$ approaches one and $\sin\theta$ approaches zero, then Formula (5) can be simplified as:

$$\begin{cases} x = (x_1 + x_2)/2 \\ y = y_1 \\ \theta = (x_2 - x_1)/(2l_0) \end{cases} \tag{6}$$

The dynamic equation of the actuator is:

$$\begin{cases} m\ddot{x} = F_{x1} + F_{x2} - F_{ex} - f_x \\ m\ddot{y} = F_y - F_{ey} - f_y \\ J\ddot{\theta} = -F_{x1} \cdot l_1 + F_{x2} \cdot l_1 - M_\theta \end{cases} \tag{7}$$

where m is mass of moving plate, J is the moment of inertia of the moving plate around the Z axis; x is the displacement of the moving plate along the X direction; y is the displacement of the moving plate along the Y direction; θ is the angle that the moving plate rotates around the Z axis; F_{x1} is the resultant force of the differential electromagnet composed of electromagnet 3 and electromagnet 4; F_{x2} is the resultant force of the differential electromagnet composed of electromagnet 1 and electromagnet 2; F_y is the resultant force of the differential electromagnet composed of electromagnet 5 and electromagnet 6; F_{ex} is the component force of the restoring force F_e generated by the annular permanent magnet in the X direction; F_{ey} is the component force of the restoring force F_e generated by the annular permanent magnet in the y direction; f_x is the friction force generated in the X direction; f_y is the friction force generated in the Y direction; M_θ is the frictional resistance torque generated around the Z axis; l_1 is the distance from the action point of F_{x1} to the central axis.

The electromagnet is a nonlinear element arranged differentially and can be linearized near the operating point. The force of the differential electromagnet is:

$$F_\lambda = \frac{4\mu_0 N^2 A}{(2\delta)^2} \cdot \frac{i_0^2}{d_0^3} \cdot \lambda + \frac{4\mu_0 N^2 A}{(2\delta)^2} \cdot \frac{i_0}{d_0^2} \cdot i_\lambda \,;(\lambda = \Delta x_1; \Delta x_2; \Delta y) \tag{8}$$

where μ_0 is magnetic permeability; N is the number of turns of the electromagnet coil; A is the cross-sectional area of the magnetic circuit in the E-type iron core; δ is the magnetic circuit reluctance coefficient; i_0 is the bias current of the electromagnet coil; d_0 is the initial air gap between the electromagnet and the armature; λ is the displacement change detected by the sensor; i_λ is control current.

Displacement stiffness coefficient k_x and the current stiffness coefficient k_i is introduced. Formula (8) is simplified as:

$$\begin{array}{c} F_\lambda = k_x \cdot \lambda + k_i \cdot i_\lambda; (\lambda = \Delta x_1, \Delta x_2, \Delta y) \\ k_x = 4 \cdot \frac{\mu_0 N^2 A}{(2\delta)^2} \cdot \frac{i_0^2}{d_0^3}; k_i = 4 \cdot \frac{\mu_0 N^2 A}{(2\delta)^2} \cdot \frac{i_0}{d_0^2}; \end{array} \tag{9}$$

λ is positive in the direction away from the sensor, so it is a unified force system coordinate system and a generalized coordinate system.

Therefore, $\Delta x_1 = -x_1$, $\Delta x_2 = -x_2$, $\Delta y = y$.

According to Kirchhoff's law and the virtual displacement method, the restoring force of the annular permanent magnet is solved, and the restoring force F_e is obtained as [42]. The F_e can be found in Appendix A.

F_e is nonlinear. The k_e of Formula (10) can be found in Appendix A. In order to reduce the complexity of the control system, Taylor series expansion is performed at $e_0 = 0$, and the high-order infinitesimal quantities are omitted to obtain:

$$F_e = k_e \cdot e \tag{10}$$

where k_e is the restoring force coefficient of the ring permanent magnet.

Formulas (6), (9) and (10) are brought into Formula (7) to solve the dynamic equation.

$$\begin{cases} m\ddot{x} = -(2k_x + k_e)x + k_i(i_{x_1} + i_{x_2}) - c\dot{x} \\ m\ddot{y} = (k_x - k_e)y + k_i i_y - c\dot{y} \\ J\ddot{\theta} = -2l_0 l_1 k_x \theta + l_1 k_i (i_{x_2} - i_{x_1}) - c_\theta \dot{\theta} \end{cases} \tag{11}$$

The Laplace transform of Formula (11) is:

$$\begin{cases} X(s) = \frac{k_i}{ms^2+cs+(2k_x+k_e)}[I_{x_1}(s) + I_{x_2}(s)] \\ Y(s) = \frac{k_i}{ms^2+cs+(k_e-k_x)} I_y(s) \\ \theta(s) = \frac{l_1 k_i}{Js^2+c_\theta s+2l_0 l_1 k_x}[I_{x_2}(s) - I_{x_1}(s)] \end{cases} \tag{12}$$

3. Mixed Sensitivity Robust Controller Design

According to Formulas (6), (9) and (10), it can be seen that the measurement value of the sensor ignores the influence of the rotation angle and the driving force, and the restoring force ignores the high-order term of the Taylor expansion. The actuator has external environmental interference and some inevitable modeling errors. A mixed sensitivity robust control strategy is adopted to ensure the system can work stably under the above uncertain conditions. The system control block diagram is shown in Figure 5.

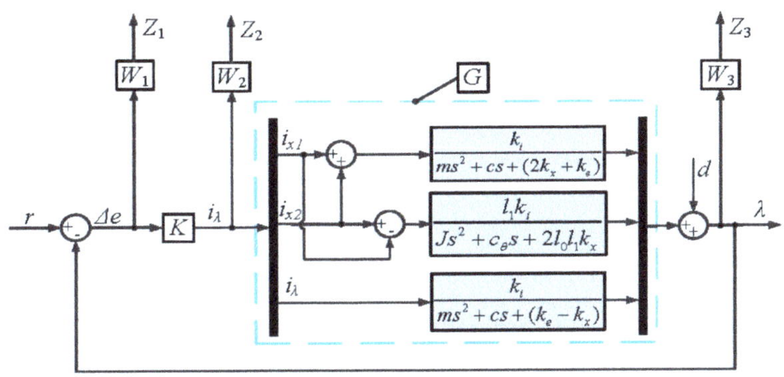

Figure 5. System control diagram. The control block diagram of the system is shown.

G is the system open-loop transfer function; K is the controller; Δe is position error; r is reference input; i_λ is control current; λ is system output; d is the system interference signal; W_1 is performance weight function; W_2 is control function; W_3 is robust weight function; Z_1, $Z_{2,}$ and Z_3 are evaluation system output.

The closed-loop transfer functions from r to Δe, $i_{\lambda,}$ and λ are:

$$\begin{cases} S = \frac{\Delta E(s)}{R(s)} = \frac{I}{I+GK} \\ R = \frac{I_\lambda(s)}{R(s)} = \frac{K}{I+GK} = KS \\ T = \frac{\lambda(s)}{R(s)} = \frac{GK}{I+GK} = I - S \end{cases} \tag{13}$$

where S is the system sensitivity function, and T is the system-supplemented sensitivity function.

The purpose of designing the mixed sensitivity $H\infty$ controller is to reasonably select W_1, W_2, and W_3, so that Formula (14) meets the requirements.

$$\|W_1 S \quad W_2 R \quad W_3 T\|_\infty^T = \gamma \leq 1 \tag{14}$$

In general, in electromagnetic drive systems, external disturbances are at low frequencies, and unmodeled disturbances are at high frequencies. Therefore, in selecting the weighting function, W_1 and W_3 should have low-pass and high-pass filtering properties, respectively.

To avoid increasing the actuator order, W_2 usually takes a constant value. According to the above analysis, W_1, W_2, and W_3 are selected as:

$$\begin{cases} W_1(s) = k_1/(1+\tau_1 s) \\ W_2(s) = k_2 \\ W_3(s) = k_3 s(1+\tau_3 s) \end{cases} \quad (15)$$

where k_1, k_2, k_3, τ_1, and τ_3 are the coefficients of the weighting function, respectively.

The values in Table 1 are brought into Formula (10) to calculate k_e, and the results are shown in Table 2.

Table 1. Parameters of ring permanent magnets.

System Parameters	Symbol	Value
Air permeability	μ_0 (H/m)	$4\pi \times 10^{-7}$
Relative permeability	μ_r	1.05
Material remanence	B_r (T)	1.18
Number of permanent magnets	n	2
Permanent magnet thickness	δ_P (mm)	4
Permanent magnet inner diameter	r_1 (mm)	15
Permanent magnet outer diameter	r_2 (mm)	20.5
Permanent magnet axial air gap	ε (mm)	4
Permanent magnet radial displacement	e (mm)	$0 < e < 4$

Table 2. Parameters of system.

System Parameters	Symbol	Value
Moving plate weight	m (kg)	1.3995
The distance between two pairs of electromagnet force centers in the X direction	l_1 (mm)	44.5
The X direction sensor measures the distance from the point to the center	l_0 (mm)	167
The rotational inertia of the moving plate	J (kg·mm^2)	0.004972
The magnetic permeability of air	μ_0 (H/m)	$4\pi \times 10^{-7}$
Number of coil turns	N	128
The cross-sectional area of the magnetic circuit	S (mm^2)	912
Magnetic circuit reluctance coefficient	δ	1.38
X direction damping	c_x	10
Y direction damping	c_y	10
θ direction damping	c_θ	10
Initial current	i_0 (A)	1.2
Initial gap	d_0 (mm)	2
Permanent magnet stiffness	k_e (N/m)	3032.7

The mixing sensitivity controller K is solved by selecting the appropriate coefficients.

The parameters in Table 2 are brought into Formula (12) to obtain the transfer function of the system as:

$$\begin{cases} G_x(s) = \frac{2.96}{1.3995s^2+10s+6582.1} \\ G_y(s) = \frac{2.96}{1.3995s^2+10s+1258} \\ G_\theta(s) = \frac{0.13}{0.00497s^2+10s+13.19} \end{cases} \quad (16)$$

The Y direction parameters are selected as $k_1 = 1500$, $k_2 = 0.002$, $k_3 = 0.01$, $\tau_1 = 100$, $\tau_3 = 0.01$, and the calculated $\gamma = 0.9338$. The controller is:

$$G_y(s) = \frac{6.0557e05(s^2+7.145s+898.9)}{(s+0.01)(s^2+1328.37s+33057.11)} \quad (17)$$

The X direction parameters are selected as $k_1 = 1500$, $k_2 = 0.0004$, $k_3 = 0.01$, $\tau_1 = 100$, $\tau_3 = 0.01$, and the calculated $\gamma = 0.9186$. The controller is:

$$G_x(s) = \frac{1.7905e06(s^2 + 7.145s + 898.9)}{(s + 876.5)(s + 66.32)(s + 0.01)} \tag{18}$$

The θ direction parameters are selected as $k_1 = 2000$, $k_2 = 0.0004$, $k_3 = 0.01$, $\tau_1 = 100$, $\tau_3 = 0.01$, and the calculated $\gamma = 0.8411$. The controller is:

$$G_\theta(s) = \frac{6.6604e05(s^2 + 2012.32s + 2654.52)}{(s + 2090)(s + 350.2)(s + 0.01)} \tag{19}$$

4. Simulation Analysis

4.1. Analysis of the Performance of the Control System

4.1.1. Analysis of the Performance of the Y Direction Control System

The amplitude-frequency characteristic curves of $S(s)$ and $W_1{}^{-1}(s)$ in the Y direction are shown in Figure 6a. The amplitude of $\overline{\sigma}(S(s))$ is extremely small in the low-frequency range, and its maximum amplitude is less than $\overline{\sigma}(W_1{}^{-1}(s))$ in the whole frequency range, which satisfies Formula (20). Therefore, the anti-interference ability and tracking characteristics of the system are ensured. The amplitude–frequency characteristic curves of $T(s)$ and $W_3{}^{-1}(s)$ are shown in Figure 6b. $\overline{\sigma}(T(s))$ has a large slope in the high-frequency range, and its maximum amplitude is smaller than $\overline{\sigma}(W_3{}^{-1}(s))$ in the whole frequency range, which satisfies Formula (21). It ensures that high-frequency noise is rapidly attenuated.

$$\overline{\sigma}(S(s)) \le \overline{\sigma}\left(W_1^{-1}(s)\right) \tag{20}$$

$$\overline{\sigma}(T(s)) \le \overline{\sigma}\left(W_3^{-1}(s)\right) \tag{21}$$

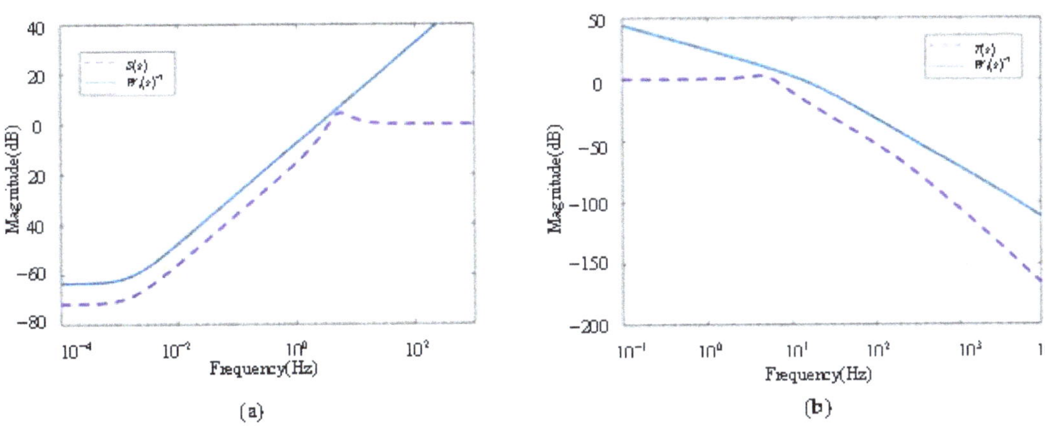

Figure 6. Amplitude–frequency curve. (**a**) Amplitude–frequency curve of $S(s)$ and $W_1{}^{-1}(s)$; (**b**) Amplitude–frequency curve of $T(s)$ and $W_3{}^{-1}(s)$.

4.1.2. Analysis of the Performance of the X Direction Control System

The amplitude–frequency characteristic curves of $S(s)$ and $W_1{}^{-1}(s)$ in the X direction are shown in Figure 7a. The amplitude of $\overline{\sigma}(S(s))$ is extremely small in the low-frequency range, and its maximum amplitude is less than $\overline{\sigma}(W_1{}^{-1}(s))$ in the whole frequency range, which satisfies Formula (20). Therefore, the anti-interference ability and tracking characteristics of the system are ensured. The amplitude-frequency characteristic curves of $T(s)$ and $W_3{}^{-1}(s)$ are shown in Figure 7b. $\overline{\sigma}(T(s))$ has a large slope in the high-frequency range, and its maximum amplitude is smaller than $\overline{\sigma}(W_3{}^{-1}(s))$ in the whole frequency range, which satisfies Formula (21). It ensures that high-frequency noise is rapidly attenuated.

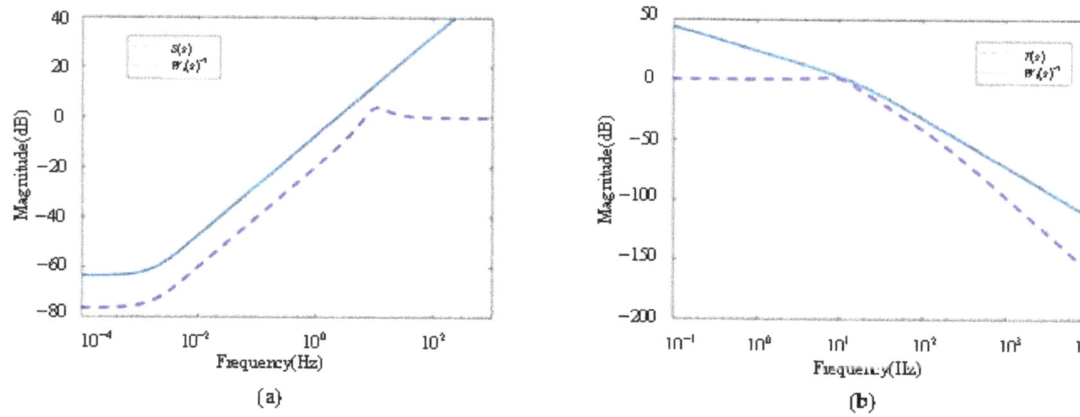

Figure 7. Amplitude–frequency curve. (**a**) Amplitude–frequency curve of $S(s)$ and $W_1{}^{-1}(s)$; (**b**) Amplitude–frequency curve of $T(s)$ and $W_3{}^{-1}(s)$.

4.1.3. Analysis of the Performance of the θ Direction Control System

The amplitude–frequency characteristic curves of $S(s)$ and $W_1{}^{-1}(s)$ in the θ direction are shown in Figure 8a. The amplitude of $\overline{\sigma}(S(s))$ is extremely small in the low-frequency range, and its maximum amplitude is less than $\overline{\sigma}(W_1{}^{-1}(s))$ in the whole frequency range, which satisfies Formula (20). Therefore, the anti-interference ability and tracking characteristics of the system are ensured. The amplitude–frequency characteristic curves of $T(s)$ and $W_3{}^{-1}(s)$ are shown in Figure 8b. $\overline{\sigma}(T(s))$ has a large slope in the high-frequency range, and its maximum amplitude is smaller than $\overline{\sigma}(W_3{}^{-1}(s))$ in the whole frequency range, which satisfies Formula (21). It ensures that high-frequency noise is rapidly attenuated.

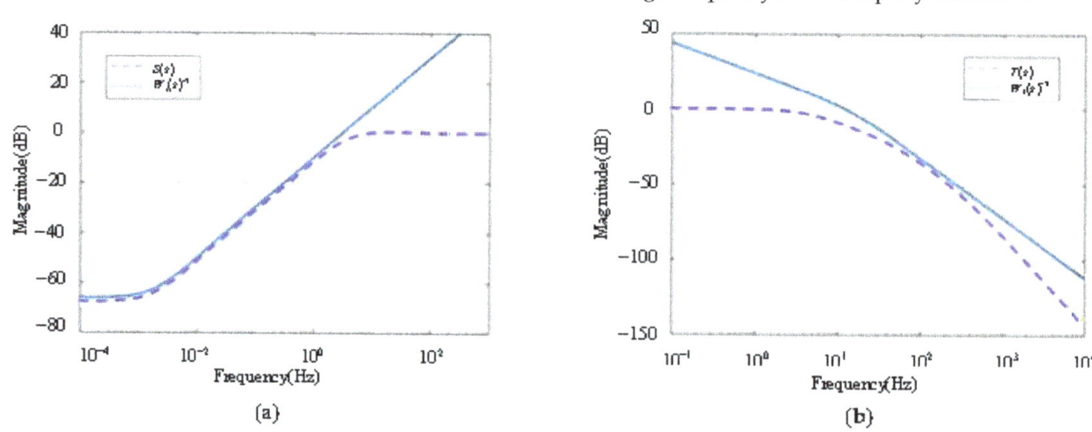

Figure 8. Amplitude–frequency curve. (**a**) Amplitude–frequency curve of $S(s)$ and $W_1{}^{-1}(s)$; (**b**) Amplitude–frequency curve of $T(s)$ and $W_3{}^{-1}(s)$.

4.2. Analysis of the System Response of the Standard Model

To verify the effect of the controller, MATLAB/Simulink is used to simulate and analyze the system.

The Y direction simulation: At 0.2 s, a signal of 0.1 mm is input in the Y direction. The system response results are shown in Figure 9a. The X direction simulation: At 0.2 s, a signal of 0.1 mm is input in the X direction. The system response results are shown in Figure 9b. The θ direction simulation: At 0.2 s, a signal of 0.1 mrad is input in the θ direction. The system response results are shown in Figure 9c.

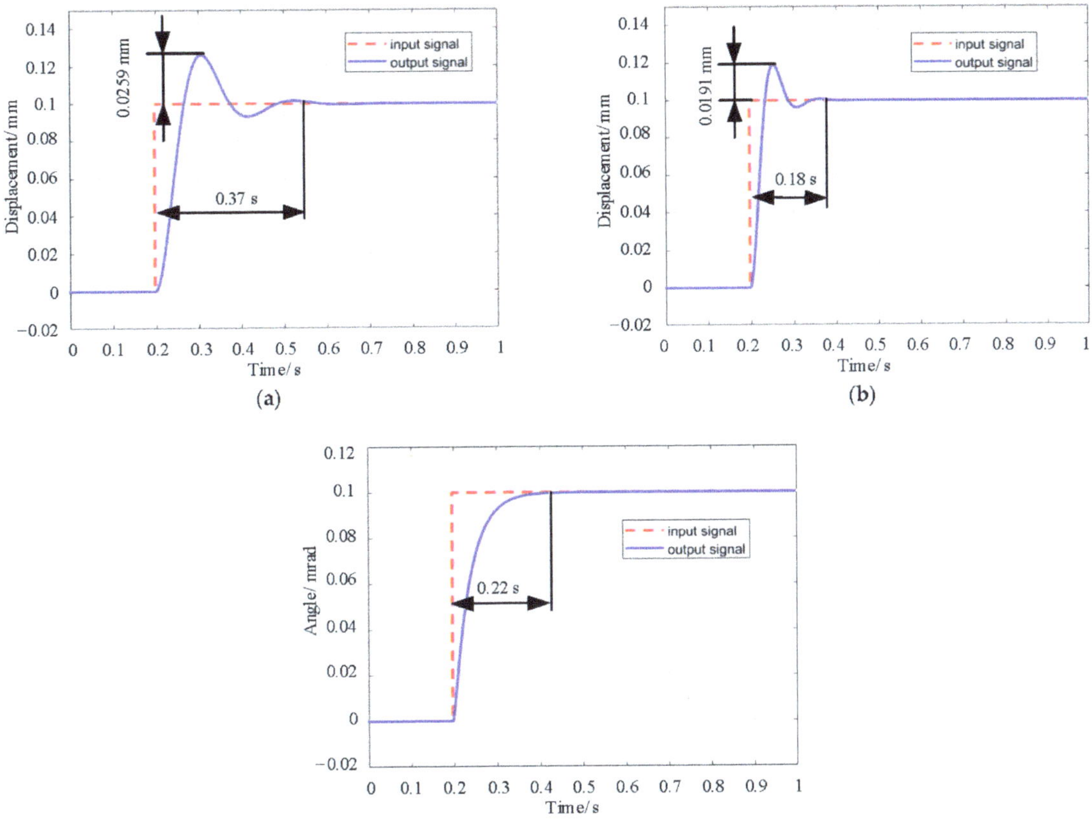

Figure 9. Step response. (**a**) Step response of Y direction; (**b**) Step response of X direction; (**c**) Step response of θ direction.

The performance indicators of the step response curve (Figure 9) in the Y, X, and θ directions are shown in Table 3.

Table 3. Performance index of each step response curve of the nominal model.

Direction	Maximum Overshoot	Regulation Time	Steady-State Error
Y direction	25.9%	0.37 s	0
X direction	19.1%	0.18 s	0
θ direction	0	0.22 s	0

According to the data in Table 3, the overshoot, adjustment time, and steady-state error in the Y direction are 25.9%, 0.37 s, and 0, respectively; in the X direction, they are 19.1%, 0.18 s, and 0; in the θ direction, they are 0, 0.22 s, and 0. Their steady-state performance meets the response requirements for laser off-axis cutting.

4.3. Analysis of System Response to Parameter Disturbance

The model of the system ignores some factors. A 20% perturbation of parameters is added to verify the system's robustness based on its standard model. Then, the parameters of the original controller were used to control the system, and its step response curve is shown in Figure 10.

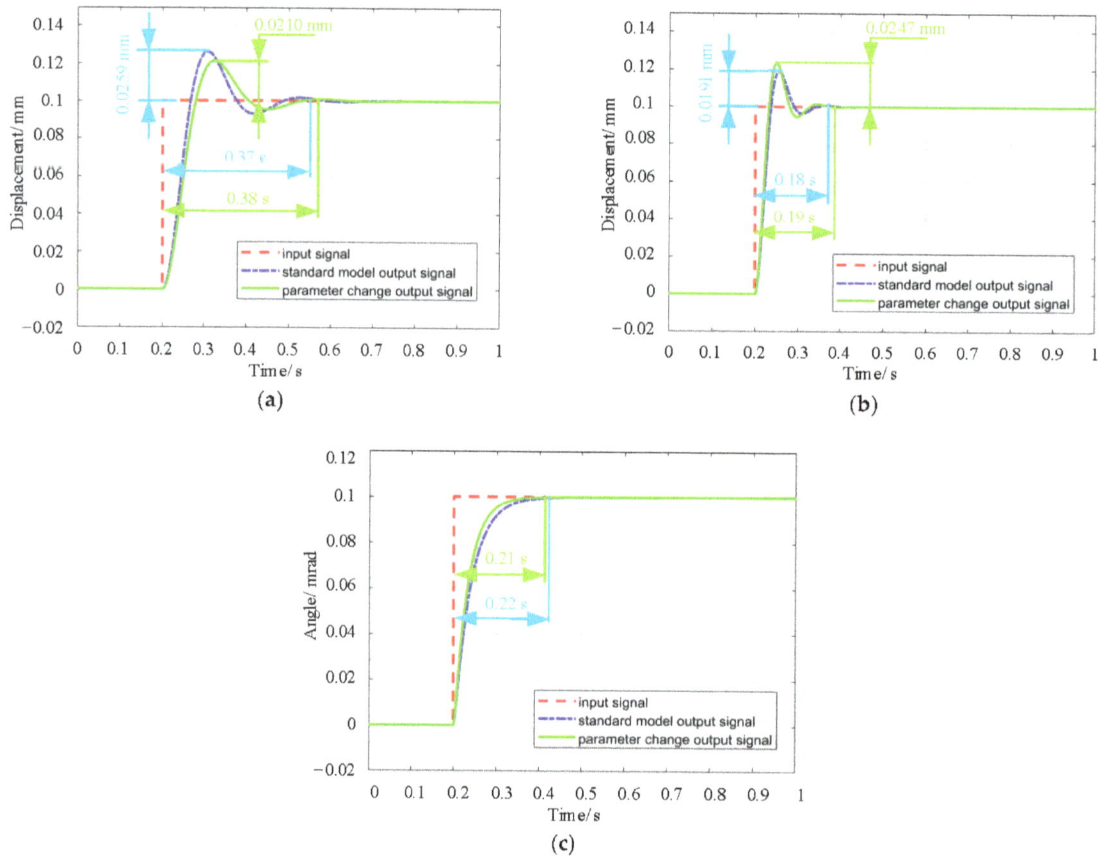

Figure 10. Step response. (**a**) Step response of Y direction with 20% parameters change; (**b**) Step response of X direction with 20% parameters change; (**c**) Step response of θ direction with 20% parameters change.

The performance indicators of the step response curve (Figure 10) in the Y, X, and θ directions are shown in Table 4.

Table 4. Performance index of each step response curve of the model with parameter variation.

Direction	Maximum Overshoot	Regulation Time	Steady-State Error
Y direction	21%	0.38 s	0
X direction	24.7%	0.19 s	0
θ direction	0	0.21 s	0

The data in Table 4 show that after adding 20% parameter disturbance, the overshoot and adjustment time in the Y direction decrease from 25.9% to 21% and increase from 0.37 s to 0.38 s, respectively. The overshoot and adjustment time in the X direction increase from 19.1% to 24.7% and 0.18 s to 0.19 s, respectively. The θ direction adjustment time reduces from 0.22 s to 0.21 s. Steady-state errors are unchanged. In summary, compared with the standard model, the steady-state performance after adding 20% parameter disturbance has no significant change. It shows that the controller can still meet the performance requirements when the model is inaccurate.

4.4. Analysis of System Response to Impulse External Disturbance

An external pulse disturbance was input to verify the system's response in the case of external disturbances, and the results are shown in Figure 11. The vertical coordinates of Figure 11 have two units, "mm" and "mrad", respectively. The unit of vertical coordinate is "mm", which indicates the disturbance signal in Y and X directions. The unit of vertical coordinate is "mrad", which means the interference signal in θ direction.

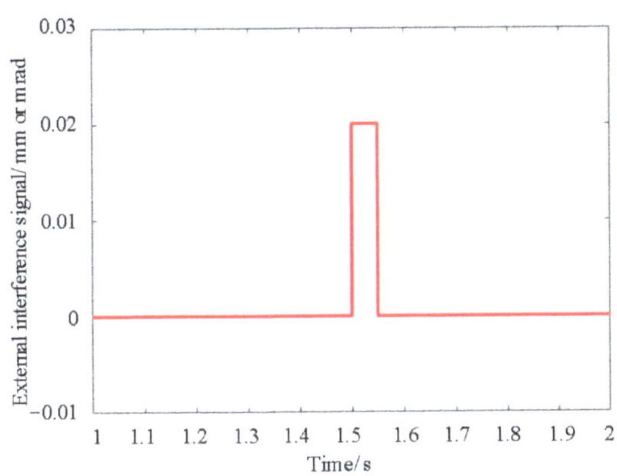

Figure 11. Pulse interference signal.

After the external disturbance signal is input, the system can quickly recover to the steady-state value, and the steady-state error is 0, as shown in Figure 12.

The performance indicators of the step response curve (Figure 12) in the Y, X, and θ directions after being subjected to interference signals are shown in Table 5.

The data in Table 5 show that after adding the interference signal, the over-shoot and adjustment time in the Y direction are 20% and 0.24 s, respectively. The overshoot and adjustment time in the X direction are 20% and 0.14 s, respectively. The overshoot and adjustment time in the θ direction are 20% and 0.16 s, respectively. The steady-state errors are unchanged. In summary, the steady-state performance after the interference signal has no significant change. It also indicates that the control method can effectively resist external interference.

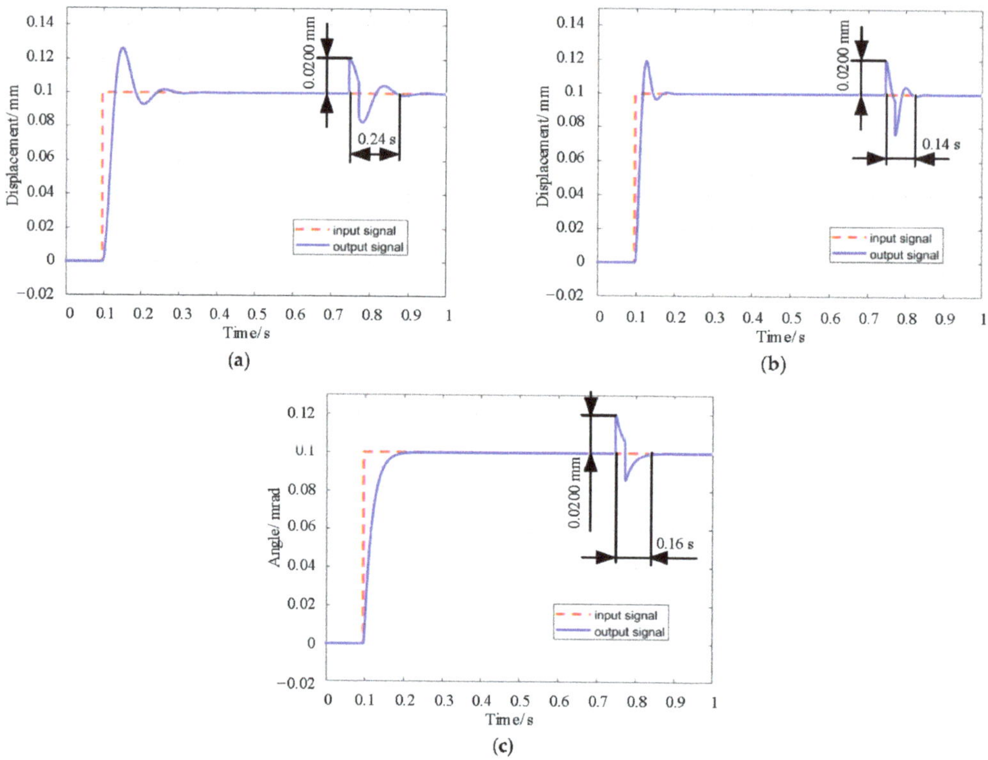

Figure 12. Step response. (**a**) Step response of Y direction with disturbance; (**b**) Step response of X direction with disturbance; (**c**) Step response of θ direction with disturbance.

Table 5. Performance index of each step response curve of the model with interference signal.

Direction	Maximum Overshoot	Regulation Time	Steady-State Error
Y direction	20%	0.24 s	0
X direction	20%	0.14 s	0
θ direction	20%	0.16 s	0

5. Experimental Verification

5.1. Experimental Equipment

Experiments were carried out to verify the controller's effectiveness in the laser focus position control. The experimental equipment mainly comprises a power amplifier (E120/06), eddy current sensor (EX-V10), dSPACE1104, actuator, etc., as shown in Figure 13. With dSPACE1104 as the core hardware, the mixed-sensitivity robust control of the actuator was performed to control its real-time position. We used dSPACE1104 to connect with MATLAB/Simulink for experiments seamlessly. First, the sensor detected the position signal and input it to the ADC module in dSPACE1104. After the control block diagram operation in Simulink, the input signal of the system was obtained. Second, the position signal detected by the sensor and the input signal of the system was subtracted to obtain the deviation signal. Then, the deviation signal was input into the controller in Simulink to obtain the control signal. Third, the control signal was input to the DAC module in dSPACE1104, and the input signal of the power amplifier was obtained. Finally, the power amplifier input the signal to three sets of differential electromagnets to realize real-time control of the actuator position.

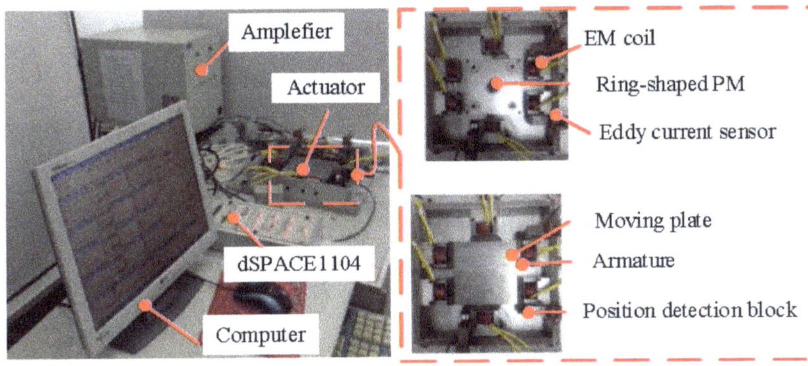

Figure 13. Experimental equipment. The equipment and actuator structures used in the experiments are shown.

5.2. Experimental Results

Step position control experiments were performed in the X and Y directions. First, an initial current of 1.2 A was fed to the three pairs of differential solenoids to make the actuator electromagnetically stiff in the X and Y directions. Then, the position compensation parameters of the system were adjusted so that the detection signal of the sensors was equal to the set value of the system.

When the system was stable, a step signal in the Y direction of 0.1 mm was input in 0.5 s, and the result is shown in Figure 14. The control current first increased and then leveled off. The system stabilized after 0.38 s with an overshoot of 0.034 mm. According to Figure 14, there was a time lag of 0.13 s, which was mainly caused by the following three reasons. First, in the actuator structure, there was only one pair of differential electromagnets in the Y direction and two pairs of differential electromagnets in the X direction, so the stiffness in the Y direction was smaller than that in the x direction. In the initial stage of current change, the driving force could not overcome the initial electromagnetic force. Second, the ring permanent magnet in the actuator was opposite the pole, so the attraction was larger, resulting in greater friction resistance. Third, to make the actuator in a wider range of stable control, the designed controller had robust conservatism. For these three reasons, the time lag was 0.13 s.

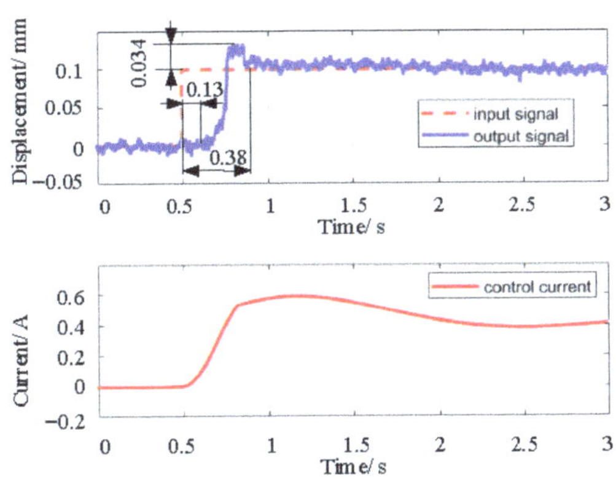

Figure 14. Step response of Y direction.

When the system is stable, we input 0.1 mm X direction step signal in 0.5 s, and the result is shown in Figure 15. According to Figure 15a, the control current of the system increases first and then becomes stable, reaching stability after 0.28 s. According to Figure 15b, after we input the X direction signal, 0.66 mrad fluctuation appears in the direction of θ. This is because the motion of the actuator in both directions X and θ is driven by X direction 2 to the differential electromagnet, and there is coupling. According to Figure 15, control current 1 and control current 2 are not equal because there is an error in the actuator installation, which leads to the different initial air gaps between the two pairs of differential electromagnets in the X direction. Different initial air gaps will lead to different driving forces provided by two pairs of differential electromagnets in the X direction. After the actuator reaches equilibrium, the control current values will differ.

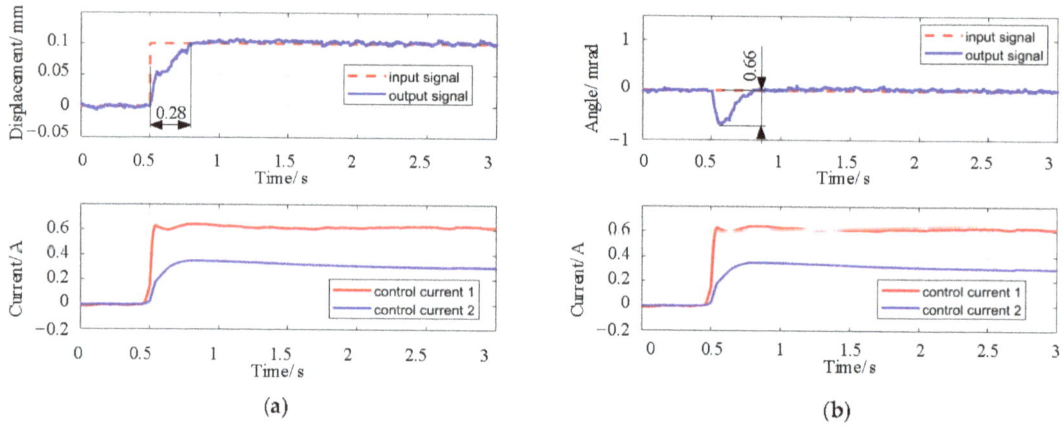

Figure 15. Step response of X direction. (**a**) The position and current signals in the X direction are shown; (**b**) The *θ* interfering position signal and current signal in the X direction are shown.

When the system is stable, a step signal in the direction of θ of 1.2 mrad is input in 0.5 s, and the result is shown in Figure 16. According to Figure 16a, control current 1 decreases first and then tends to be stable, while control current 2 increases first and then tends to be stable, and the output signal in the θ direction reaches the steady value in 0.29 s. According to Figure 16b, after we input θ signal, a fluctuation of 0.059 mm appears in the X direction. This is because the motion of the actuator in both X and θ directions is driven by X direction 2 to the differential electromagnet, and there is coupling.

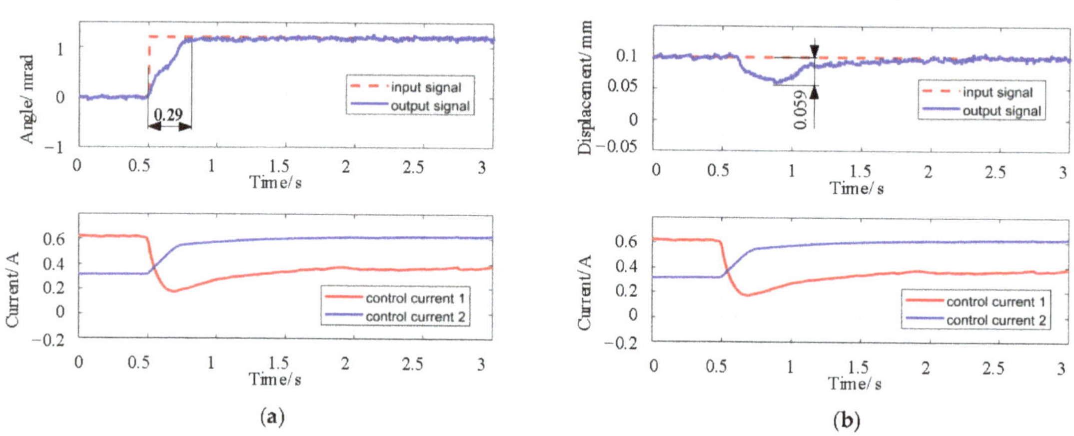

Figure 16. Step response. (**a**) The position and current signals in the θ direction are shown; (**b**) The X direction interfering position signal and current signal in the θ direction are shown.

The maximum overshoot in the Y direction is calculated as the maximum value between 0.5–1 s adjustment time, and the steady-state error is calculated as the average value from 2–3 s. Since there is no obvious overshoot in the X direction and θ direction, the average value of 0.2 s after stabilization is taken as the maximum overshoot, and the steady-state error is taken as the average value of 2–3 s. The values are shown in Table 6.

Table 6. Performance index of the step response experiment in each direction.

Direction	Maximum Overshoot	Regulation Time	Steady-State Error
Y direction	34%	0.38 s	0.39%
X direction	3.1%	0.28 s	1.6%
θ direction	0%	0.29 s	0.45%

According to Table 6, the steady-state errors in the 3 directions are all less than 2.5%. That is, the system's error after stability can be made within 0.025 mm of the machine tool's accuracy in the position step response, which meets the performance requirements.

In order to be able to analyze the experimental data quantitatively. The simulated and experimental data are included in Table 7 for comparative analysis.

Table 7. Performance index of step response experiment in each direction.

Direction		Maximum Overshoot	Regulation Time	Steady-State Error
Y direction	Simulation	25.9%	0.37 s	0
	Experiment	34%	0.38 s	0.39%
	ǀ Difference value ǀ	8.1%	0.01 s	0.39%
X direction	Simulation	19.1%	0.18 s	0
	Experiment	3.1%	0.28 s	1.60%
	ǀ Difference value ǀ	16%	0.10 s	1.60%
θ direction	Simulation	0%	0.22 s	0
	Experiment	0%	0.29 s	0.45%
	ǀ Difference value ǀ	0%	0.07 s	0.45%

According to Table 7, the results of the standard model simulation are compared and analyzed with the experimental result data. The overshoot differences value in the Y, X, and θ directions are 8.1%, 16%, and 0%, respectively. The differences in adjustment time are 0.01 s, 0.10 s, and 0.07 s, respectively. The differences in steady-state errors are 0.39%, 1.60%, and 0.45%, respectively. The experimental and simulation data are slightly different because there are some non-artificial errors in the experiments, including the machining errors of the parts, assembly errors, etc. In addition, there is noise interference in the experiment, so it leads to some steady-state errors in the experiment.

6. Conclusions

A mixed sensitivity robust controller for a 3-DOF laser focal magnetic actuator is designed and simulated by MATLAB/Simulink. The following conclusions are obtained through simulation analysis and experimental verification.

1. The mixed sensitivity controller of the system is calculated by a reasonable selection of the weighting function, and the simulation analysis of the position step response in Y, X, and θ directions is carried out, and the steady-state errors are all 0. It shows that the designed controller has good position response characteristics.
2. Based on the standard model, 20% parameter variation is introduced, and the original controller simulates and analyzes the system. In the simulation of position step response in Y, X, and θ directions, the response characteristics are consistent with the results of the standard model. There are slight differences in overshoot and adjustment time between the above, and the steady-state error is 0. It shows that the system has better position response characteristics when the model is inaccurate.

3. When the external pulse disturbance is added to the system, the steady-state values of Y, X, and θ can be quickly restored in the position step experiment. It shows that the controller has good anti-interference properties.
4. There is a coupling phenomenon in the direction of X and θ, so both show certain coupling characteristics in the step response experiment. Still, the moving plate can be quickly adjusted to the steady-state value. The steady-state errors in the three directions are all less than 25%, which meets the requirement of machine tool accuracy of 0.025 mm.

In the future, we will optimize the structure of the actuator and design an actuator that can be mounted on the laser cutting head. Then, we will install it on the laser cutting machine for in-machine experimental research in order to explore the actuator's performance during the operation of the laser cutting head.

Author Contributions: Methodology, L.W.; validation, L.W. and Q.Z.; formal analysis, L.W.; resources, F.S.; data curation, L.W.; writing—original draft preparation, L.W.; writing—review and editing, L.T., G.Y., F.X., J.J., X.Z. and F.S. All authors have read and agreed to the published version of the manuscript.

Funding: This research is supported by National Natural Science Fund of China (No. 52005345, No. 52005344), National Key Research and Development Project (No. 2020YFC2006701), Scientific Research Fund Project of Liaoning Provincial Department of Education (No. LJKMZ20220506, No. LJKMZ20220460), Major Project of the Ministry of Science and Technology of Liaoning Province (No. 2022JH1/10400027).

Data Availability Statement: Not applicable.

Acknowledgments: We thank Zhou R., Zhao C., and Liu X. for their suggestions and recommendations.

Conflicts of Interest: The authors declare no conflict of interest.

Appendix A

$$
\begin{aligned}
F_e = &\left\{4\pi^2\mu_0 B_r n\delta_p\varepsilon(r_2^2-r_1^2)[\pi r_2^2(1+\sqrt{\varepsilon^2+e^2})+2(r_1+r_2)\delta_p^2][4\pi(r_2^2-r_1^2)\varepsilon+1.632(r_1+r_2)(\varepsilon^2+e^2)]\right\}2\\
&\cdot e/\left\{\left\{\mu_r\pi(r_2^2-r_1^2)\left\{\delta_p(1+\sqrt{\varepsilon^2+e^2})[4\pi(r_2^2-r_1^2)\varepsilon+1.632(r_1+r_2)(\varepsilon^2+e^2)]+(\varepsilon^2+e^2)[\pi r_2^2(1+\sqrt{\varepsilon^2+e^2})\right.\right.\right.\\
&\left.\left.+2(r_1+r_2)\delta_p^2]\right\}+n\delta_p[\pi r_2^2(1+\sqrt{\varepsilon^2+e^2})+2(r_1+r_2)\delta_p^2][4\pi(r_2^2-r_1^2)\varepsilon+1.632(r_1+r_2)(\varepsilon^2+e^2)]\right\}[4\pi\mu_0(r_2^2\\
&-r_1^2)\varepsilon+1.632\mu_0(r_1+r_2)(\varepsilon^2+e^2)]\}^2;\\
e = &\sqrt{x^2+y^2}
\end{aligned}
\tag{A1}
$$

where ε is an axial air gap, e is radial displacement, t is the width of the ring permanent magnet, B_r is the remanence of permanent magnet material, n is the number of permanent magnets, δ_P is the thickness of permanent magnet, μ_r is the relative permeability of the magnetic ring, μ_0 is air permeability, r_1 is the inner radius of the annular permanent magnet, r_2 is the outer radius of the ring permanent magnet.

$$
\begin{aligned}
k_e = &4\mu_0\varepsilon^{-3}(r_2^2-r_1^2)\left\{B_r n\delta_p\pi(r_2^2-r_1^2)[\pi r_2^2(1+\varepsilon)+2(r_1+r_2)\delta_p^2][4\pi(r_2^2-r_1^2)\varepsilon^{-1}+1.632(r_1+r_2)]\right\}2/\{\{\mu_r\pi(r_2^2\\
&-r_1^2)\left\{\delta_p(1+\varepsilon)[4\pi(r_2^2-r_1^2)\varepsilon^{-1}+1.632(r_1+r_2)]+[\pi r_2^2(1+\varepsilon)+2(r_1+r_2)\delta_p^2]\right\}+n\delta_p[\pi r_2^2(1+\varepsilon)+2(r_1+r_2)\delta_p^2]\\
&[4\pi(r_2^2-r_1^2)\varepsilon^{-1}+1.632(r_1+r_2)]\}[4\pi\mu_0(r_2^2-r_1^2)\varepsilon^{-1}+1.632\mu_0(r_1+r_2)]\}^2
\end{aligned}
\tag{A2}
$$

References

1. He, Y.; Xie, H.; Ge, Y.; Lin, Y.; Yao, Z.; Wang, B.; Jin, M.; Liu, J.; Chen, X.; Sun, Y. Laser Cutting Technologies and Corresponding Pollution Control Strategy. *Processes* **2022**, *10*, 732. [CrossRef]
2. Singh, Y.; Singh, J.; Sharma, S.; Sharma, A.; Chohan, J. Process parameter optimization in laser cutting of Coir fiber reinforced Epoxy composite—A review. *Mater. Today Proc.* **2021**, *48*, 1021–1027. [CrossRef]
3. Sim, M.; Lee, C. Determination of optimal laser power according to the tool path inclination angle of a titanium alloy workpiece in laser-assisted machining. *Int. J. Adv. Manuf. Technol.* **2016**, *83*, 1717–1724. [CrossRef]

4. Jain, A.; Agarwal, A.; Chauhan, P.S. Optimization of laser cutting quality characteristics for basalt–glass hybrid composite using hybrid approach. *Mater. Today Proc.* **2022**, *62*, 12. [CrossRef]
5. Wang, H.; Yang, T. A review on laser drilling and cutting of silicon. *J. Eur. Ceram. Soc.* **2021**, *41*, 4997–5015. [CrossRef]
6. Marimuthu, S.; Dunleavey, J.; Liu, Y.; Antar, M.; Smith, B. Laser cutting of aluminium-alumina metal matrix composite. *Opt. Laser Technol.* **2019**, *117*, 251–259. [CrossRef]
7. Sharifi, M.; Akbari, M. Experimental investigation of the effect of process parameters on cutting region temperature and cutting edge quality in laser cutting of AL6061T6 alloy. *Optik* **2019**, *184*, 457–463. [CrossRef]
8. Mushtaq, R.T.; Wang, Y.; Rehman, M.; Khan, A.M.; Mia, M. State-Of-The-Art and Trends in CO2 Laser Cutting of Polymeric Materials—A Review. *Materials* **2020**, *13*, 3839. [CrossRef]
9. Li, M.; Han, H.; Jiang, X.; Zhang, X.; Chen, Y. Surface morphology and defect characterization during high-power fiber laser cutting of SiC particles reinforced aluminum metal matrix composite. *Opt. Laser Technol.* **2022**, *155*, 108419. [CrossRef]
10. Li, M.; Li, S.; Yang, X.; Zhang, Y.; Liang, Z. Effect of lay-up configuration and processing parameters on surface quality during fiber laser cutting of CFRP laminates. *Int. J. Adv. Manuf. Technol.* **2019**, *100*, 623–635. [CrossRef]
11. Elsheikh, A.; Deng, W.; Showaib, E. Improving laser cutting quality of polymethylmethacrylate sheet: Experimental investigation and optimization. *J. Mater. Res. Technol.* **2020**, *9*, 1325–1339. [CrossRef]
12. Vora, J.; Chaudhari, R.; Patel, C.; Pimenov, D.; Patel, V.; Giasin, K.; Sharma, S. Experimental investigations and Pareto optimization of fiber laser cutting process of Ti6Al4V. *Metals* **2021**, *11*, 1461. [CrossRef]
13. Hong, L.; Zhang, Y.; Mi, C. Technological study of laser cutting silicon steel controlled by rotating gas flow. *Opt. Laser Technol.* **2009**, *41*, 328–333. [CrossRef]
14. Quintero, F.; Pou, J.; Fernandez, J.L.; Doval, A.F.; Lusquiños, F.; Boutinguiza, M.; Soto, R.; Pérez-Amor, M. Optimization of an off-axis nozzle for assist gas injection in laser fusion cutting. *Opt. Lasers Eng.* **2006**, *44*, 1158–1171. [CrossRef]
15. Riveiro, A.; Quintero, F.; Boutinguiza, M.; Del Val, J.; Comesaña, R.; Lusquiños, F.; Pou, J. Laser cutting: A review on the influence of assist gas. *Materials* **2019**, *12*, 157. [CrossRef]
16. Riveiro, A.; Quintero, F.; Lusquiños, F.; Comesaña, R.; Pou, J. Effects of processing parameters on laser cutting of aluminium-copper alloys using off-axial supersonic nozzles. *Appl. Surf. Sci.* **2011**, *257*, 5393–5397. [CrossRef]
17. Riveiro, A.; Quintero, F.; del Val, J.; Boutinguiza, M.; Comesaña, R.; Lusquiños, F.; Pou, J. Laser cutting using off-axial supersonic rectangular nozzles. *Precis. Eng.* **2018**, *51*, 78–87. [CrossRef]
18. Yagi, A.; Kadonaga, S.; Okamoto, Y.; Ishiguro, H.; Ito, R.; Sugiyama, A.; Okawa, H.; Fujta, R.; Okada, A. Fundamental study on reduction of dross in fiber laser cutting of steel by shifting nozzle axis. *J. Laser Appl.* **2021**, *33*, 012022. [CrossRef]
19. Murakami, I.; Zhao, Y.; Tashiro, T. Stabilization of a Magnetic Repulsive Levitation Flywheel System Using a High-Efficiency Superconducting Magnetic Bearing. *Actuators* **2022**, *11*, 180. [CrossRef]
20. Soni, T.; Dutt, J.; Das, A.S. Parametric stability analysis of active magnetic bearing supported rotor system with a novel control law subject to periodic base motion. *IEEE Trans. Ind. Electron.* **2019**, *67*, 1160–1170. [CrossRef]
21. Sillanpää, T.; Smirnov, A.; Jaatinen, P.; Vuojolainen, J.; Nevaranta, N.; Jastrzebski, R.; Pyrhönen, O. Three-axis inductive displacement sensor using phase-sensitive digital signal processing for industrial magnetic bearing applications. *Actuators* **2021**, *10*, 115. [CrossRef]
22. Cao, S.; Niu, P.; Wang, W.; Zhao, T.; Liu, Q.; Bai, J.; Sheng, S. Novel Magnetic Suspension Platform with Three Types of Magnetic Bearings for Mass Transfer. *Energies* **2022**, *15*, 5691. [CrossRef]
23. Javed, A.; Mizuno, T.; Takasaki, M.; Ishino, Y.; Hara, M.; Yamaguchi, D. Lateral Vibration Suppression by Varying Stiffness Control in a Vertically Active Magnetic Suspension System. *Actuators* **2018**, *7*, 21. [CrossRef]
24. Ren, G.P.; Chen, Z.; Zhang, H.; Wu, Y.; Meng, H.; Wu, D.; Ding, H. Design of interval type-2 fuzzy controllers for active magnetic bearing systems. *IEEE/ASME Trans. Mechatron.* **2020**, *25*, 2449–2459. [CrossRef]
25. Fan, X.; Yin, J.; Lu, Q. Design and Analysis of a Novel Composited Electromagnetic Linear Actuator. *Actuators* **2022**, *11*, 6. [CrossRef]
26. Xu, Y.; Jiang, Q.; Yang, K.; Zhou, J.; Guo, Q. A novel ultra-high-resolution inclination sensor based on diamagnetic levitation. *Sens. Actuators A Phys.* **2022**, *343*, 113686. [CrossRef]
27. Zhang, X.; Shinshi, T.; Fukuoka, T.; Nakai, T. Development of a 2-DOF Controlled Magnetic Drive Actuator for Laser Beam Cutting. *J. Adv. Mech. Des. Syst. Manuf.* **2013**, *7*, 448–457. [CrossRef]
28. He, D.; Shinshi, T.; Nakai, T. Development of a maglev lens driving actuator for off-axis control and adjustment of the focal point in laser beam machining. *Precis. Eng.* **2013**, *37*, 255–264. [CrossRef]
29. He, D.; Shinshi, T.; Nakai, T. Development of a Lens Driving Maglev Actuator for Laser Beam Off-Axis Cutting and Deep Piercing. In *Key Engineering Materials*; Trans Tech Publications Ltd.: Bäch, Switzerland, 2012; Volume 523, pp. 774–779. [CrossRef]
30. Morimoto, Y.; Shinshi, T.; Nakai, T. A two-DOF controlled lens drive actuator for off-axis laser beam cutting. *J. Adv. Mech. Des. Syst. Manuf.* **2012**, *6*, 875–884. [CrossRef]
31. Gasmi, N.; Boutayeb, M.; Thabet, A.; Bel Haj Frej, G.; Aoun, M. Robust Control of a Class of Nonlinear Discrete-Time Systems: Design and Experimental Results on a Real-Time Emulator. *Actuators* **2021**, *10*, 303. [CrossRef]
32. Seyfi, N.; Khalaji, A. Robust control of a cable-driven rehabilitation robot for lower and upper limbs. *ISA Trans.* **2022**, *125*, 268–289. [CrossRef] [PubMed]

33. Zhang, X.; Ma, K.; Wei, Y.; Han, Y. Finite-time robust H∞ control for high-speed underwater vehicles subject to parametric uncertainties and disturbances. *J. Mar. Sci. Technol.* **2017**, *22*, 201–218. [CrossRef]
34. Bu, X.; Wu, X.; Wei, D.; Huang, J. Neural-approximation-based robust adaptive control of flexible air-breathing hypersonic vehicles with parametric uncertainties and control input constraints. *Inf. Sci.* **2016**, *346*, 29–43. [CrossRef]
35. Gosiewski, Z.; Mystkowski, A. Robust control of active magnetic suspension: Analytical and experimental results. *Mech. Syst. Signal Process.* **2008**, *22*, 1297–1303. [CrossRef]
36. Zhang, Q.; Yu, R.; Li, C.; Chen, Y.-H.; Gu, J. Servo Robust Control of Uncertain Mechanical Systems: Application in a Compressor/PMSM System. *Actuators* **2022**, *11*, 42. [CrossRef]
37. Zhang, K.; Scorletti, G.; Ichchou, M.; Mieyeville, F. Quantitative robust linear parameter varying H∞ vibration control of flexible structures for saving the control energy. *J. Intell. Mater. Syst. Struct.* **2015**, *26*, 1006–1027. [CrossRef]
38. Ko, Y.; Kim, T. Feedforward Plus Feedback Control of an Electro-Hydraulic Valve System Using a Proportional Control Valve. *Actuators* **2020**, *9*, 45. [CrossRef]
39. Isbeih, Y.; El Moursi, M.; Xiao, W.; El-Saadany, E. H∞ mixed-sensitivity robust control design for damping low-frequency oscillations with DFIG wind power generation. *IET Gener. Transm. Distrib.* **2019**, *13*, 4274–4286. [CrossRef]
40. Wu, X.; Xu, C.; Wei, B.; Xia, C.; Li, X. H∞ mixed sensitivity robust control method of relay ICPT system for output voltage regulation. *Electr. Eng.* **2021**, *103*, 781–792. [CrossRef]
41. Xu, Z.; Zhang, Q.; Sun, F. Analysis of Wave Characteristics of Laser Focus Magnetic Drive Platform. *Modul. Mach. Tool Autom. Manuf. Tech.* **2020**, *3*, 13–15. [CrossRef]
42. Tian, L.; Ai, X.P.; Tian, Y. Analytical model of magnetic force for axial stack permanent-magnet bearings. *IEEE Trans. Magn.* **2012**, *48*, 2592–2599. [CrossRef]

actuators

Article

Analysis of the Notch Filter Insertion Position for Natural Frequency Vibration Suppression in a Magnetic Suspended Flywheel Energy Storage System

Hongjin Hu [1,†], Jingbo Wei [1,†], Haoze Wang [2], Peng Xiao [1], Yuan Zeng [1] and Kun Liu [1,*]

1 School of Aeronautics and Astronautics, Sun Yat-Sen University, Shenzhen 518107, China
2 School of Civil Aviation, Northwestern Polytechnical University, Suzhou 215400, China
* Correspondence: liukun6@mail.sysu.edu.cn
† These authors contributed equally to this work.

Abstract: The composite material flywheel rotor of a flywheel energy storage system (FESS) has a low natural frequency. When the system suffers from noise interference, the magnetic bearing generates a force with the same frequency as the natural frequency and causes vibration to occur. Thus, it is necessary to suppress the natural vibration of the magnetic suspended (MS) FESS. The LMS adaptive notch filter is generally adopted for vibration suppression. The vibration suppression performance of the system is different when the insertion position of the notch filter is different. This paper analyzes the influence of the notch filter in different insertion positions of the control system. Through the transfer function from noise to magnetic bearing force, theoretical analysis of the influence of different positions of the notch filter is performed. Corresponding experiments are performed in a 500 kW MS FESS prototype. The theoretical analysis is verified experimentally.

Keywords: magnetic suspension flywheel energy storage; natural frequency; vibration suppression; insertion position of the notch filter

Citation: Hu, H.; Wei, J.; Wang, H.; Xiao, P.; Zeng, Y.; Liu, K. Analysis of the Notch Filter Insertion Position for Natural Frequency Vibration Suppression in a Magnetic Suspended Flywheel Energy Storage System. *Actuators* **2023**, *12*, 22. https://doi.org/10.3390/act12010022

Academic Editor: Kirill Poletkin

Received: 8 December 2022
Revised: 28 December 2022
Accepted: 30 December 2022
Published: 4 January 2023

1. Introduction

Energy storage systems (ESSs) play increasingly important roles in modern I ndustrial applications. A flywheel energy storage system (FESS) is a type of ESS that has the advantages of high efficiency, fast response, instantaneous high power, low maintenance, and long life [1–3]. Thus, FESSs have been applied in electric vehicles [4,5], hydraulic excavators [6], wind power generation [7,8], microgrids [9], photovoltaic generation [10], uninterruptible power supply (UPS) [11,12], and aerospace applications [13]. In an FESS, energy is stored in a high-speed flywheel rotor. The FESS generally adopts a composite material flywheel rotor [14,15]. By using a composite flywheel rotor, the FESS can obtain a higher linear velocity and energy storage density [15]. However, the stiffness of the composite material is small, resulting in a relatively low natural frequency of the composite flywheel, which is generally a few hundred hertz. In the magnetic suspended (MS) FESS, the composite material flywheel rotor is supported by magnetic bearings (MBs), which have the advantages of low loss, frictionlessness, and silent operation [16,17]. The MB controller generates noise interference of the same frequency as the natural frequency of the flywheel rotor. Therefore, the MB generates an electromagnetic force that has the same frequency as the natural frequency. This electromagnetic force becomes the excitation source of the flywheel rotor. When the frequency of the force is the same as the natural frequency, the flywheel rotor vibrates violently, causing system instability [18].

The notch filter can filter out signals with a certain frequency. Therefore, the vibration of the MB system can be suppressed. Vibration suppression methods based on notch filters have been widely applied in MB systems. The main notch filters applied in MB systems include the generalized notch filter [19,20], the least mean square (LMS) adaptive notch filter [21–23], and the phase-shift notch filter [24,25]. In [19], a generalized narrow-band notch filter is proposed for an MB system. This notch filter is inserted into the multivariable feedback without destabilizing the closed loop. In [20], a generalized notch filter based on synchronous rotating frame (SRF) transformation is proposed for autobalancing of the magnetically suspended rotor. In [21], a harmonic based on a frequency-domain adaptive LMS is proposed to suppress the vibration force. This method introduces a convergence factor in the frequency domain that improves the updating strategy of the step size and the convergence rate of the algorithm. In [22], a double-channel adaptive LMS error algorithm is proposed to suppress the imbalanced vibration. A modified adaptive notch filter with phase shift is proposed in [23] to suppress the unbalanced vibration. This notch filter does not require angular velocity information. In [24], a phase-shift notch filter is proposed to ensure stable operations over the entire speed range. In [25], a control system scheme with cascaded mode phase-shift notch filters is proposed for the MB rotor control system.

Notch filters have been widely applied in the vibration suppression of MB systems and show satisfactory vibration suppression performance. However, research on the notch filters in MB control systems is mainly focused on the notch filter algorithm. The influence of the insertion positions of the notch filters is not sufficiently discussed. The MB control system is a double closed-loop system. The outer loop is a position control loop, and the inner loop is a current control loop. The notch filter can be inserted into the feedback path of the outer loop, the forward path of the outer loop, the feedback path of the inner loop, and the forward path of the inner loop. In general, the notch filter is inserted into the forward path of the position control loop. In this paper, the influence of the notch filter access position is analyzed, and the control model of the MB system is clarified. The LMS adaptive notch filter is adopted for vibration suppression, and the influence of the insertion position of the notch filter is analyzed through the Bode plot.

The remainder of this paper is organized as follows. In Section 2, the structure of the MS FESS is presented. In Section 3, the influence of the access position of the notch filter is analyzed. In Section 4, the experiment is performed on the FESS prototype to verify the theoretical analysis. Finally, the conclusion is drawn in Section 5.

2. Structure of the FESS

An FESS is a type of mechanical energy storage device in which mechanical energy is stored in a high-rotation-speed flywheel rotor. The structure of the MS FESS is shown in Figure 1. The MS FESS is installed vertically, which mainly includes the flywheel rotor, MBs, high-speed motor, corresponding position sensors, and base. The flywheel rotor is supported by MBs and installed in a vacuum chamber. Thus, the flywheel rotor can operate at a high speed without mechanical friction and with low wind resistance. FESS achieves energy conversion through a high-speed motor. The motor rotor connects to the flywheel rotor. The flywheel rotor has the same rotation speed as the rotor of the motor. The FESS has three MBs as follows: the top radial MB, the bottom radial MB, and the top axial MB. The power amplifier (PA) should generate currents to the MBs of the FESS. By applying the appropriate current, the MB system can generate the corresponding electromagnetic force to suspend the flywheel rotor.

Stable MS control is important to the MS FESS. Natural frequency vibration is one of the issues that cause system instability. To avoid natural frequency vibration, the force with the same frequency as the natural frequency should be filtered out.

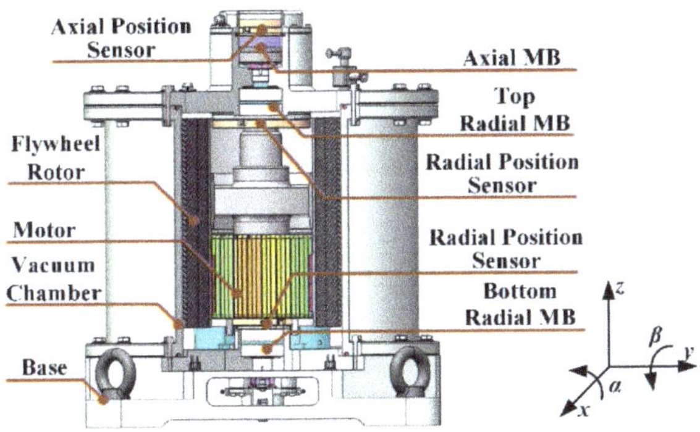

Figure 1. The structure of the FESS.

3. Analysis of the Insertion Position of the Notch Filter

In this section, the influence of the insertion position of the notch filter is analyzed. The model of the MB control system and the LMS adaptive notch filter are presented. Based on the control model and the notch filter, the influence of the insertion position of the notch filter is analyzed through a Bode plot.

3.1. Model of the Magnetic Bearing Control System

The MS FESS of Figure 1 adopts a permanent-magnet-biased MB. The structure diagram of the hybrid MB is shown in Figure 2. The electromagnetic force generated by the hybrid MB is deduced as Equation (1):

$$f_{mag} = \frac{\mu_0 S N^2}{2}\left[\left(\frac{I_0 - I}{D_0 - x}\right)^2 - \left(\frac{I_0 + I}{D_0 + x}\right)^2\right] \tag{1}$$

where f_{mag} is the electromagnetic force of the MB, N is the number of turns of the MB coil, μ_0 is the vacuum permeability, S is the area directly opposite to the magnetic pole, D_0 is the MB air gap, x is the flywheel rotor position, and I is the coil current.

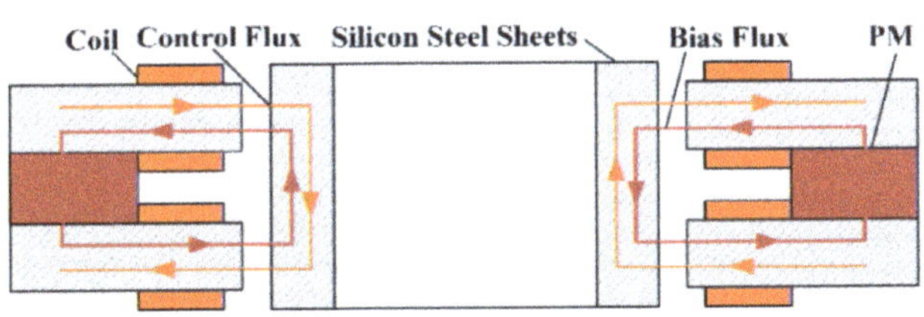

Figure 2. The structure diagram of the hybrid MB.

The rotor is suspended at the position where x = 0. According to the Taylor series expansion, Equation (1) can be linearized as follows in Equation (2):

$$f_{mag} = K_x x + K_i I \tag{2}$$

where K_x is the displacement stiffness of the MB and K_i is the current stiffness of the MB.

According to Newton's second law, the kinematic equation can be deduced as follows in Equation (3):

$$f_{mag} = m\ddot{x} = K_x x + K_i I \tag{3}$$

The transfer function of Equation (3) is presented as follows:

$$G_m(s) = \frac{K_i}{ms^2 - K_x}. \tag{4}$$

The MS control system is a double closed-loop control system. The control block diagram is shown in Figure 3. A proportional–derivative (PD) controller is adopted for MB position control, and the transfer function of the PD controller is presented as follows in Equation (5):

$$G_{pd}(s) = K_{px} + K_{dx}s \tag{5}$$

where K_{px} is the proportional coefficient of the PD controller and K_{dx} is the differential coefficient of the PD controller.

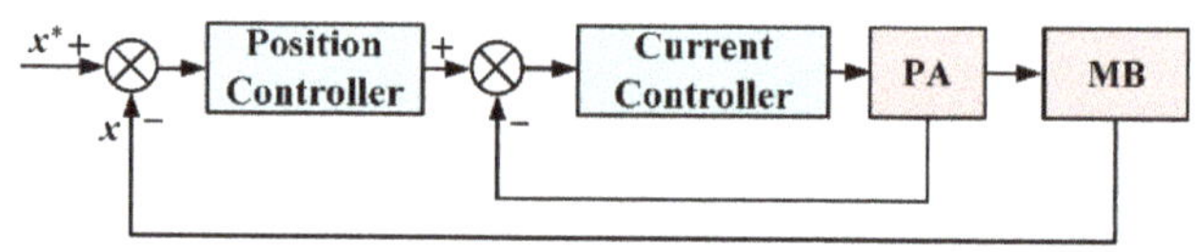

Figure 3. Control block diagram of the MS control.

For current control, the coil of the MB is equivalent to the resistance inductance series circuit. Thus, the model of the coil can be deduced as follows in Equation (6):

$$G_c(s) = \frac{1}{Ls + R} \tag{6}$$

where L is the inductance of the MB coil and R is the resistance of the MB coil.

A (proportional integral) PI controller is generally used for current control, and the transfer function of the PI controller is deduced as follows in Equation (7):

$$G_{pi}(s) = K_{pi} + \frac{K_{ii}}{s} \tag{7}$$

The influence of the insertion position of the notch filter is analyzed based on the above control system model.

3.2. LMS Adaptive Notch Filter Algorithm

The vibration is suppressed by the LMS adaptive notch filter. The LMS adaptive notch filter is based on the Wiener filter and minimum mean square error (MMSE) criteria, which uses the gradient descent method to minimize the mean square error of the signal [26]. Thus, the filter can output the best signal estimate. The block diagram of the LMS adaptive filter is shown in Figure 4. In Figure 4, T_s is the sampling time of the discrete system, ω_0 is the center frequency of the LMS notch filter, the input vector $\mathbf{X}(kT_s)$ is the reference signal of the algorithm, $d(kT_s)$ is the expected signal, and w is the weighting coefficient of the filter.

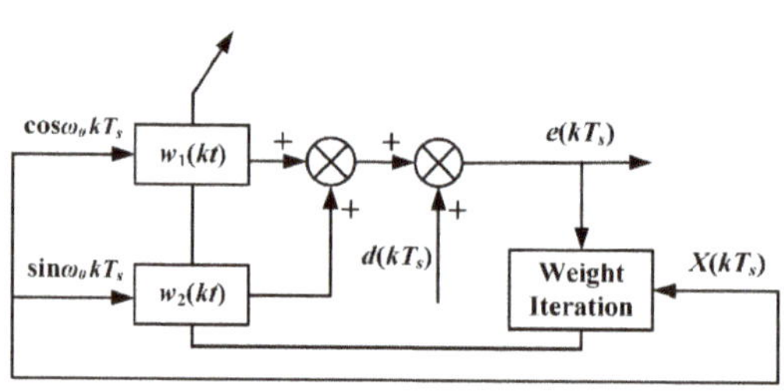

Figure 4. The block diagram of the LMS notch filter.

The weighting coefficient is adjusted by the gradient descent method to minimize the mean square error. The mean square error is presented as follows in Equation (8):

$$MSE = E\left[e^2(kT_s)\right] \tag{8}$$

The error of the signal can be deduced as follows in Equation (9):

$$e(kT_s) = d(kT_s) - \mathbf{X}^{\mathrm{T}}(kT_s)\mathbf{W}(kT_s)\mathbf{X}(kT_s) \tag{9}$$

where **W** is the filter weighting coefficient matrix.

Combining (8) and (9), the mean square error can be further deduced as follows in Equation (10).

$$MSE = E\left(d^2\right) + \mathbf{W}^{\mathrm{T}}\mathbf{R}\mathbf{W} - 2\mathbf{W}^{\mathrm{T}}\mathbf{P} \tag{10}$$

where **R** is the autocorrelation matrix of the input signal, **P** is the cross-correlation matrix of the expected signal $d(kT_s)$, and $\mathbf{X}(kT_s)$ is the input signal.

The partial differential of the mean square error can be deduced as follows in Equation (11).

$$\frac{\partial MSE}{\partial \mathbf{W}} = 2\mathbf{R}\mathbf{W} - 2\mathbf{P} \tag{11}$$

According to Equation (11), the filter coefficient matrix **W** can be directly calculated. However, the inverse matrix of **R** needs to be obtained. In practical applications, matrix inversion requires massive computing resources. Thus, the descent method is adopted to obtain the filter weighting matrix. The weighting matrix is adjusted along the negative gradient direction. The recursive equations of the LMS adaptive notch filter algorithm can be deduced as follows in Equation (12) [26]:

$$\begin{cases} e(kT_s) = d(kT_s) - y(kT_s) \\ y(kT_s) = w_1(kT_s)\cos\omega_0 kT_s + w_2(kT_s)\sin\omega_0 kT_s \\ W[(k+1)T]_s = W(kT_s) + 2\mu e(kT_s)X(kT_s) \end{cases} \tag{12}$$

where μ is the step length.

The step length can be used to adjust the convergence speed and steady-state accuracy of the filter. According to Equation (12), the transfer function of the LMS adaptive notch filter can be deduced as follows in Equation (13):

$$G_f(z) = \frac{1 - 2z^{-1}\cos\omega_0 T_s + z^{-2}}{1 - 2(1-\mu)z^{-1}\cos\omega_0 T_s + (1-2\mu)z^{-2}} \tag{13}$$

In this paper, the LMS adaptive notch filter of Equation (13) is used to analyze the influence of the notch filter at different insertion positions of the control system on vibration suppression.

3.3. Insertion Position of the Notch Filter

The MS control system is a double closed-loop system. The LMS adaptive notch filter has four insertion positions. The four insertion positions are the forward path of the outer loop, the feedback path of the outer loop, the forward path of the inner loop, and the feedback path of the inner loop, which are shown in Figure 5.

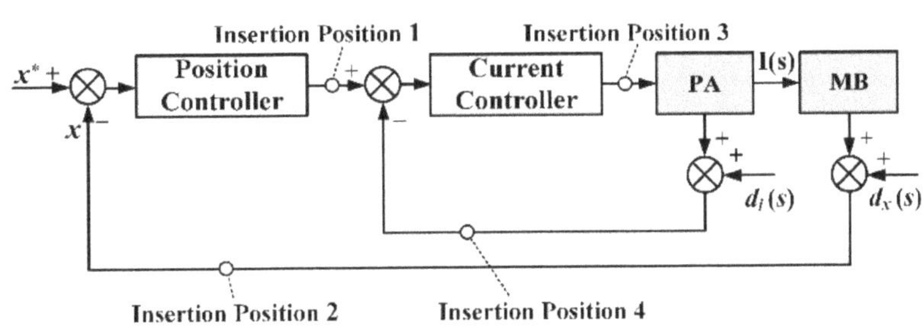

Figure 5. Control block diagram with four notch filter insertion positions.

According to Equation (2), the electromagnetic force of the MB is proportional to the current of the PA. When the PA generates a current that has the same frequency as the natural frequency of the flywheel rotor and has a sufficiently large amplitude, a violent vibration occurs in the flywheel rotor. To suppress the vibration, the PA should not generate a current that has the same frequency as the natural frequency. In MS control, signals with the same frequency as the natural frequency mainly come from the noise of the position sensors and the current sensor. To analyze the influence of different insertion points on vibration suppression, the transfer function from the sensor noise to the PA current is deduced. Based on the transfer function, this paper uses a Bode plot to qualitatively analyze the influence of different insertion positions.

When the notch filter is inserted at position 1, i.e., the forward path of the outer loop, the transfer function from the sensor noise to the current can be deduced as follows in Equations (14) and (15):

$$G_{1i}(s) = \frac{I(s)}{d_i(s)} = \frac{G_{pi}(s)G_c(s)}{1 + G_{pd}(s)G_{pi}(s)G_c(s)G_m(s)G_f(s) + G_{pi}(s)G_c(s)} \tag{14}$$

$$G_{1x}(s) = \frac{I(s)}{d_x(s)} = \frac{G_f(s)G_{pd}(s)G_{pi}(s)G_c(s)}{1 + G_{pd}(s)G_{pi}(s)G_c(s)G_m(s)G_f(s) + G_{pi}(s)G_c(s)} \tag{15}$$

where $G_f(s)$ is the continuous transfer function of the LMS adaptive notch filter, $d_i(s)$ is the Laplace transform of the current noise signal, $d_x(s)$ is the Laplace transform of position noise, and $I(s)$ is the Laplace transform of the current.

The Bode plots of Equations (14) and (15) are shown in Figure 6. According to Figure 6, the natural frequency noise of the position sensor can be filtered out when the notch filter is inserted at position 1. However, the natural frequency noise of the current sensor cannot be filtered out.

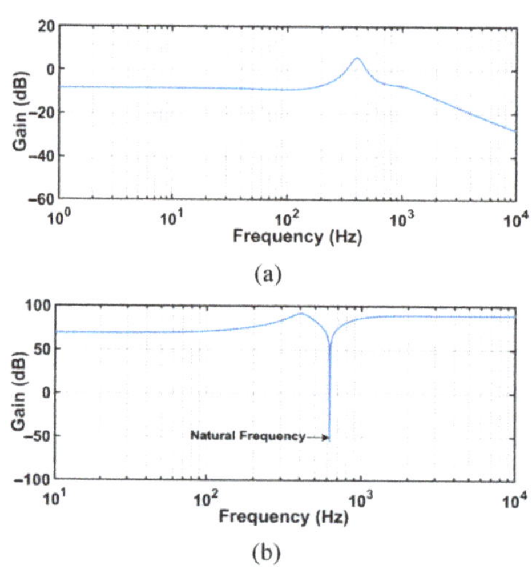

Figure 6. Bode plot of the transfer function from the noise to the currents of insertion positions 1 and 2: (**a**) transfer function from the current noise to the current; (**b**) transfer function from the position noise to the current.

When the notch filter is inserted into position 2, the transfer function from the sensor noise to the current is deduced as follows in Equations (16) and (17).

$$G_{2i}(s) = \frac{I(s)}{d_i(s)} = \frac{G_{pi}(s)G_c(s)}{1 + G_{pd}(s)G_{pi}(s)G_c(s)G_m(s)G_f(s) + G_{pi}(s)G_c(s)} \tag{16}$$

$$G_{2x}(s) = \frac{I(s)}{d_x(s)} = \frac{G_f(s)G_{pd}(s)G_{pi}(s)G_c(s)}{1 + G_{pd}(s)G_{pi}(s)G_c(s)G_m(s)G_f(s) + G_{pi}(s)G_c(s)} \quad (17)$$

According to Equations (14)–(17), the transfer functions of insertion position 1 and insertion position 2 are the same. Thus, the Bode plots of insertion position 1 and insertion position 2 are the same.

When the notch filter is inserted at position 3, the transfer function from the sensor noise to the current is deduced as follows in Equations (18) and (19).

$$G_{3i}(s) = \frac{I(s)}{d_i(s)} = \frac{G_f(s)G_{pi}(s)G_c(s)}{1 + G_{pd}(s)G_{pi}(s)G_c(s)G_m(s) + G_{pi}(s)G_c(s)G_f(s)} \quad (18)$$

$$G_{3x}(s) = \frac{I(s)}{d_x(s)} = \frac{G_f(s)G_{pd}(s)G_{pi}(s)G_c(s)}{1 + G_{pd}(s)G_{pi}(s)G_c(s)G_m(s) + G_{pi}(s)G_c(s)G_f(s)} \quad (19)$$

The Bode plots of Equations (18) and (19) are shown in Figure 7. When the notch filter is inserted at position 3, the natural frequency noise of the position sensor and the current sensor can both be filtered out.

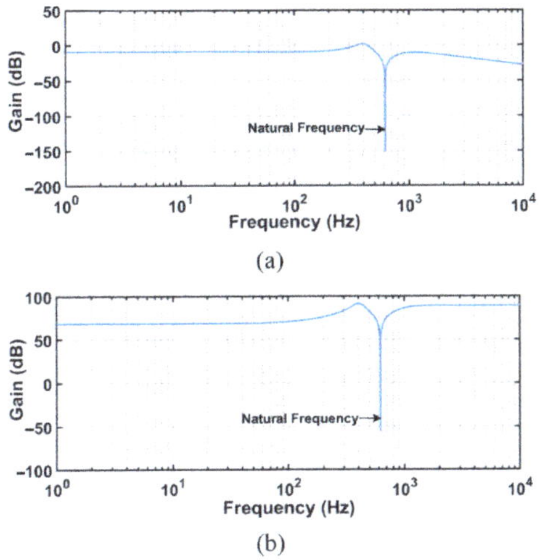

Figure 7. Bode plot of the transfer function from the noise to the current of insertion position 3: (**a**) transfer function from the current noise to the current; (**b**) transfer function from the position noise to the current.

When the notch filter inserts to position 4, the transfer function from the sensor noise to the current is deduced as follows in Equations (20) and (21):

$$G_{4i}(s) = \frac{I(s)}{d_i(s)} = \frac{G_f(s)G_{pi}(s)G_c(s)}{1 + G_{pd}(s)G_{pi}(s)G_c(s)G_m(s) + G_{pi}(s)G_c(s)G_f(s)} \quad (20)$$

$$G_{4x}(s)\frac{I(s)}{d_x(s)} = \frac{G_{pd}(s)G_{pi}(s)G_c(s)}{1 + G_{pd}(s)G_{pi}(s)G_c(s)G_m(s)G_f(s) + G_{pi}(s)G_c(s)G_f(s)} \quad (21)$$

The Bode plots of Equations (20) and (21) are shown in Figure 8. According to Figure 8, the natural frequency noise of the current sensor can be filtered out when the notch filter is inserted at position 4. However, the natural frequency noise of the position sensor cannot be filtered out.

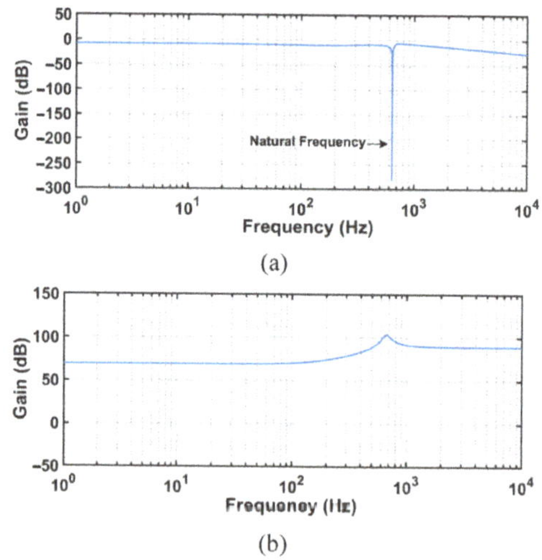

Figure 8. Bode plot of the transfer function from the noise to the current of insertion position 4: (**a**) transfer function from the current noise to the current; (**b**) transfer function from the position noise to the current.

Based on the above analysis, the notch filter is inserted at position 3, i.e., the forward path of the inner loop is the most effective to filter out the natural frequency noise of the sensors.

4. Experimental Results

The influence of the insertion position of the LMS adaptive notch filter is experimentally analyzed on a 500-kW MS FESS prototype. The natural frequency is tested through a laser vibrometer. The position responses of different insertion positions of the notch filter are tested.

4.1. Experimental Platform

The experimental platform is shown in Figure 9. The experimental platform includes a 500-kW MS FESS prototype, Tektronix oscilloscope, Rohde & Schwarz FSH4 dynamic signal analyzer, and power supply. The MS FESS prototype is shown in Figure 10, and its main parameters are listed in Table 1.

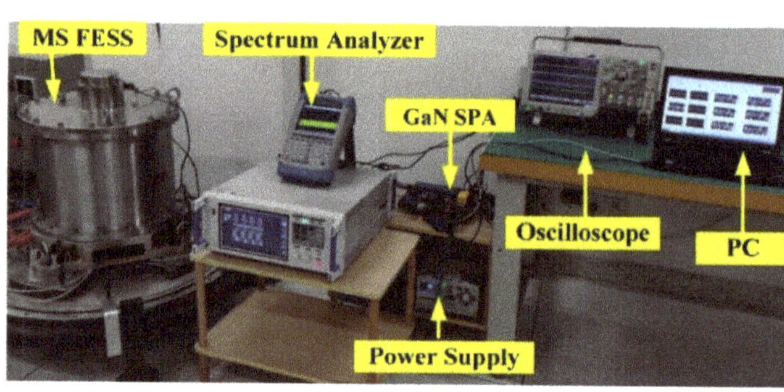

Figure 9. The experimental platform of the MS FESS.

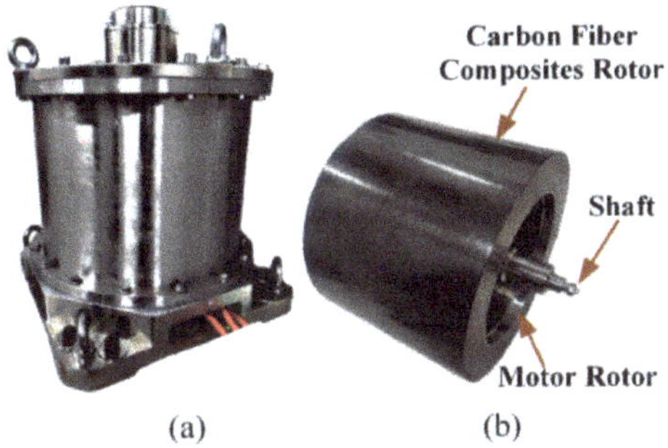

Figure 10. (**a**) The MS FESS prototype. (**b**) The flywheel rotor.

Table 1. Main Parameter Specification of the MS.

Symbol	Description	Value
V_{DC}	Maximum voltage	500 V
P_r	Maximum power	500 kW
E_m	Stored energy	1.2 kWh
ω_m	Maximum design speed	27,000 rpm
M_r	Rotor weight	58.0 kg
S_m	Prototype Size	Φ500 × 590 mm

4.2. Experiment A: Test of the Natural Frequency of the Flywheel Rotor

The natural frequency test device is shown in Figure 11. The flywheel rotor is suspended by a rope. A Polytecpsv-500 laser vibrometer is used to test the natural frequency. The tested natural frequency is shown in Figure 12. According to Figure 12, the natural frequency of the flywheel rotor is 635 Hz. Thus, the selected center frequency of the notch filter is 635 Hz.

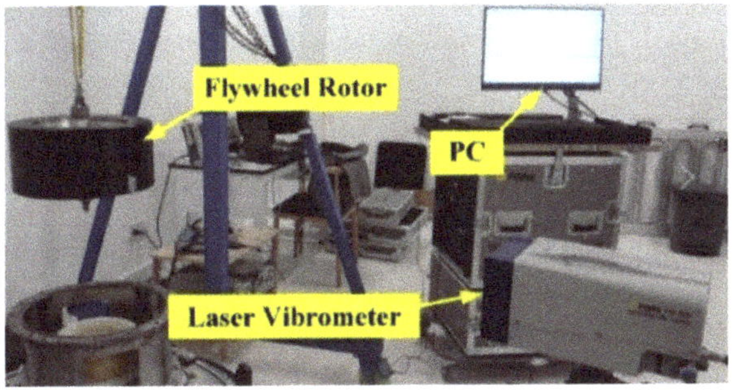

Figure 11. The device for testing the natural frequency of the flywheel rotor.

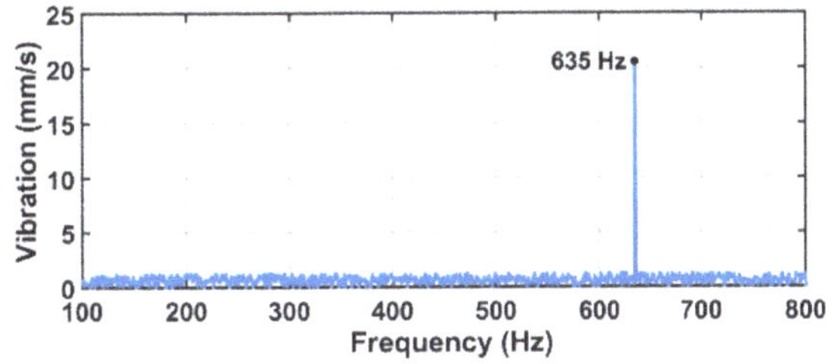

Figure 12. The tested natural frequency of the flywheel rotor.

The frequency analysis of the current sensor signal and position sensor signal is presented in Figure 13. The frequency analysis shows that the current sensor signal and position sensor signal include a noise signal that is the same as the natural frequency. To avoid natural frequency vibration, the noise signal that is the same as the natural frequency should be filtered out.

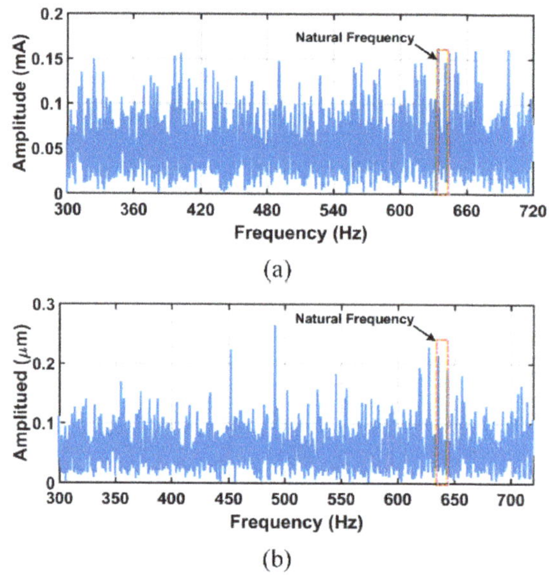

Figure 13. Frequency analysis of the sensor signal: (**a**) current sensor; (**b**) position sensor.

4.3. Experiment B: Position Responses of Different Insertion Positions

The experimental analysis of different insertion positions of the notch filter is performed. The parameters of the notch filter are listed in Table 2. Figure 14 shows the experimental results when the notch filter is inserted into the forward path of the outer loop, i.e., insertion position 1. According to Figure 14a, the amplitude of the vibration without a notch filter is larger than 25 μm. With a notch filter, the amplitude of the vibration can be smaller than 4 μm. The experimental result illustrates that the vibration can be suppressed when the notch filter is inserted at position 1. However, the notch filter cannot filter out the natural frequency noise of the current sensor. When the gain of the current controller increases, the current noise is amplified. Figure 14b shows the position response curve when the gain of the current controller is increased by 30%. The amplitude of the vibration is increased and is larger than 10 μm.

Table 2. Main Parameters of the LMS filter.

Symbol	Description	Value
ω_0	Center frequency	635 Hz
μ	Step length	0.01
f_s	Sampling frequency	5 kHz

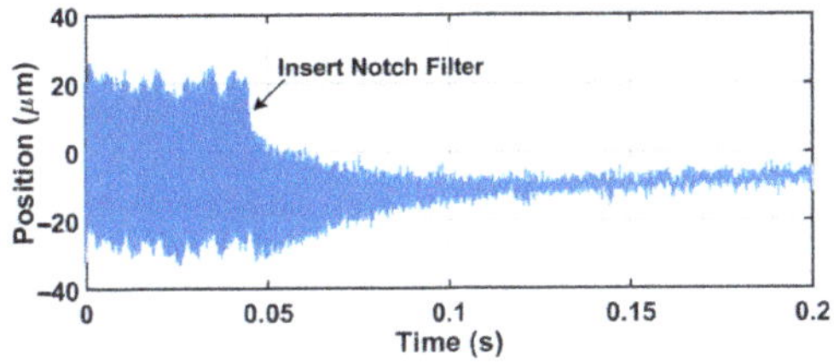

(a)

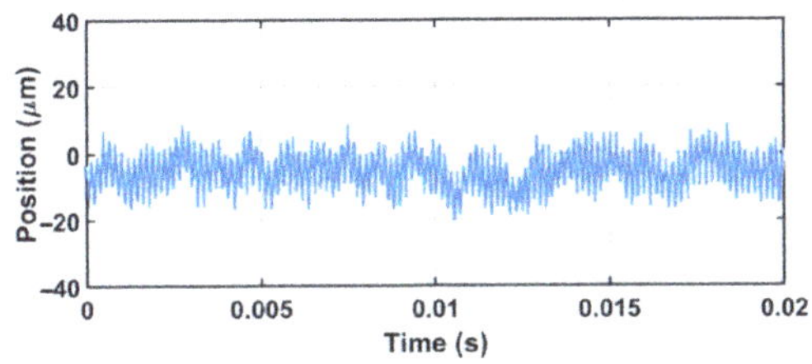

(b)

Figure 14. The experimental results when the notch filter is inserted at position 1: (**a**) position signal when the notch filter is inserted; (**b**) position signal when the gain of the current controller increases.

Figure 15 shows the experimental results when the notch filter is inserted into the feedback path of the outer loop, i.e., insertion position 2. With this notch filter, the vibration can be suppressed, and the amplitude of the vibration is smaller than 4 μm. Similar to insertion at position 1, the notch filter cannot filter out the natural frequency noise of the current sensor. Figure 15b shows the position response curve when the gain of the current controller is increased by 30%. The amplitude of the vibration is increased and is larger than 12 μm.

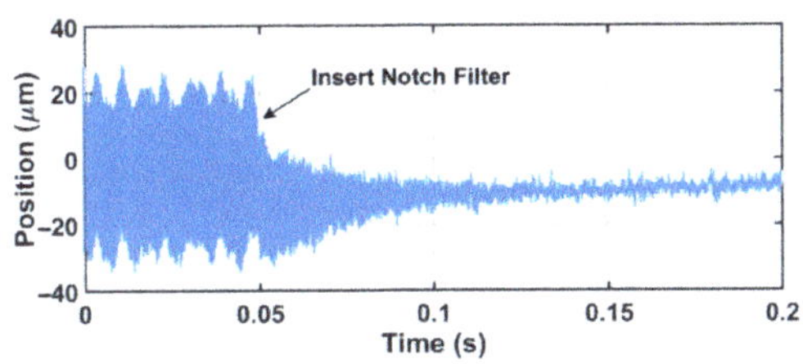

(a)

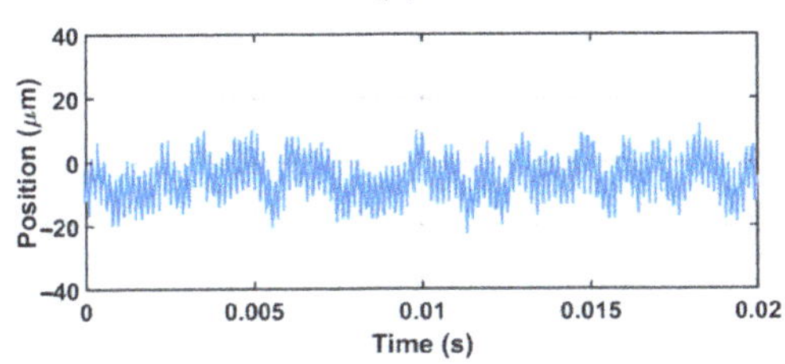

(b)

Figure 15. The experimental results when the notch filter is inserted at position 2: (**a**) position signal when the notch filter is inserted; (**b**) position signal when the gain of the current controller increases.

Figure 16 shows the experimental results when the notch filter is inserted into the forward path of the inner loop, i.e., insertion position 3. With a notch filter, the amplitude of the vibration can be suppressed and is smaller than 4 μm. Based on the theoretical analysis, the notch filter can filter out the natural frequency noise of the current sensor and the position sensor. Figure 16b shows the position response curve when the gain of the position controller is increased by 30%. The amplitude of the vibration does not significantly increase and is smaller than 6 μm.

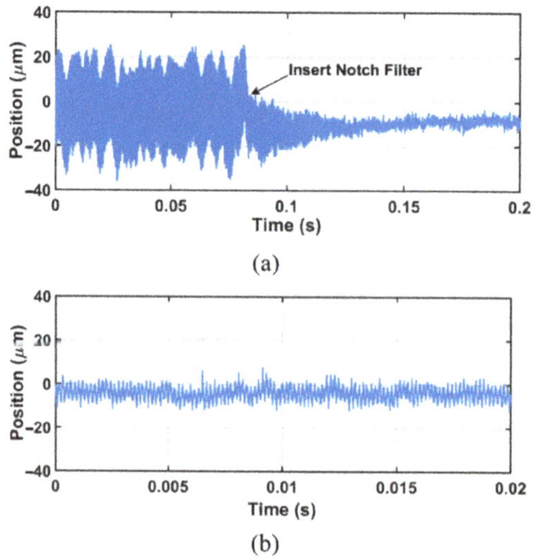

Figure 16. The experimental results when the notch filter is inserted at position 3: (**a**) position signal when the notch filter is inserted; (**b**) position signal when the gain of the position controller increases.

Figure 17 shows the experimental results when the notch filter is inserted into the feedback path of the inner loop, i.e., insertion position 4. With a notch filter, the vibration can be suppressed, and the amplitude of the vibration is smaller than 4 μm. The notch filter cannot filter out the natural frequency noise of the position sensor. Figure 17b shows the position response curve when the gain of the position controller is increased by 30%. The amplitude of the vibration is increased and is larger than 18 μm.

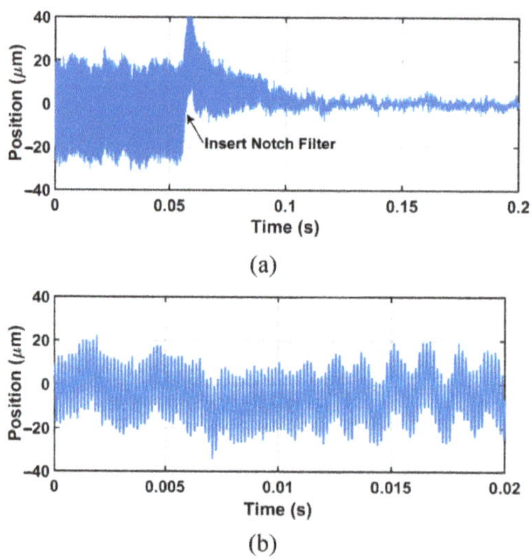

Figure 17. The experimental results when the notch filter is inserted at position 2: (**a**) position signal when the notch filter is inserted; (**b**) position signal when the gain of the position controller increases.

The experimental results show that the vibration can be suppressed to a certain extent by inserting the notch filter into the four positions of Figure 5. When the notch filter is inserted at position 3, the MS system shows the most appropriate vibration suppression performance. The experimental results match the theoretical analysis.

5. Conclusions

In this paper, the influence of the insertion position of the notch filter in an MS FESS is analyzed for natural frequency vibration suppression. The MS control model and LMS adaptive notch filter are given. Based on the control mode and notch filter, the influence of the insertion position is analyzed by the Bode plot. The experimental analysis is carried out on a 500 kW MS FESS prototype. The main conclusions are presented as follows.

(1) The vibration can be suppressed to some extent by inserting the notch filter into the forward and feedback path of the outer loop and the forward and feedback path of the inner loop.
(2) When the notch filter is inserted into the forward and feedback path of the outer loop, the natural frequency noise of the position sensor can be filtered out; nevertheless, the natural frequency noise of the current sensor cannot be filtered out. When the notch filter is inserted into the feedback path of the inner loop, the natural frequency noise of the position sensor can be filtered out; nevertheless, the natural frequency noise of the current sensor cannot be filtered out.
(3) When the notch filter is inserted into the forward path of the inner loop, the natural frequency noise of the position sensor and current sensor can be filtered out. Inserting the notch filter into the forward path of the inner loop is the most appropriate way to suppress the natural vibration.

Author Contributions: Conceptualization, H.H.; Data curation, P.X.; Funding acquisition, K.L.; Investigation, H.H. and H.W.; Methodology, H.H.; Software, J.W.; Validation, H.W. and H.H.; Writing—original draft, H.H.; Writing—review and editing, H.W. and Y.Z. All authors have read and agreed to the published version of the manuscript.

Funding: Technology program of Shenzhen, China, under grants JCYJ20200109142205924 and 202001093000459.

Data Availability Statement: The data presented in this study are available upon request from the corresponding author. The data are not publicly available due to the privacy agreement among coauthors.

Conflicts of Interest: The authors declare no conflict of interest.

References

1. Komori, M.; Kato, H.; Asami, K.-I. Suspension-Type of Flywheel Energy Storage System Using High Tc Superconducting Magnetic Bearing (SMB). *Actuators* **2022**, *11*, 215. [CrossRef]
2. Hu, H.; Liu, K.; Wei, J.; Wang, H. Multirate model predictive current control of a permanent magnet synchronous machine for a flywheel energy storage system. *Energy Rep.* **2022**, *8*, 11579–11591. [CrossRef]
3. Shadnam Zarbil, M.; Vahedi, A.; Azizi Moghaddam, H.; Khlyupin, P.A. Design and Sizing of Electric Bus Flash Charger Based on a Flywheel Energy Storage System: A Case Study. *Energies* **2022**, *15*, 8032. [CrossRef]
4. Farrokhzad Ershad, N.; Tafazzoli Mehrjardi, R.; Ehsani, M. High-Performance 4WD Electric Powertrain with Flywheel Kinetic Energy Recovery. *IEEE Trans. Power Electron.* **2021**, *36*, 772–784. [CrossRef]
5. Masouleh, M.I.; Limebeer, D.J.N. Fuel Minimization for a Vehicle Equipped With a Flywheel and Battery on a Three-Dimensional Track. *IEEE Trans. Intell. Veh.* **2017**, *2*, 161–174. [CrossRef]
6. Li, J.; Han, Y.; Li, S. Flywheel-Based Boom Energy Recovery System for Hydraulic Excavators with Load Sensing System. *Actuators* **2021**, *10*, 126. [CrossRef]
7. Hutchinson, A.; Gladwin, D.T. Optimisation of a wind power site through utilisation of flywheel energy storage technology. *Energy Rep.* **2020**, *6*, 259–265. [CrossRef]
8. Adhikari, S.; Karki, R.; Piya, P. Recovery Risk Mitigation of Wind Integrated Bulk Power System With Flywheel Energy Storage. *IEEE Trans. Power Syst.* **2019**, *34*, 3484–3493. [CrossRef]
9. Ghanaatian, M.; Lotfifard, S. Control of Flywheel Energy Storage Systems in the Presence of Uncertainties. *IEEE Trans. Sustain. Energy* **2019**, *10*, 36–45. [CrossRef]

10. Mukoyama, S.; Nakao, K.; Sakamoto, H.; Matsuoka, T.; Nagashima, K.; Ogata, M.; Yamashita, T.; Miyazaki, Y.; Miyazaki, K.; Maeda, T.; et al. Development of Superconducting Magnetic Bearing for 300 kW Flywheel Energy Storage System. *IEEE Trans. Appl. Supercond.* **2017**, *27*, 3600804. [CrossRef]
11. Werfel, F.N.; Floegel-Delor, U.; Riedel, T.; Rothfeld, R.; Wippich, D.; Goebel, B.; Reiner, G.; Wehlau, N. A Compact HTS 5 kWh/250 kW Flywheel Energy Storage System. *IEEE Trans. Appl. Supercond.* **2007**, *17*, 2138–2141. [CrossRef]
12. Gengji, W.; Ping, W. Rotor Loss Analysis of PMSM in Flywheel Energy Storage System as Uninterruptable Power Supply. *IEEE Trans. Appl. Supercond.* **2016**, *26*, 0609905. [CrossRef]
13. Nagorny, A.S.; Dravid, N.V.; Jansen, R.H.; Kenny, B.H. Design Aspects of a High Speed Permanent Magnet Synchronous Motor/Generator for Flywheel Applications. In Proceedings of the IEEE International Conference on Electric Machines & Drives, San Antonio, TX, USA, 15 May 2005.
14. Wang, P.; Gu, T.; Sun, B.; Liu, R.; Zhang, T.; Yang, J. Design and Performance Analysis of Super Highspeed Flywheel Rotor for Electric Vehicle. *World Electr. Veh. J.* **2022**, *13*, 147. [CrossRef]
15. Kale, V.; Secanell, M. A comparative study between optimal metal and composite rotors for flywheel energy storage systems. *Energy Rep.* **2018**, *4*, 576–585. [CrossRef]
16. Wang, H.; Wu, Z.; Liu, K.; Wei, J.; Hu, H. Modeling and Control Strategies of a Novel Axial Hybrid Magnetic Bearing for Flywheel Energy Storage System. *IEEE/ASME Trans. Mechatron.* **2022**, *27*, 3819–3829. [CrossRef]
17. Hu, H.; Liu, K.; Wang, H.; Wei, J. A Wide Bandwidth GaN Switching Power Amplifier of Active Magnetic Bearing for a Flywheel Energy Storage System. *IEEE Trans. Power Electron.* **2022**, *38*, 2589–2605. [CrossRef]
18. Wu, D.; Zhang, D.; Han, Q.; Wang, H. Study on nonlinear force transmissibility of flywheel rotor system considering periodic base motions. *J. Sound Vib.* **2022**, *529*, 116953. [CrossRef]
19. Herzog, R.; Buhler, P.; Gahler, C.; Larsonneur, R. Unbalance compensation using generalized notch filters in the multivariable feedback of magnetic bearings. *IEEE Trans. Control. Syst. Technol.* **1996**, *4*, 580–586. [CrossRef]
20. Zheng, S.; Han, B.; Feng, R.; Jiang, Y. Vibration Suppression Control for AMB-Supported Motor Driveline System Using Synchronous Rotating Frame Transformation. *IEEE Trans. Ind. Electron.* **2015**, *62*, 5700–5708. [CrossRef]
21. Cui, P.; Liu, Z.; Xu, H.; Zheng, S.; Han, B.; Zhang, D. Harmonic Vibration Force Suppression of Magnetically Suspended Rotor With Frequency-Domain Adaptive LMS. *IEEE Sens. J.* **2020**, *20*, 1166–1175. [CrossRef]
22. Li, J.; Liu, G.; Zheng, S.; Cui, P.; Chen, Q. Micro-Jitter Control of Magnetically Suspended Control Moment Gyro Using Adaptive LMS Algorithm. *IEEE/ASME Trans. Mechatron.* **2022**, *27*, 327–335. [CrossRef]
23. Chen, Q.; Liu, G.; Han, B. Unbalance vibration suppression for AMBs system using adaptive notch filter. *Mech. Syst. Signal Process.* **2017**, *93*, 136–150. [CrossRef]
24. Cui, P.; Li, S.; Wang, Q.; Gao, Q.; Cui, J.; Zhang, H. Harmonic Current Suppression of an AMB Rotor System at Variable Rotation Speed Based on Multiple Phase-Shift Notch Filters. *IEEE Trans. Ind. Electron.* **2016**, *63*, 6962–6969. [CrossRef]
25. He, J.; Deng, Z.; Peng, C.; Li, K. Reduction of the High-Speed Magnetically Suspended Centrifugal Compressor Harmonic Vibration Using Cascaded Phase-Shifted Notch Filters. *IEEE Sens. J.* **2021**, *21*, 1315–1323. [CrossRef]
26. Moir Tom, J. *Rudiments of Signal Processing and Systems*; Springer: Berlin/Heidelberg, Germany, 2022.

actuators

MDPI

Article

Design and Implementation of Permanent and Electromagnet Composite Vibration Isolation System Based on Negative Stiffness Theory

Mingda Zhai, Bo Zhang, Xiaolong Li * and Zhiqiang Long

College of Intelligence Science and Technology, National University of Defense Technology, Changsha 410073, China

* Correspondence: 13787786254@163.com

Abstract: In order to decrease the transmission of vibration and achieve the attenuation of the vibration magnitude of an isolated object, a new type of permanent and electromagnet composite vibration isolation system is designed based on negative stiffness theory. Firstly, according to the characteristic analysis, the design of a permanent and electromagnet hybrid actuator is accomplished; secondly, the vibration isolation system model is established, and the active control strategy based on the fuzzy PID algorithm is designed. Finally, a test platform is built to verify the vibration isolation effect. The results indicate that the developed permanent and electromagnet composite vibration isolation system renders the sharp attenuation of external vibration in multiple frequency bands. When the external vibration frequency is within the frequency range of 20 Hz to 100 Hz, the vibration attenuation is greater than 80%; when the external vibration frequency is within the frequency range of 100 Hz to 500 Hz, the vibration attenuation rate is greater than 90%.

Keywords: vibration; negative stiffness; permanent and electromagnet composite vibration isolation; vibration attenuation rate

Citation: Zhai, M.; Zhang, B.; Li, X.; Long, Z. Design and Implementation of Permanent and Electromagnet Composite Vibration Isolation System Based on Negative Stiffness Theory. *Actuators* **2023**, *12*, 44. https://doi.org/10.3390/act12010044

Academic Editor: Eihab M. Abdel-Rahman

Received: 29 November 2022
Revised: 21 December 2022
Accepted: 4 January 2023
Published: 16 January 2023

1. Introduction

Vibration, which is a critical bottleneck restricting accuracy and performance, runs through the entire life cycle of weapons and equipment. Additionally, it also seriously affects the reliability of equipment and reduces normal service life [1–4]. In order to decrease the transmission of vibration and realize the vibration level attenuation of the object to be isolated, various vibration isolation systems and vibration control technologies have emerged. In recent years, research of negative stiffness technology has become a focus in the field of vibration isolation. Saitama University in Japan proposed a three-degrees-of-freedom modular vibration isolation system based on springs and ferromagnetic materials which can realize the dynamic adjustment of negative stiffness and has a noticeable vibration isolation effect on the disturbance of the load [5,6]. The Naval Engineering University designed a positive and negative stiffness parallel active and passive vibration isolation controller to be applied to ships. Permanent and electromagnet hybrid actuators are used in the air spring for significant vibration isolation [7–9].

In order to reduce the damage of external vibration to a certain type of equipment, this paper studies and designs a new type of permanent and electromagnet composite vibration isolation system (PECVIS) based on the negative stiffness theory, taking permanent magnets as negative stiffness components. This system can dynamically adjust the stiffness and damping of the system to achieve sharp attenuation of external vibration in multiple frequency bands.

2. Design of PECVIS Based on Negative Stiffness Theory

The stiffness of the traditional vibration isolation device is positive. Hence, there is a contradiction between the system stiffness, the vibration isolation effect and the carrying capacity. What accounts for it is that the greater the force of the positive stiffness element, the greater the deformation. On the contrary, the negative stiffness element is less than zero, and the direction of the force increment is opposite to that of the deformation increment [10–14]. Adding negative stiffness components to the traditional vibration isolation device can effectively decrease the overall stiffness of the system without damaging the bearing capacity of the system and efficiently resolve the conflicts of the traditional passive vibration isolation system, as well as achieve a relatively better low-frequency vibration isolation effect [15–17]. Therefore, based on the negative stiffness theory, a new type of PECVIS is designed in this paper. The overall structure of the PECVIS is illustrated in Figure 1.

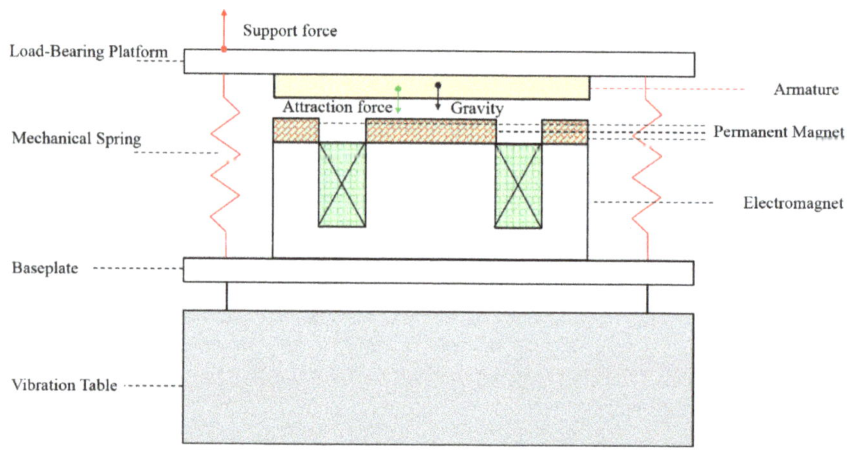

Figure 1. The overall structure of the PECVIS.

The mechanical spring provides the support force for the vibration-isolated object through pre-compression. Under the condition of no current, the vibration-isolated object is always near the initial balance position by balancing the electromagnetic attraction between the permanent and electromagnet hybrid actuator and the armature. In the permanent and electromagnet hybrid actuator, the permanent magnet is mainly used to provide negative stiffness, reduce or balance the positive stiffness of the system brought by the mechanical spring and then make the total stiffness of the system reach the quasi-zero-stiffness state near the balance point. The electromagnet in the permanent and electromagnet hybrid actuator can change the magnetic field strength of the permanent and electromagnet hybrid actuator by changing the size and direction of the voltage and current and finally achieve the goal of dynamically adjusting the electromagnetic attraction. The directions of the electromagnetic attraction force, support force and gravity have been marked in Figure 1.

The designed PECVIS in this paper can not only reduce the total stiffness of the system through permanent magnets, but also dynamically adjust the equivalent stiffness and equivalent damping of the system. Therefore, the designed PECVIS is able to achieve a good low-frequency vibration isolation effect and broaden the vibration isolation frequency band.

The PECVIS is designed according to modularization, which is divided into two parts: the passive vibration isolation device and the permanent and electromagnet hybrid actuator.

2.1. Design of Passive Vibration Isolation Device

The selected springs need high strength and good performance so as to meet the design requirements. Therefore, the equal-pitch cylindrical coil spring is selected, which can not only simplify the design of the vibration isolation system but can also achieve excellent passive vibration isolation function. There are four mechanical springs in the whole vibration isolation system, and their installation positions are shown in Figure 2.

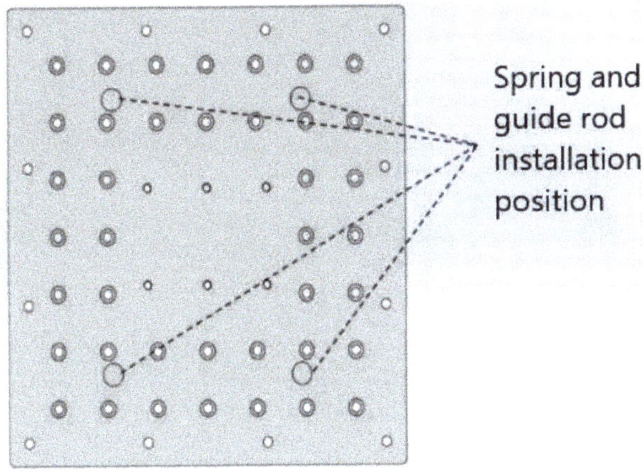

Figure 2. Installation position of spring and guide rod.

The spring material utilized in this paper is carbon spring steel wire with a grade of 65 Mn, and its specific parameters are shown in Table 1.

Table 1. Spring specific parameters.

Parameter	Magnitude	Unit
inner diameter of material	5	mm
inner diameter of spring	23	mm
stiffness	86,461	$N \cdot m^{-1}$
damping	2120	$N \cdot s \cdot m^{-1}$

2.2. Design of Electromagnet for Permanent and Electromagnet Hybrid Actuator

The electromagnet is the active control module of the permanent and electromagnet hybrid actuator, as well as the active vibration isolation device. This module changes the input current of the electromagnet by controlling the voltage of the electromagnet, thus, changing the magnetic field strength of the permanent and electromagnet hybrid actuator and achieving the goal of dynamically adjusting the attraction force.

According to experience of electromagnet design and installation, the E-type electromagnet has a good spatial structure which is highly in line with the design requirements of the PECVIS. Therefore, the spatial structure of the electromagnet is designed as an E-type electromagnet, as shown in Figure 3. Given that the electromagnet faces different vibration environments and heat problems, the silicon steel disc of model EI-180 is used as the core and armature of the electromagnet. The enameled copper round wire with low resistance is made into the coil of the electromagnet. The attraction of the E-type electromagnet is proportional to the pole area of the iron core. The larger the pole area of the iron core, the stronger the attraction of the E-type electromagnet. However, due to the requirements of the installation space of the E-type electromagnet, the pole surface of the middle iron core is square, that is, the length and width is 60 mm, respectively. The pole surfaces of the left and right iron cores are the same rectangle, and the length and width are 60 mm and 30 mm, respectively.

The armature is bolted to the bottom of the load platform. The armature is made of silicon steel discs. The length, width and height of the armature are 180 mm, 60 mm and 30 mm, respectively.

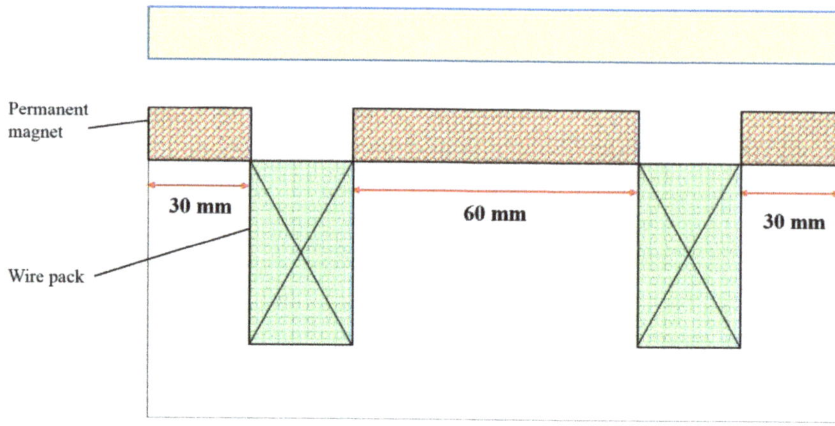

Figure 3. Structure of permanent and electromagnet hybrid actuator.

2.3. Design of Permanent Magnet for Permanent and Electromagnet Hybrid Actuator

The design of a permanent magnet mainly includes the selection of material, the geometric parameters and the analysis of the attraction force. Considering the extraordinary energy density, great coercivity and magnetic energy product of NdFeB materials, the permanent and electromagnet hybrid actuator designed in this paper selects N35 NdFeB material. The specific parameters are shown in Table 2.

Table 2. Detailed parameters of permanent magnet.

Parameter	Symbol	Value
residual magnetic induction	Br	1.2 T
coercivity	Hc	896 KA/m
relative permeability	μ_r	1.05
maximum magnetic energy product	$(BH)_{max}$	29MGOe

The permanent magnet is assembled on the upper end of the electromagnet core, the thickness of which directly determines the magnitude of its attraction force. Considering that lots of factors are often ignored in theoretical calculation, this paper uses the finite element analysis method to numerically simulate the electromagnetic field of the permanent and electromagnet hybrid actuator. As shown in Figure 4, a three-dimensional finite element model is established by Maxwell software, and the attraction force generated by the permanent magnet is numerically calculated and analyzed. According to the different thicknesses of the permanent magnets in the permanent and electromagnet hybrid actuator, the numerical analysis of the force is carried out, and the calculation results are shown in Table 3. The magnetic flux density of the permanent and electromagnet hybrid actuator is shown in Figure 5.

Table 3. Attraction force of permanent magnet near working point.

Permanent Magnet Thickness/mm	Air Gap/mm		
	9.5	10	10.5
7	465.1 N	430.4 N	399.0 N
8	532.5 N	493.9 N	459.3 N
9	595.4 N	553.1 N	515.6 N
10	652.2 N	608.9 N	567.8 N
11	705.5 N	658.7 N	615.3 N
12	753.2 N	704.2 N	659.3 N
13	797.1 N	747.0 N	699.9 N

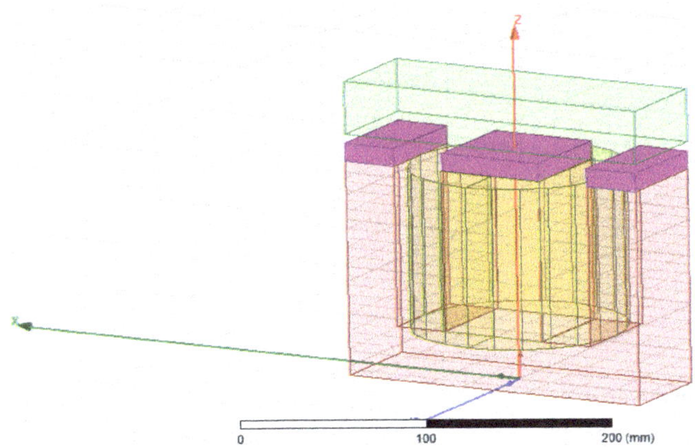

Figure 4. The three-dimensional model of permanent and electromagnet hybrid actuator.

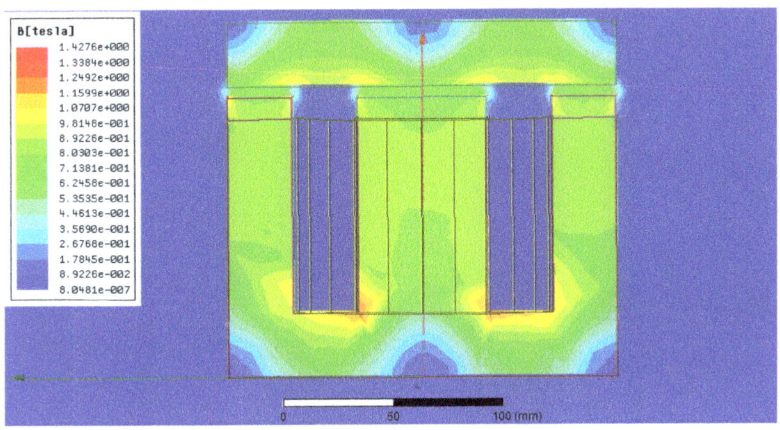

Figure 5. Magnetic flux density of permanent and electromagnet hybrid actuator.

The stiffness of the passive vibration isolation device composed of the spring is $k_s = 86.461\text{N/mm}$. As a negative stiffness element, the permanent magnet needs to enable the whole system to work normally under the circumstances of nearly zero stiffness; so, the stiffness of the permanent magnet should be approximately -86.461N/mm. This means that the permanent magnet is required to produce an attraction force of 600 N.

Since the air gap between the permanent magnet and the armature does not vary significantly, it is considered that the attraction force of the permanent magnet changes linearly near the working point. Therefore, the stiffness of the permanent magnet can be calculated between the 9.5 mm and 10.5 mm air gaps. The stiffness of the permanent magnet is:

$$k_m = \frac{F_{m10.5} - F_{m9.5}}{\Delta z} = -84.4\text{N/mm} \quad (1)$$

Based on the above analysis, the optimal thickness of the permanent magnet is determined to be 10 mm. The overall stiffness of the PECVIS can be approximated as:

$$k = k_s - k_m = 2.061\text{N/mm} \quad (2)$$

By reasonably designing the size of the permanent magnet, the positive and negative stiffness near the equilibrium point is roughly equal, and the total stiffness tends to be zero, reaching a quasi-zero-stiffness state.

3. Modeling and Nominal Controller Design of PECVIS

In order to dynamically adjust the stiffness and damping of the system, combined with the dynamic characteristics of the permanent and electromagnet hybrid actuator, the model of the PECVIS is established, and the nominal controller is designed.

3.1. Modeling of PECVIS

As to the complex electromechanical system of the PECVIS, many factors are ignored in order to reduce the difficulty of solving problems [18,19]. The magnetic flux leakage and edge effects are also ignored in this study, and it is considered that the intensity distribution of the permanent magnetic–electromagnetic hybrid magnetic field is uniform. According to the theory of the permanent magnet electromagnetic hybrid magnetic field, the equivalent magnetic circuit of the permanent and electromagnet hybrid actuator is shown in Figure 5.

According to Ohm's law, the entire circuit satisfies the following relationship:

$$\begin{cases} Ni + H_c z_m - R_{m1}\Phi_1 - R_{m2}\Phi_2 + H_c z_m = 0 \\ Ni + H_c z_m - R_{m1}\Phi_1 - R_{m3}\Phi_3 + H_c z_m - 0 \end{cases} \quad (3)$$

In the formula, N. is the number of turns of the electromagnetic coil, i is the current passing through the coil of the electromagnet, H_c is the coercive force of the NdFeB permanent magnet, z_m is the thickness of the permanent magnet, and Φ_1 is the magnetic flux of the intermediate magnetic circuit of the E-type electromagnet. In addition, Φ_2 and Φ_3 are the magnetic fluxes of the left and right magnetic circuits, respectively, R_{m1} is the total magnetic resistance of the middle air gap and permanent magnet, R_{m2} is the total magnetic resistance of the left air gap and permanent magnet and R_{m3} is the total magnetic resistance of the right air gap and permanent magnet.

The magnetic flux of the middle magnetic circuit of the E-type electromagnet is:

$$\Phi_1 = 2\Phi_2 = \frac{(Ni + 2H_c z_m)\mu_0 \mu_r s_1}{2(z_m + z\mu_r)} \quad (4)$$

In this formula, μ_0 and μ_r are the vacuum permeability and the relative permeability of NdFeB permanent magnet respectively, s_1 is the magnetic pole area of the middle permanent magnet and z is the gap between the permanent magnet and the armature.

The magnetic force generated by the permanent and electromagnet hybrid actuator is:

$$F_m = \frac{\Phi_1^2}{\mu_0 s_1} + \frac{\Phi_2^2}{\mu_0 s_2} + \frac{\Phi_3^2}{\mu_0 s_3} = \frac{(Ni + 2H_c z_m)^2 \mu_0 \mu_r^2 s_1}{2(z\mu_r + z_m)^2} \quad (5)$$

In the formula, F_m is the magnetic force generated by the permanent and electromagnet hybrid actuator, s_2 is the magnetic pole area of the upper surface of the left permanent magnet and s_3 is the magnetic pole area of the upper surface of the right permanent magnet.

The relationship between the electromagnetic attraction force generated by the permanent and electromagnet hybrid actuator and the current and air gap can also be listed in the following form:

$$F_m = \frac{\alpha(\beta i + 1)^2}{(z + \lambda)^2} \quad (6)$$

The specific values of α, β and λ are fitted by the numerical results of the electromagnetic attraction force. The electromagnetic attraction force generated by the permanent and electromagnet hybrid actuator under different currents and different air gaps is shown in Table 4.

Table 4. Force of permanent and electromagnet hybrid actuator under different currents and air gaps.

Air Gap/mm	Current/A			
	0	2	4	6
8	813.9 N	899.2 N	988.0 N	1079.5 N
9	701.8 N	775.9 N	853.5 N	934.4 N
10	608.9 N	672.9 N	741.1 N	813.1 N
11	530.2 N	587.7 N	648.4 N	711.1 N
12	464.2 N	515.2 N	569.2 N	625.1 N

According to the above Table 4, the relationship between electromagnetic attraction force and the gap is as shown in Figure 6.

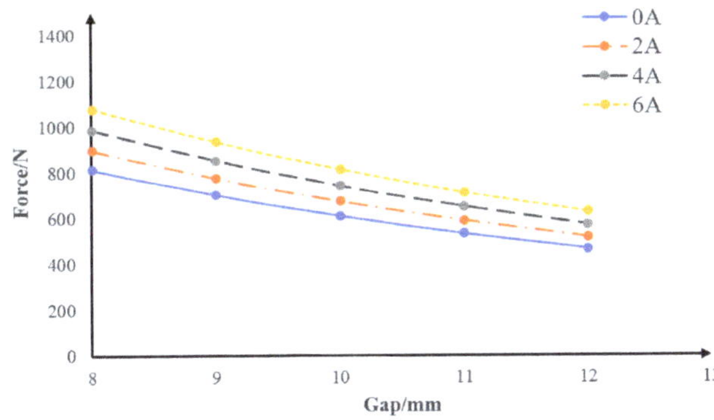

Figure 6. The relationship between electromagnetic attraction force and gap.

According to the above table, the following is obtained:

$\alpha = 131500$, $\beta = 0.02574$, $\lambda = 4.696$

The electromagnet is an inductive element, so the relationship between its voltage and current is:

$$u = Ri + \frac{N^2\mu_0\mu_r s_1}{2(z\mu_r + z_m)}\frac{di}{dt} - \frac{N(Ni + 2H_c z_m)\mu_0\mu_r^2 s_1}{2(z\mu_r + z_m)^2}\frac{dz}{dt} \tag{7}$$

The schematic diagram of the model of the PECVIS is shown in Figure 7. k_s and c represent the stiffness and damping of the spring, respectively, and k_m represents the stiffness of the permanent magnet–electromagnetic hybrid actuator, which includes the negative stiffness caused by the permanent magnet and the variable stiffness caused by the electromagnet.

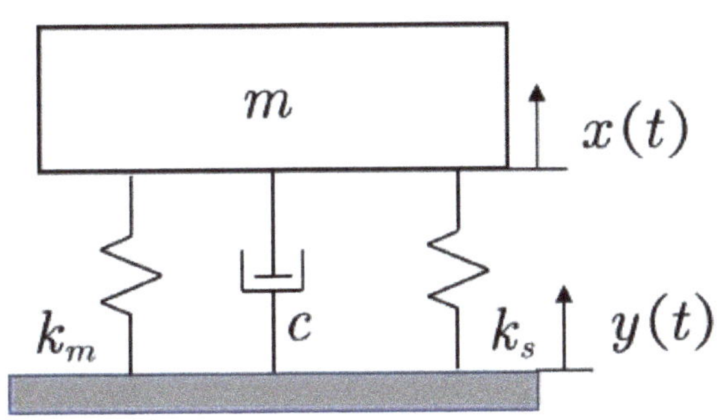

Figure 7. Diagrammatic sketch of PECVIS.

Taking the vertical upward as the positive direction, the external excitation $y(t)$ forces the vibration-isolated object of the system to vibrate in the vertical direction. $x(t)$ is the vibration.

The dynamic model of the system can be expressed as:

$$m\frac{d^2x}{dt^2} = k_s(y - x + \Delta l) + c\left(\frac{dy}{dt} - \frac{dx}{dt}\right) - mg - F_m \tag{8}$$

where Δl is the compression amount of the spring at the equilibrium point.

The system parameter values are referred to in Table 5.

Table 5. The system parameters.

Symbolic	Numerical	Unit
k_s	8646.1	N/m
c	2120	Ns/m
m	10	kg
z_m	0.01	m
μ_0	$4\pi 10^{-7}$	N/A^2
N	378	
R	0.62	Ω.

3.2. Design of Nominal Controller

Given that the working point of the PECVIS is stable, and the operating range is concentrated around the working point, the linear analysis method can be used to analyze the systems. It is assumed that, when the system is static, the initial distance z_0 satisfies:

$$z = z_0 - y + x \tag{9}$$

Taking the state variables $X = [X_1\ X_2\ X_3] = [x\ \dot{x}\ i]$, the linearized model is:

$$\begin{cases} \dot{X}_1 = X_2 \\ \dot{X}_2 = \left(\frac{k_s}{m} - \frac{2\alpha(\beta i_0+1)^2}{m(z_0+\lambda)^3}\right)(y - X_1) \\ \quad + \frac{c(\dot{y}-X_2)}{m} - \frac{2\alpha\beta(\beta i_0+1)}{m(z_0+\lambda)^2}X_3 \\ \dot{X}_3 = \frac{2(z_0\mu_r+z_m)}{N^2\mu_0\mu_r s_1}(u - RX_3) \end{cases} \tag{10}$$

Taking the input variable $U = [y\ \dot{y}\ u]$, the output variable of the system is $Y = x$. Then, the system model can be represented by the following state-space equation:

$$\begin{aligned} \dot{X} = & \begin{bmatrix} 0 & 1 & 0 \\ -\frac{k_s}{m} + \frac{2\alpha(\beta i_0+1)^2}{m(z_0+\lambda)^3} & -\frac{c}{m} & -\frac{2\alpha\beta(\beta i_0+1)}{m(z_0+\lambda)^2} \\ 0 & 0 & -\frac{2(z_0\mu_r+z_m)}{N^2\mu_0\mu_r s_1}R \end{bmatrix} X \\ & + \begin{bmatrix} 0 & 0 & 0 \\ \frac{k_s}{m} - \frac{2\alpha(\beta u_0+1)^2}{m(z_0+\lambda)^3} & \frac{c}{m} & 0 \\ 0 & 0 & \frac{2(z_0\mu_r+z_m)}{N^2\mu_0\mu_r s_1} \end{bmatrix} U \end{aligned} \tag{11}$$

The characteristic polynomial of the PECVIS does not satisfy the Hurwitz stability criterion. According to the Hartman–Grobman theorem, the system is unstable. Further analysis of the controllability of the system shows that the controllability matrix rank of the system is 3, so the PECVIS is completely controllable.

Based on the thought of cascade control, the control algorithm is usually divided into two parts, including the inner loop (also called the current loop) and the outer loop (also called the position loop) [20–22]. When the electromagnet works in the equilibrium position, due to the small variation range of inductance, its inductance can be considered as a constant. Therefore, close to the operating point, the transfer function of the relationship between the voltage and the current of the levitating electromagnet is:

$$G_i(s) = \frac{i(s)}{u(s)} = \frac{\frac{1}{R}}{\frac{N^2\mu_0\mu_r s_1}{2R(z_0\mu_r+z_m)}s+1} \tag{12}$$

The transfer functions of the electromagnet voltage and current show that the electromagnet is a typical inertial link, and the response speed of the output current seriously lags behind the changing speed of the input voltage. Therefore, current negative feedback is introduced in the design of the control algorithm. k_{c1} is the preamplification factor, and k_{c2} is the negative feedback coefficient, as shown in Figure 8.

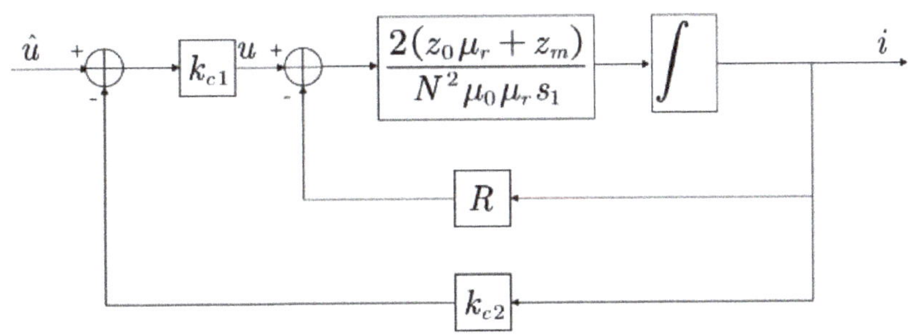

Figure 8. Current negative feedback dynamic structure.

The input voltage before the introduction of current feedback is $\hat{u}$, then the voltage directly acting on the electromagnet is:

$$u = k_{c1}(\hat{u} - k_{c2}i) \tag{13}$$

When the input voltage is a unit step signal, the response curve after introducing the current loop is as shown in Figure 9.

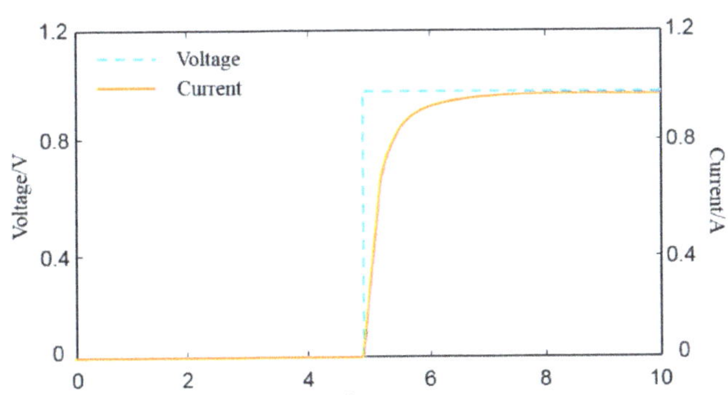

Figure 9. Step response curve of suspended solenoid.

After introducing the current negative feedback, the output current can track the change of the input voltage in time, and the order of the system is decreased from 3 to 2.

After the design of the current loop is completed, the design of the position loop adopts the PID control algorithm:

$$u(t) = K_P e(t) + K_I \int_0^t e(t)dt + K_D \frac{de(t)}{dt} \tag{14}$$

where K_p is the proportional coefficient, K_I is the integral coefficient and K_D is the differential coefficient. In the PECVIS, the $e(t)$ deviation is the displacement $x(t)$ of the vibration-isolated object in the vertical direction. Initially, the vibration-isolated object is at the equilibrium point, and its initial value is $x_0 = 0$. Therefore, the PID control law of the system is:

$$u(t) = -\left(K_P x(t) + K_I \int_0^t x(t)dt + K_D \frac{dx(t)}{dt} \right) \tag{15}$$

In summary, after the control law design is completed, it is brought into the PECVIS, which satisfies:

$$\begin{aligned} m\tfrac{d^2x}{dt^2} = &\left(k_s - \tfrac{2\alpha(\beta u_0+1)^2}{(z_0+\lambda)^3} \right)(y-x) + c(\dot{y}-\dot{x}) \\ &+ \tfrac{2\alpha\beta(\beta u_0+1)}{m(z_0+\lambda)^2}\left(K_P x + K_I \textstyle\int_0^t x dt + K_D \tfrac{dx}{dt} \right) \end{aligned} \tag{16}$$

At this time, the equivalent stiffness and equivalent damping of the PECVIS are:

$$\widetilde{k} = k_s - \frac{2\alpha(\beta u_0 + 1)^2}{(z_0 + \lambda)^3} - \frac{2\alpha\beta K_P(\beta u_0 + 1)}{m(z_0 + \lambda)^2} \tag{17}$$

$$\widetilde{c} = c - \frac{2\alpha\beta K_D(\beta u_0 + 1)}{m(z_0 + \lambda)^2} \tag{18}$$

The PECVIS designed in this paper can realize the dynamic adjustment of the stiffness and damping of the system by adjusting the three parameters K_p, K_I and K_D.

4. Active Control Strategy Based on Fuzzy PID Algorithm

In order to realize the vibration attenuation of the isolated object in multi-frequency bands, it is necessary to complete the dynamic adjustment and control parameter optimization of the existing nominal controller. Therefore, this paper adopts the fuzzy PID algorithm, takes the error ***e*** and the error variation ***ec*** as the input signals of the input fuzzification module and uses the fuzzy rules and the defuzzification process to automatically update the correction values ΔK_P, ΔK_I and ΔK_D in real time. The block diagram of fuzzy PID controller is shown in in Figure 10.

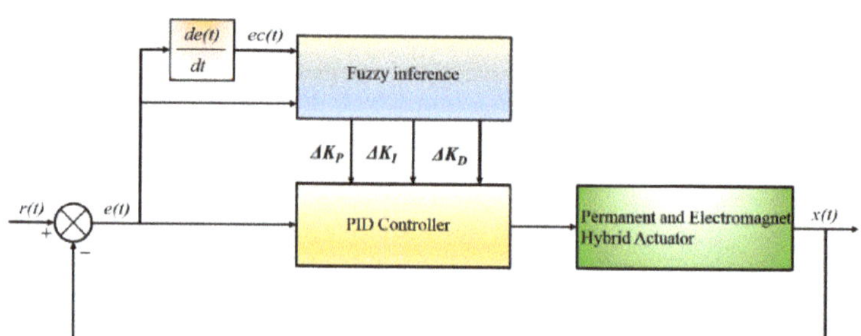

Figure 10. The block diagram of fuzzy PID controller.

4.1. Determining Input and Output Variables

The vibration system analysis determines *e* and *ec* controller output ΔK_P, ΔK_I and ΔK_D and then the one-to-one mapping relationship with fuzzy domain is established according to this value range [23–26]. Table 6 is the parameter table of the established fuzzy controller.

Table 6. Parameter table of the established fuzzy controller.

Fuzzy Variable	Basic Domain	Fuzzy Domain	Mapping Coefficient
e	[−0.0025,0.0025]	[−1,1]	400
ec	[−0.05,0.05]	[−1,1]	20
ΔK_P	[−5000,5000]	[−1,1]	0.0002
ΔK_I	[−10,10]	[−1,1]	0.1
ΔK_D	[−50,50]	[−1,1]	0.02

4.2. Select Membership Function

In fuzzy control, the size of input and output variables is usually described by human experience [27,28]. The three levels of description are usually 'big', 'middle' and 'small', and the positive and negative direction and zero state are added so there are seven levels to describe the size of input and output variables in total. Therefore, the fuzzy subset of input and output variables is taken as 'positive big (PB)', 'positive middle (PM)', 'positive small (PS)', 'zero state (ZE)', 'negative small (NS)', 'negative middle (NM)' or 'negative large (NL)'. Secondly, the input and output variables are fuzzified to determine the membership degree. In this paper, the Gaussian membership function is used to calculate the membership of input and output variables, as shown in Figure 11.

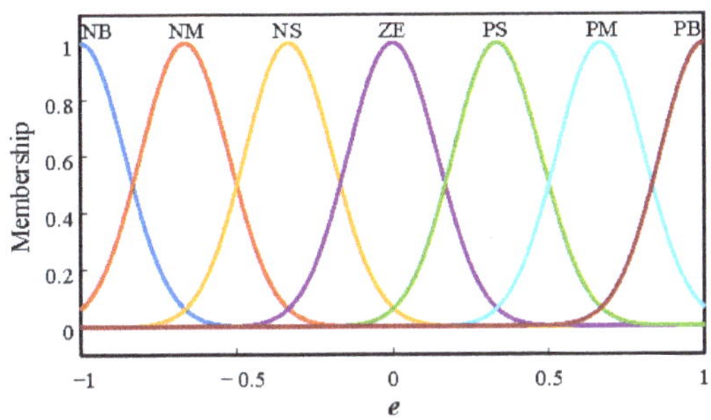

Figure 11. Gaussian membership function.

4.3. Fuzzy Reasoning

The fuzzy reasoning process is built on fuzzy rules, and the establishment of fuzzy rules is based on vibration characteristics. The principle of ΔK_P, ΔK_I and ΔK_D setting is shown in Tables 7–9.

Table 7. Setting rules of ΔK_P.

e	*ec*						
	NB	**NM**	**NS**	**ZE**	**PS**	**PM**	**PB**
NB	NB	NB	NM	NM	NM	ZE	ZE
NM	NB	NM	NM	NM	NS	ZE	PS
NS	NM	NM	NS	NS	ZE	PS	PS
ZE	NM	NS	NS	ZE	PS	PS	PM
PS	NS	NS	ZE	PS	PS	PM	PM
PM	NS	ZE	PS	PM	PM	PM	PB
PB	ZE	ZE	PM	PM	PM	PB	PB

Table 8. Setting rules of ΔK_I.

e	*ec*						
	NB	**NM**	**NS**	**ZE**	**PS**	**PM**	**PB**
NB	ZE	ZE	ZE	ZE	ZE	ZE	ZE
NM	ZE	ZE	ZE	ZE	ZE	ZE	ZE
NS	PS	PS	PS	PS	PS	PS	PS
ZE	PS	PS	PS	PS	PS	PS	PS
PS	PS	PS	PS	PS	PS	PS	PS
PM	ZE	ZE	ZE	ZE	ZE	ZE	ZE
PB	ZE	ZE	ZE	ZE	ZE	ZE	ZE

Table 9. Setting rules of ΔK_D.

e	*ec*						
	NB	**NM**	**NS**	**ZE**	**PS**	**PM**	**PB**
NB	NB	NB	PM	PM	PS	ZE	ZE
NM	NB	NB	PM	PS	PS	ZE	ZE
NS	NB	NM	PS	PS	ZE	NS	NS
ZE	NM	NS	PS	ZE	NS	NS	NM
PS	NS	NS	ZE	NS	NS	NM	NB
PM	ZE	ZE	NS	NS	NM	NB	NB
PB	ZE	ZE	NS	NM	NM	NB	NB

4.4. Defuzzification

The maximum membership method and the weighted average method are the most commonly used defuzzification methods. The weighted average method calculates multiple membership degrees through a parameter and regards the membership degree as the weight coefficient multiplied by the value of the degree. The fuzzy domain value of the output variables is able to become the actual outputs ΔK_P, ΔK_I and ΔK_D. Based on the above fuzzy PID rules, the active control strategy based on fuzzy PID control is designed.

5. Simulation Verification

The model of the system is built by Simulink, and the designed active control strategy based on the fuzzy PID algorithm is substituted into the model shown in Figure 12 for simulation verification.

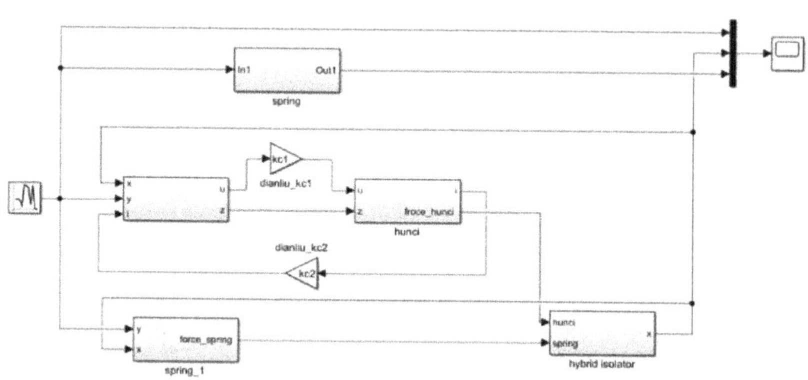

Figure 12. The Simulink model.

When the external vibration frequency is 20 Hz, 50 Hz, 100 Hz, 200 Hz, 300 Hz and 500 Hz, the results are as presented in Figure 13.

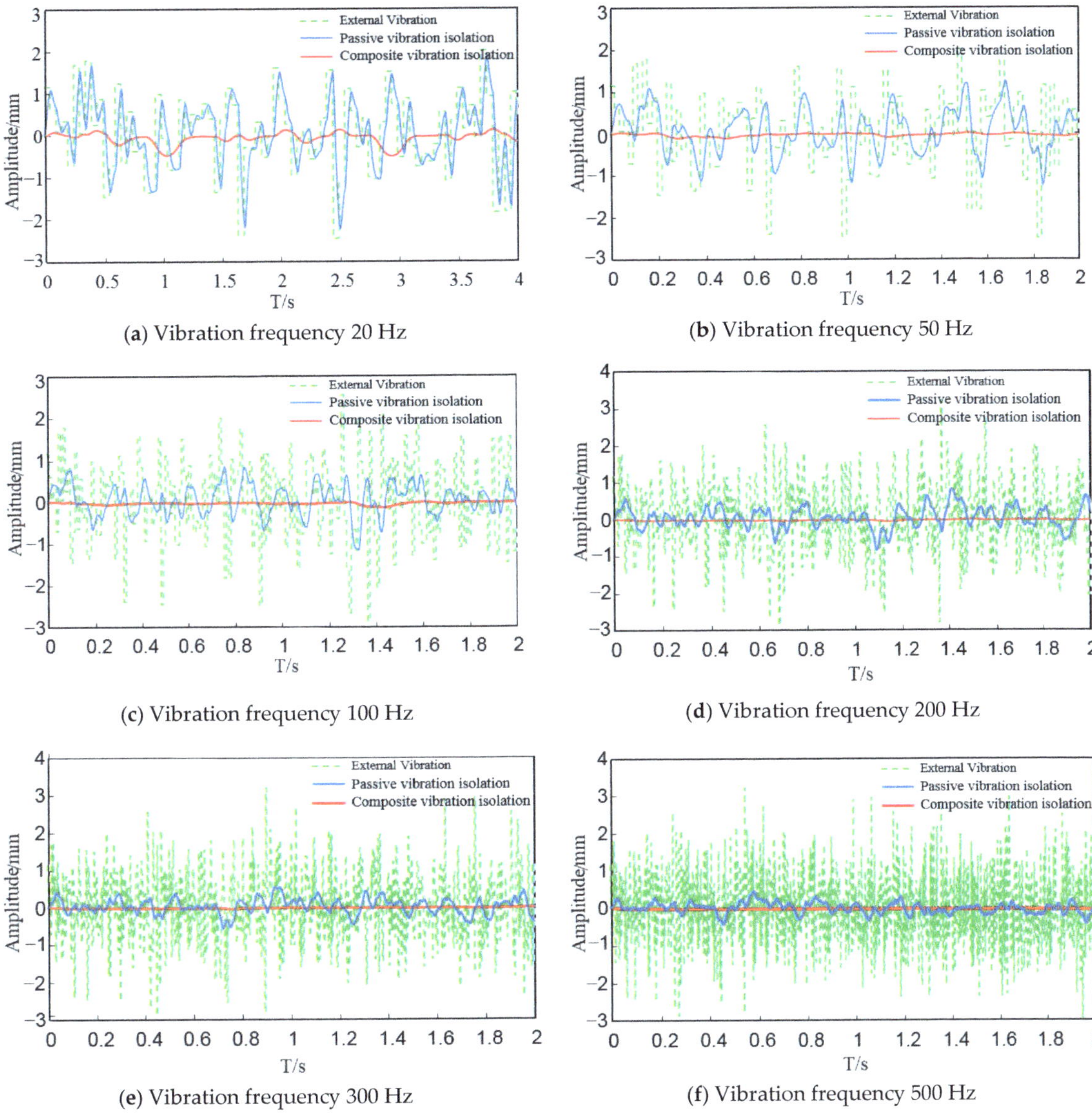

(**a**) Vibration frequency 20 Hz

(**b**) Vibration frequency 50 Hz

(**c**) Vibration frequency 100 Hz

(**d**) Vibration frequency 200 Hz

(**e**) Vibration frequency 300 Hz

(**f**) Vibration frequency 500 Hz

Figure 13. Vibration isolation effect.

As shown in Figure 14, the green curve is the external random vibration, the blue curve is the isolated object vibration only caused by the passive vibration isolation system and the red curve is the isolated object of the PECVIS. In order to evaluate the vibration isolation ability of the PECVIS, the vibration attenuation rate of the system is calculated with the different external vibration frequencies, as shown in Table 10. The attenuation rate of the designed PECVIS for a 20–50 Hz external random vibration is greater than 86.5% and, for a 100–500 Hz external random vibration, is greater than 95%. The simulation results show that the designed PECVIS achieves a good low-frequency vibration isolation effect and broadens the vibration isolation frequency band.

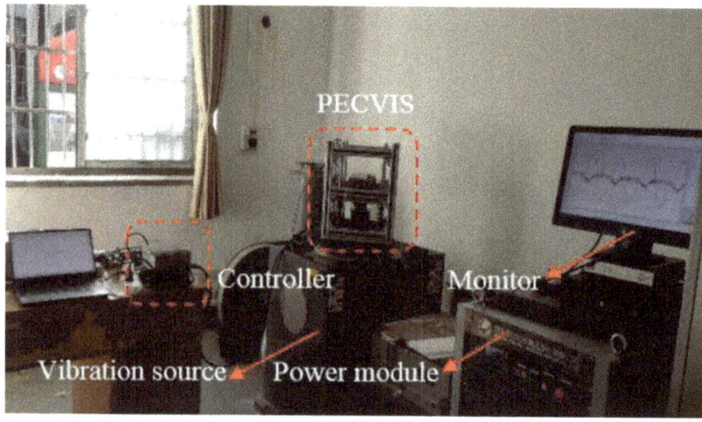

Figure 14. Photograph of PECVIS test platform.

Table 10. Vibration isolation effect of mechanical spring and PECVIS.

External Vibration Frequency (Hz)	Vibration Attenuation Rate of Mechanical Spring	Vibration Attenuation Rate of PECVIS
20	18.02%	86.53%
50	44.83%	91.96%
100	59.45%	95.57%
200	74.55%	97.29%
300	79.59%	97.97%
500	85.84%	98.48%

6. Realization and Testing of PECVIS

A PECVIS test platform is built, which is used to test the vibration isolation effect of the PECVIS.

6.1. The Construction of the Test Platform of PECVIS

This paper uses the Baseline series real-time target machine from Speedgoat to develop the controller of the PECVIS.

Considering that the vibration-isolated object is in the process of vibration, the displacement sensor adopts the non-contact TR81 series sensor, and the acceleration sensor adopts the YA23-type ICP high-sensitivity acceleration sensor. The Sushi DC-600-6 vibration table is selected as the external vibration source device to carry out the test of the PECVIS in this paper. The test site is presented in Figure 14.

6.2. Vibration Isolation Tests

In the actual experiment process, the vibration table can only provide a pseudo-sinusoidal random vibration excitation source at present. For the vibration isolation requirements of multiple frequency bands, the same external vibration frequency setting as the simulation is used to verify the vibration isolation effect. The vibration table set the vibration frequency to 20 Hz, 50 Hz, 100 Hz, 200 Hz, 300 Hz and 500 Hz and the maximum amplitude to 2.5 mm. The vibration isolation effect is illustrated in Figure 15, where the red and green curves correspond to the vibration acceleration of the isolated object and the external vibration, respectively.

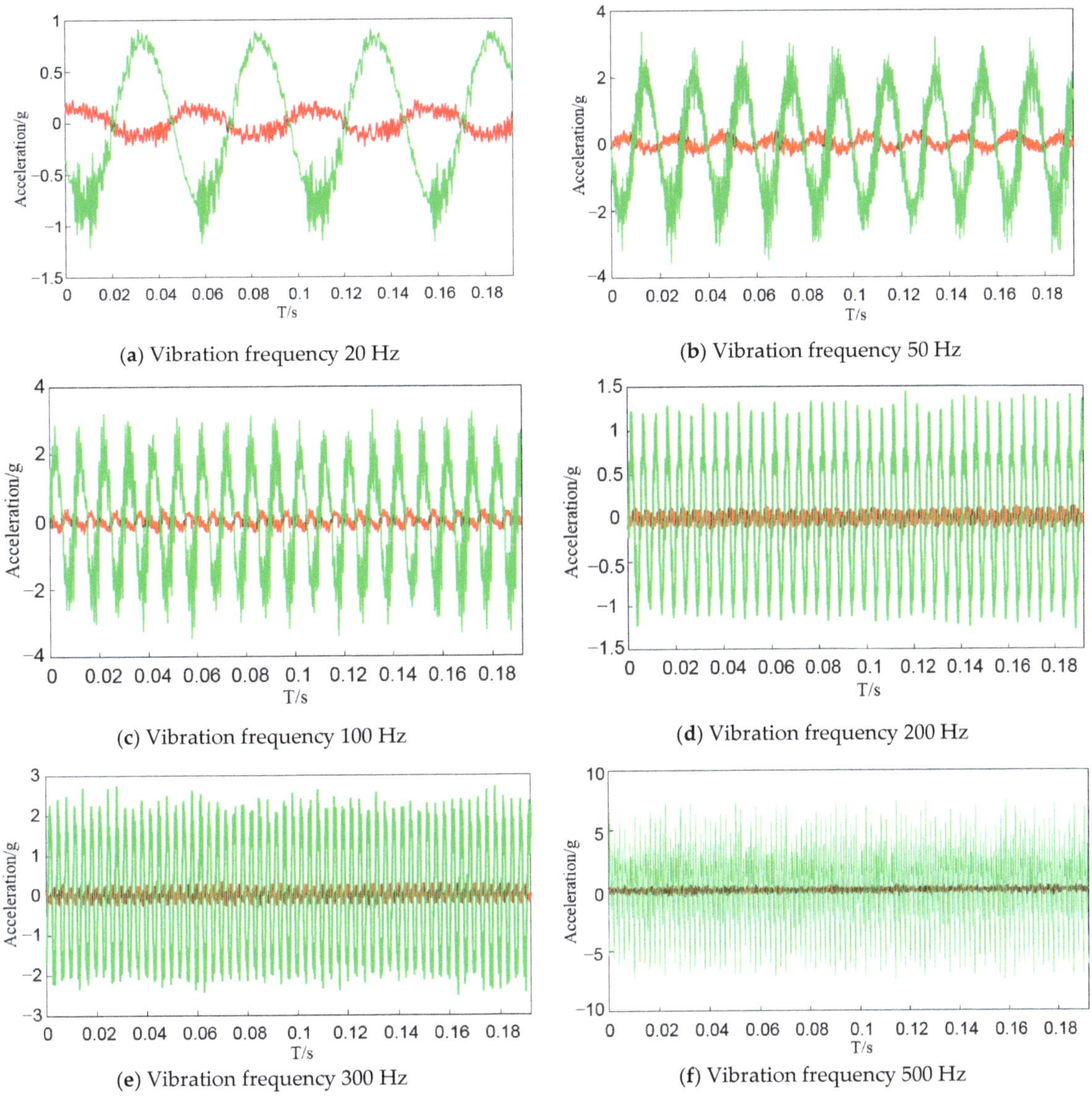

(**a**) Vibration frequency 20 Hz

(**b**) Vibration frequency 50 Hz

(**c**) Vibration frequency 100 Hz

(**d**) Vibration frequency 200 Hz

(**e**) Vibration frequency 300 Hz

(**f**) Vibration frequency 500 Hz

Figure 15. Vibration isolation effect of PECVIS.

The vibration attenuation rate and vibration transmissibility of the system are calculated with the vibration level difference method to evaluate the vibration isolation ability of the PECVIS. The vibration attenuation rate and vibration transmissibility of the PECVIS at frequencies of 20 Hz, 50 Hz, 100 Hz, 200 Hz, 300 Hz and 500 Hz are shown in Table 11.

Table 11. Vibration attenuation rate and vibration transmission rate.

External Vibration Frequency/Hz	Vibration Attenuation Rate	Vibration Transmissibility/dB
20 Hz	83.64%	−15.72
50 Hz	90.55%	−19.83
100 Hz	91.91%	−21.84
200 Hz	92.60%	−22.61
300 Hz	93.52%	−23.77
500 Hz	95.35%	−26.66

7. Conclusions

In this paper, a permanent and electromagnet composite vibration isolation system is designed based on negative stiffness theory. The mathematical model of the PECVIS is established, and an active control strategy based on the fuzzy PID algorithm is designed to dynamically adjust the stiffness and damping of the PECVIS successfully. A test platform is built to verify the vibration isolation effect. The results indicate that the PECVIS is able to decrease the transmission of vibration and achieve the attenuation of the vibration magnitude of the isolated object. The designed vibration isolation system has a vibration attenuation rate of 83.64% when the external vibration frequency is 20 Hz; the vibration attenuation rate is 91.91% at 50 Hz; the vibration attenuation rate is 90.55% at 100 Hz; the vibration attenuation rate is 92.60% at 200 Hz; the vibration attenuation rate is 93.52% at 300 Hz; and the vibration attenuation rate is 95.35% at 500 Hz.

The study demonstrates that the designed PECVIS can not only reduce the total stiffness of the system through permanent magnets, but also dynamically adjust the equivalent stiffness and equivalent damping of the system. Meanwhile, the designed PECVIS is able to achieve a good low-frequency vibration isolation effect and broaden the vibration isolation frequency band.

Author Contributions: Conceptualization, M.Z. and X.L.; methodology, M.Z. and B.Z. software, B.Z.; validation, M.Z., B.Z. and X.L.; formal analysis, Z.L.; investigation, X.L.; resources, X.L.; data curation, B.Z.; writing—original draft preparation, M.Z. and B.Z.; writing—review and editing, M.Z.; visualization, B.Z.; supervision, Z.L.; project administration, Z.L.; funding acquisition, X.L. All authors have read and agreed to the published version of the manuscript.

Funding: This research was funded by National Key R & D Program of China, grant number 2016YFB1200602.

Conflicts of Interest: The authors declare no conflict of interest.

References

1. Li, H.; Li, Y.; Li, J. Negative stiffness devices for vibration isolation applications: A review. *Adv. Struct. Eng.* **2020**, *23*, 1739–1755. [CrossRef]
2. Ren, C.; Yang, D.; Qin, H. Mechanical performance of multidirectional buckling-based negative stiffness metamaterials: An analytical and numerical study. *Materials* **2018**, *11*, 1078. [CrossRef] [PubMed]
3. Yan, B.; Yu, N.; Wu, C. A state-of-the-art review on low-frequency nonlinear vibration isolation with electromagnetic mechanisms. *Appl. Math. Mech.* **2022**, *43*, 1045–1062. [CrossRef]
4. Zhang, F.; Shao, S.; Tian, Z.; Xu, M.; Xie, S. Active-passive hybrid vibration isolation with magnetic negative stiffness isolator based on Maxwell normal stress. *Mech. Syst. Signal Process.* **2019**, *123*, 244–263. [CrossRef]
5. Zhou, N.; Liu, K. A tunable high-static-low-dynamic stiffness vibration isolator. *J. Sound Vib.* **2010**, *329*, 1254–1273. [CrossRef]
6. Li, C.; Jiang, T.; He, Q.; Peng, Z. Stiffness-mass-coding metamaterial with broadband tunability for low-frequency vibration isolation. *J. Sound Vib.* **2020**, *489*, 115685. [CrossRef]
7. Zeng, P.; Yang, Y.; Huang, L.; Yin, L.; Liu, B. An Origami-Inspired Quasi-zero Stiffness Structure for Low-Frequency Vibration Isolation. *J. Vib. Eng. Technol.* **2022**, 1–13. [CrossRef]
8. Ma, J.; Shuai, C.; Li, Y. Dynamic characteristics of maglev-air bag active-passive hybrid vibration isolators and active vibration isolation tests. *J. Vib. Shock* **2018**, *37*, 206–212.
9. Baek, S.W.; Yoon, K.Y. Improving the Hybrid Electromagnetic Clamping System by Reducing the Leakage Flux and Enhancing the Effective Flux. *Energies* **2019**, *12*, 3762. [CrossRef]

10. Henderson, J.P.; Plummer, A.; Johnston, N. An electro-hydrostatic actuator for hybrid active-passive vibration isolation. *Int. J. Hydromechatronics* **2018**, *1*, 47–71. [CrossRef]
11. Zhao, Y.; Cui, J.; Zou, L.; Cheng, Z. Modeling and dynamics of magnetically repulsive negative stiffness permanent magnetic array for precision air/magnetic composite vibration isolation. *Int. J. Struct. Stab. Dyn.* **2022**, *22*, 2250031. [CrossRef]
12. Wu, J.; Zeng, L.; Han, B.; Zhou, Y.; Luo, X.; Li, X.; Chen, X.; Jiang, W. Analysis and design of a novel arrayed magnetic spring with high negative stiffness for low-frequency vibration isolation. *Int. J. Mech. Sci.* **2022**, *216*, 106980. [CrossRef]
13. Tu, L.; Du, H.; Dong, M.; Ning, D.; Wu, Y.; Li, W.; Huang, H. Semiactively controllable vehicle seat suspension system with negative stiffness magnetic spring. *IEEE/ASME Trans. Mechatron.* **2020**, *26*, 156–167.
14. Chi, W.; Ma, S.J.; Sun, J.Q. A hybrid multi-degree-of-freedom vibration isolation platform for spacecrafts by the linear active disturbance rejection control. *Appl. Math. Mech.* **2020**, *41*, 805–818. [CrossRef]
15. Braun, D.J.; Chalvet, V.; Dahiya, A. Positive–negative stiffness actuators. *IEEE Trans. Robot.* **2018**, *35*, 162–173. [CrossRef]
16. Wang, R.; Yang, B.; Gao, H. Nonlinear Feedback Control of the Inductrack System Based on a Transient Model. *J. Dyn. Syst. Meas. Control* **2021**, *143*, 081003. [CrossRef]
17. Kim, J.; Kim, J.W.; Kim, H.C.; Zhai, L.; Ko, H.U.; Muthoka, R.M. Review of soft actuator materials. *Int. J. Precis. Eng. Manuf.* **2019**, *20*, 2221–2241. [CrossRef]
18. Sun, Y.; Xu, J.; Wu, H.; Lin, G.; Mumtaz, S. Deep learning based semi-supervised control for vertical security of maglev vehicle with guaranteed bounded airgap. *IEEE Trans. Intell. Transp. Syst.* **2021**, *22*, 4431–4442. [CrossRef]
19. Sun, Y.; Xu, J.; Lin, G.; Ji, W.; Wang, L. RBF Neural Network-Based Supervisor Control for Maglev Vehicles on an Elastic Track With Network Time Delay. *IEEE Trans. Ind. Inform.* **2022**, *18*, 509–519. [CrossRef]
20. Hu, Y.; Chen, C.; Wu, H.; Song, C. Study on structural optimization design and cascade PID control of maglev actuator for active vibration isolation system. *J. Vib. Control* **2018**, *24*, 1829–1847. [CrossRef]
21. Anantachaisilp, P.; Lin, Z. Fractional Order PID Control of Rotor Suspension by Active Magnetic Bearings. *Actuators* **2017**, *6*, 4. [CrossRef]
22. Du, H.; Cui, Q.; Liu, P.; Ma, X.; Wang, H. PID controller enhanced with artificial bee colony algorithm for active magnetic bearing. *Syst. Sci. Control Eng.* **2022**, *10*, 686–697. [CrossRef]
23. Eltag, K.; Aslamx, M.S.; Ullah, R. Dynamic stability enhancement using fuzzy PID control technology for power system. *Int. J. Control Autom. Syst.* **2019**, *17*, 234–242. [CrossRef]
24. Jin, X.; Chen, K.; Zhao, Y.; Ji, J.; Jing, P. Simulation of hydraulic transplanting robot control system based on fuzzy PID controller. *Measurement* **2020**, *164*, 108023. [CrossRef]
25. Zeng, W.; Jiang, Q.; Liu, Y.; Yan, S.; Zhang, G.; Yu, T.; Xie, J. Core power control of a space nuclear reactor based on a nonlinear model and fuzzy-PID controller. *Prog. Nucl. Energy* **2021**, *132*, 103564. [CrossRef]
26. Mitra, P.; Dey, C.; Mudi, R.K. Fuzzy rule-based set point weighting for fuzzy PID controller. *SN Appl. Sci.* **2021**, *3*, 651. [CrossRef]
27. Yang, T.; Sun, N.; Fang, Y. Adaptive fuzzy control for a class of MIMO underactuated systems with plant uncertainties and actuator deadzones: Design and experiments. *IEEE Trans. Cybern.* **2021**, *52*, 8213–8226. [CrossRef]
28. Wang, H.; Liu, P.X.; Zhao, X.; Liu, X. Adaptive fuzzy finite-time control of nonlinear systems with actuator faults. *IEEE Trans. Cybern.* **2019**, *50*, 1786–1797. [CrossRef]

actuators

MDPI

Article

Research on the Residual Vibration Suppression of Delta Robots Based on the Dual-Modal Input Shaping Method

Zhongfeng Guo, Jianqiang Zhang * and Peisen Zhang

Liaoning Provincial Key Laboratory of Intelligent Manufacturing and Industrial Robots, Shenyang University of Technology, Shenyang 110870, China
* Correspondence: zhang_jq@smail.sut.edu.cn

Abstract: The Delta robot is a high-speed and high-precision parallel robot. When it is in function, the end effector generates residual vibration, which reduces the repeat positioning accuracy and positioning efficiency. The input shaping method has previously been shown to suppress the residual vibration of the robot, but the vibration suppression effect of the single-modal input shaper is not good for the delta robot, which has multiple dominant modes for the residual vibration. To solve this problem, this paper proposes an effective method for residual vibration suppression of Delta robots based on dual-modal input shaping technology. Firstly, the modal analysis of the Delta robot is performed using finite element software, and the dominant modal of its residual vibration is determined. Secondly, six dual-modal input shapers are designed according to the obtained modal parameters. Finally, Simulink is used for simulation analysis to verify the robustness and vibration suppression performance of the designed six dual-modal input shapers and traditional single-modal input shapers. The simulation results show that the designed ZVD-EI dual-modal input shaper has good robustness, can effectively suppress the residual vibration of the Delta robot, and can effectively improve the repetitive positioning accuracy and work efficiency of the Delta robot when it is running at high speed.

Keywords: Delta robot; vibration suppression; modal analysis; multimodal; input shaping

Citation: Guo, Z.; Zhang, J.; Zhang, P. Research on the Residual Vibration Suppression of Delta Robots Based on the Dual-Modal Input Shaping Method. *Actuators* **2023**, *12*, 84. https://doi.org/10.3390/act12020084

Academic Editor: André Preumont

Received: 15 December 2022
Revised: 11 February 2023
Accepted: 12 February 2023
Published: 15 February 2023

1. Introduction

Parallel robots are widely used in electronics, food, pharmaceuticals, and other industrial applications due to their high speed, high accuracy, and excellent motion performance. Delta robots are a typical example, first introduced by Clavel [1] in the 1980s. Since then, researchers have focused on DELTA parallel robots, and many valuable studies have been proposed [2–5]. However, key components must be lightweight to achieve high-speed motion, which can lead to residual vibrations at high speeds and reduce the repeatability and positioning efficiency of the robot's end effector. Therefore, residual vibration suppression in Delta robots has become one of the hot topics of research.

At present, research on vibration suppression for Delta robots mainly includes the use of trajectory planning methods, adaptive robust control methods, and input shaping methods. In the literature [6], to reduce mechanism vibration and speed fluctuations and ensure smooth transitions in operation, the feasibility of the real-time path smoothing method is verified through experiments. In the literature [7], for the requirements of dynamic pick-and-place high-speed stability of Delta robots, multi-segment polynomial designs such as 4-3-4, 3-5-3, and 5-3-5 are used to establish a non-linear motion trajectory planning model, and the optimal solution of trajectory planning is obtained to verify the effectiveness in controlling mechanism vibration. In the literature [8], a trajectory planning method is proposed to consider both the motion smoothness and dynamic stress of the Delta robot. A modified fifth-order b-sample method is used for sensitivity analysis and normalized time factor optimization, and the results show that this method can improve the motion smoothness while reducing the dynamic stress. In the literature [9], an adaptive robust control method based on a fuzzy dynamics model is established for the Delta robot with unknown dynamics parameters, residual vibration disturbances, and other factors, using the uncertainty of the fuzzy description; the uncertainty information is estimated through an adaptive mechanism; and finally, the effectiveness of the control method is verified through simulation.

However, the use of trajectory planning methods or adaptive robust control can only suppress the process vibration of Delta robots to a certain extent when the robot is moving at high speeds and with high acceleration. To improve the repetitive positioning accuracy and positioning efficiency of the robot's end effector, the residual robot vibration must be effectively suppressed. The input shaping method [10,11] is a commonly used feed-forward-based vibration suppression method, where the actual input model of the controlled system is obtained by convolving the input signal with a series of pulse sequences to eliminate the residual vibrations generated by the system. The method was originally applied to overhead cranes [12], where the application of input shaping algorithms to the control system was effective in reducing residual oscillations in the crane boom and improving efficiency. The technique has now been used in various industrial robots [13,14].

The most fundamental component of input shapers is the ZV (Zero-Vibration) input shaper. In order to design an effective ZV input shaper, it is usually necessary to derive the exact intrinsic frequency and damping ratio of the system; however, due to the uncertainties in modeling and simulation, it is not possible to obtain completely accurate modal parameters, so it is necessary to ensure the robustness of the designed input shaper. Researchers have proposed different input shapers, such as the ZVD (Zero-Vibration and Derivative) input shaper and EI (Extra-Insensitive) input shaper, based on the ZV input shaper [15]. In the case of multimodal-dominated systems, multiple unimodal input shapers are used for convolution to obtain multi-modal input shapers for residual vibration suppression of multi-modal systems [16–18]. Initially, the input-shaping method was mainly used in linear constant systems, but its application has gradually been extended to non-linear systems and has been more extensively studied in residual vibration suppression of parallel robots [19–21].

The Delta robot is a typical multi-modal dominated non-linear system with residual-vibration-dominated modal parameters that vary greatly from position to position in the workspace. This makes the traditional single-modal input shaper described above less effective in suppressing the residual vibration of the Delta robot. Therefore, an effective input shaper needs to be designed to provide good robustness and vibration suppression performance of the Delta robot throughout the working space. Inspired by the existing results, this paper proposes a dual-modal input shaper-based residual vibration suppression method for Delta robots based on the multi-dominant modal characteristics of Delta robotic systems to improve the robustness and residual vibration suppression performance of conventional single-modal input shapers in Delta robotic applications.

The main contributions of this paper can be summarized as follows:(a) Using ANSYS modal analysis, the two-order modes that play a dominant role in the residual vibration of the Delta robot were identified and the modal parameters were obtained for the subsequent design of the input shaper. (b) The six dual-modal input shapers were designed for the system characteristics of the Delta robot, and the input shapers were built in Simulink. (c) The robustness of the six bimodal input shapers was evaluated by simulation because the dominant modal parameters of the Delta robot vary considerably at different locations in the workspace. After giving certain error values to the modal parameters, the robustness and residual vibration suppression performance of the six dual-modal input shapers were analyzed and it was found that the established ZVD-EI dual-modal input shaper could guarantee both a short adjustment time and good vibration suppression performance.

2. Model of the Delta Robot

The 3D model of the 4-DOF Delta robot is shown in Figure 1. This robot is mainly composed of a fixed platform, a moving platform, an intermediate shaft, three driving arms, and three driven arms. Each driven arm contains a set of parallelogram branch chains. Driven by three driving arms, the moving platform can realize high-speed movement of the X–Y–Z axis. Driven by the motor connected to the intermediate shaft, the end effector under the moving platform can realize Z-axis rotation. The specific structure of each component is shown in Figure 2.

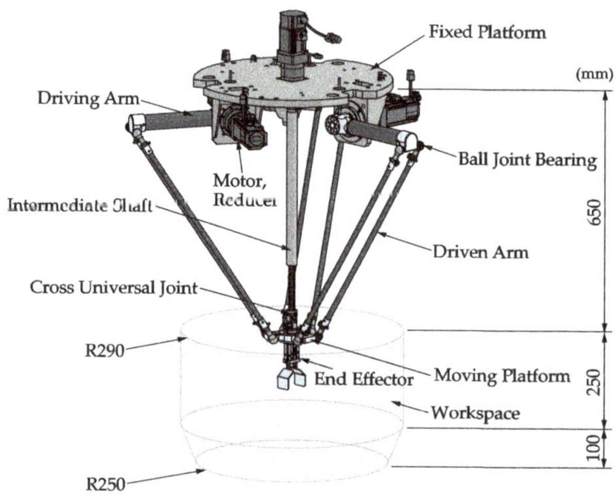

Figure 1. Delta robot 3D model.

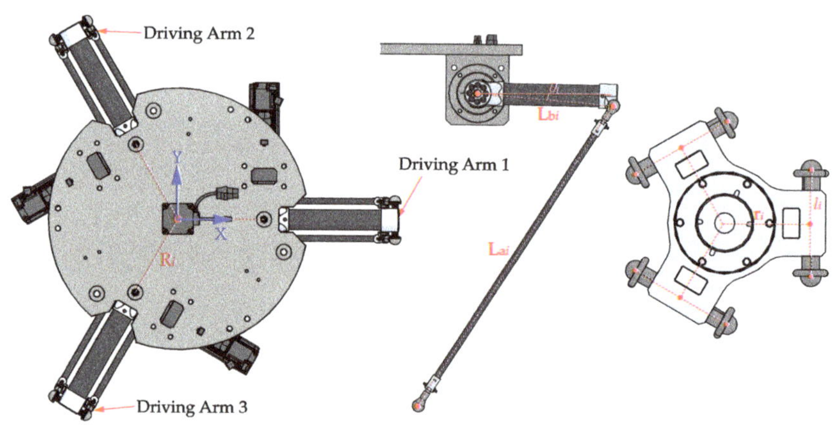

Figure 2. Notation used in the model of the Delta robot.

3. Elastic Dynamics Modeling and Modal Analysis

The Delta robot is a time-varying system with multiple modals, and its system modal parameters change according to the position of the robot. In order to design a suitable input shaper, it is first necessary to establish an elastodynamic model that accurately estimates the inherent characteristics of the Delta robot and then determine the modal that plays a dominant role in the residual vibration of the end effector [21].

3.1. *Elastic Dynamic Modeling*

Considering that the structure of the Delta robot is complex and there are many joints and hinges, to accurately estimate the modal parameters of the Delta robot, a 3D model of the Delta robot is established in Solidworks, and the overall finite element model is established by using ANSYS software after proper simplification. The geometrical and dynamic parameters of the Delta robot used in this paper are reported in Tables 1 and 2. Known system parameters: both the main body of the driving arm and the driven arm are made of carbon fiber ($\rho = 1800$ kg/m^{-3}), and the connection part is made of aluminum alloy ($\rho = 2770$ kg/m^{-3}). The diameter of the driving arm is 40 mm, the wall thickness is 5 mm, the diameter of the driven arm is 12 mm, and the wall thickness is 3 mm. The moving platform and the intermediate shaft are made of aluminum alloy.

Table 1. Geometrical parameters of the Delta robot.

Description	Notation	Value
Radius of the fixed platform	R_i	0.17 m
Radius of the moving platform	r_i	0.06 m
Length of the *i*-th driven arm	L_{ai}	0.68 m
Length of the *i*-th driving arm	L_{bi}	0.26 m
Width of the branch chain of the *i*-th driven arm	l_i	0.074 m
Angle of the *i*-th active joint	θ_i	$-35°$~$75°$

Table 2. Dynamic parameters of the Delta robot.

Description	Value
Shear modulus of elasticity of the driving arm and driven arm	9.0×10^9 N/m^2
Young's modulus of the driving arm and driven arm	2.3×10^{10} N/m^2
Poisson coefficient	0.2
Friction coefficient	0.11
Mass of the *i*-th driving arm	0.693 kg
Mass of the *i*-th driven arm	0.175 kg
Mass of moving platform	0.52 kg
Load mass	1 kg

Due to the different positions and attitudes of the Delta robot when it is in function, its mass matrix and stiffness matrix are also different. To facilitate the modal analysis, a specific trajectory is selected to analyze the modal changes of the robot. Given the following trajectory of the end effector, the position points P_0, P_1, P_2, P_3, P_4, and P_5 are selected as shown in Figure 3. The position coordinates of its position point in the global coordinate system are given in Table 3. The origin o of the global coordinate system is the center of the plane formed by the output axes of the three sets of reducers of the robot, where P_1P_2 and P_4P_5 are circular arc segments, and the rest of the trajectories are straight-line segments. The motion trajectory in Figure 3 is calculated by the quintic polynomial trajectory planning method, the total length of the motion trajectory is 0.6048 m, and the motion time is 0.25 s. The position, velocity and acceleration curves of Delta robot in joint space and Cartesian space can be obtained, as shown in Figure 4.

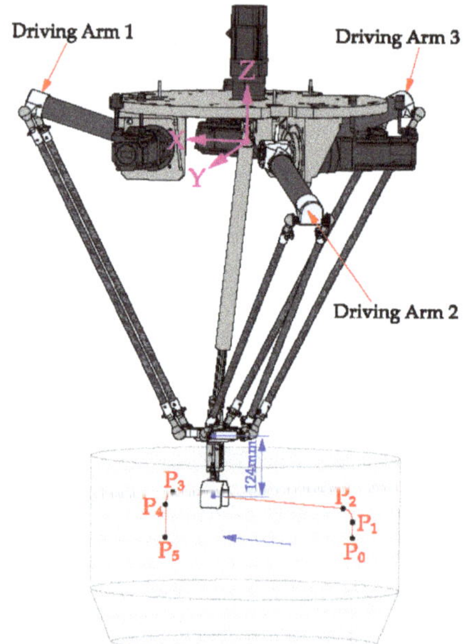

Figure 3. Trajectory of the Delta robot.

Table 3. Position coordinates.

Position Point	X (mm)	Y (mm)	Z (mm)
P_0	−200	0	−740
P_1	−200	0	−710
P_2	−182	−9	−670
P_3	182	−191	−670
P_4	200	−200	−710
P_5	200	−200	−780

The simulation steps are as follows:

1. Establish a parametric geometric model of the Delta robot according to actual needs, as shown in Figure 1.
2. Take the Delta robot model shown in Figure 1 as the verification object, simplify its geometric model, and import it into ANSYS workbench in the x_t standard format.
3. Import the component features (material and unit type) and connection features (joint contact type/fixed support) into the finite element software, perform mesh division, and establish the finite element model of the robot.

3.2. Delta Robot Modal Analysis

When analyzing the elastic dynamics of the Delta robot system, the fixed platform and the kinematic pair are regarded as rigid elements, and the deformation during the movement is not considered; the driving arm, driven arm, and moving platform are regarded as flexible elements. Therefore, the Delta robot is meshed in the ANSYS workbench, and the first six order modals of the center position coordinates of the Delta robot end effector at the above six position points are calculated. The natural frequencies of each order of the system are shown in Figure 5. It can be seen that the natural frequency of the robot gradually increases with the increase in the modal order in any position. Table 4 describes the modal shape of the robot at each position point. In the first two order modals, the modal shape of the robot is the torsion of the moving platform along the XY plane. Obviously, the torsion along the XY plane has the greatest impact on the robot and is most likely to cause the end effector vibration. In the design of the input shaper below, the influence of the first two order modals on the residual vibration of the Delta robot should be considered at the same time. Therefore, the first two order modes are taken as the dominant modal of the residual vibration of the Delta robot under this specific trajectory. Figures 6 and 7 show the first two order modals' shape contours of each location point.

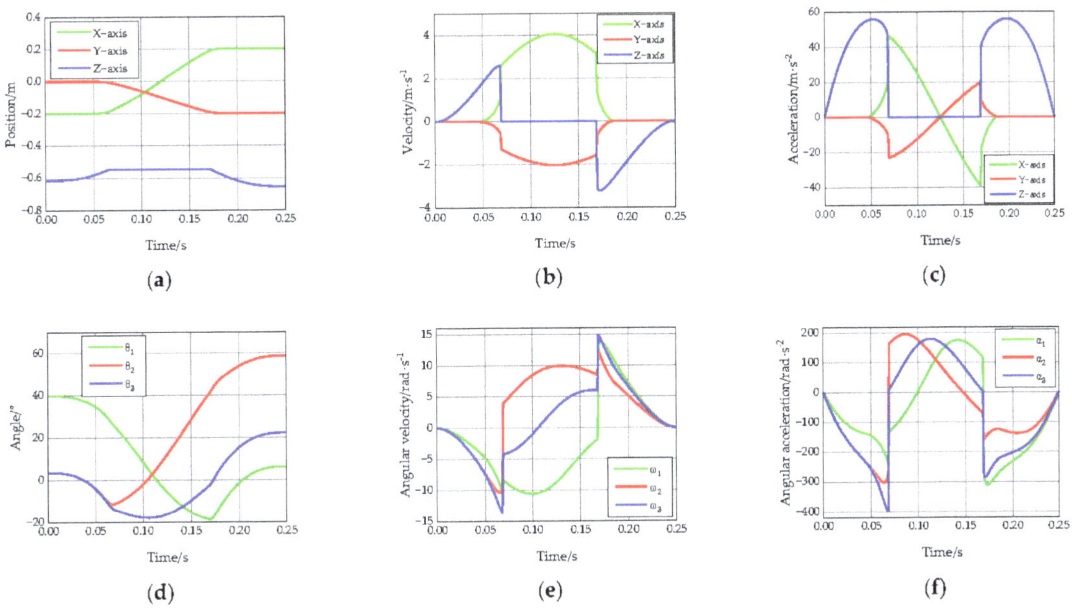

Figure 4. Physical quantity of the Delta robot under the motion trajectory: (**a**) position curve of the robot in the Cartesian space; (**b**) velocity curve of the robot in the Cartesian space; (**c**) acceleration curve of the robot in the Cartesian space; (**d**) angle curve of the robot in the joint space; (**e**) angular velocity curve of the robot in the joint space; (**f**) angular acceleration curve of the robot in the joint space.

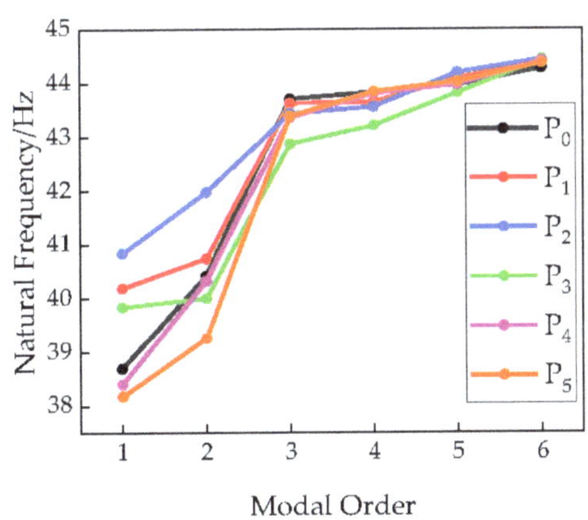

Figure 5. First six modals of each position point.

Table 4. Description of point vibration shapes of the robot at each position.

Order	Maximum Frequency/Hz	Minimum Frequency/Hz	Modal Shape Description
1	40.832	38.177	The moving platform is twisted along the XY plane
2	41.963	39.253	The moving platform is twisted along the XY plane
3	43.706	42.871	Deflection of a driven arm chain along its parallelogram plane
4	43.836	43.208	The deflection deformation of two driven arm chains along their parallelogram planes
5	44.19	43.813	The deflection deformation of two driven arm chains perpendicular to their parallelogram plane
6	44.437	44.268	Two driven arm chains deflect along its parallelogram plane, and one driven arm chain deflects perpendicularly to its parallelogram plane

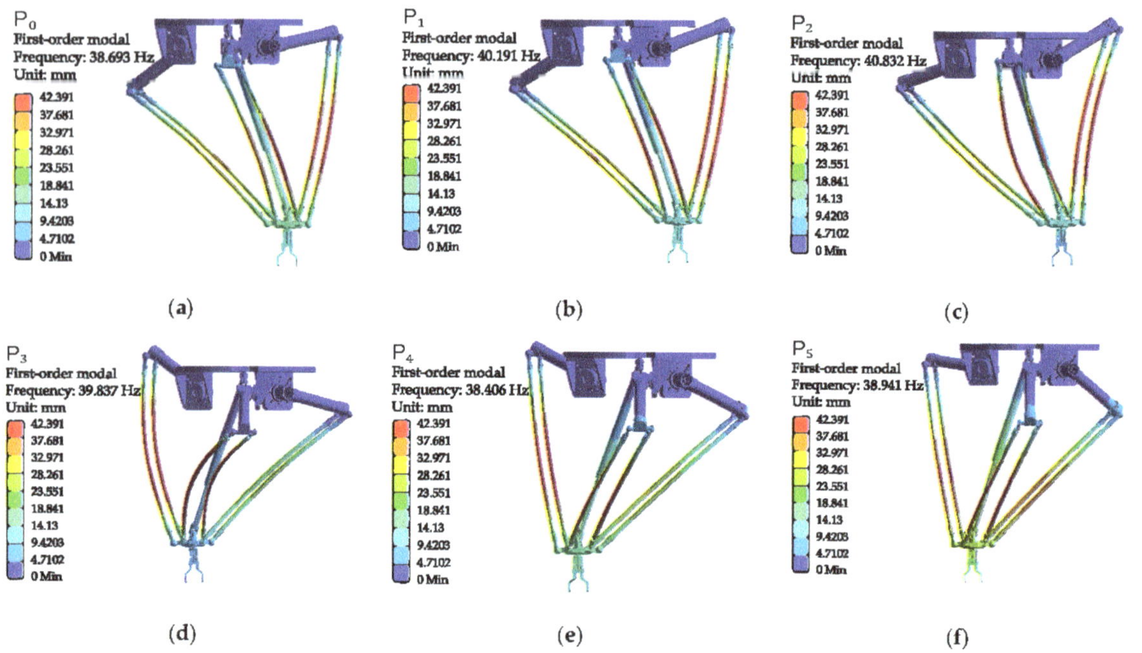

Figure 6. First-order modal of each position point: (**a**) mode shape contour at point P_0; (**b**) mode shape contour at point P_1; (**c**) mode shape contour at point P_2; (**d**) mode shape contour at point P_3; (**e**) mode shape contour at point P_4; (**f**) mode shape contour at point P_5.

According to the cloud diagram analysis of the first six order modals at each point of the Delta robot, the deformation is mainly located on the driven arm. This is because the driven arm is mainly made of carbon fiber material. The driven arm is a slender rod, and its aspect ratio is the same as that of the driving arm; it is more flexible and is more likely to cause vibration deformation. Therefore, when optimizing the vibration reduction in the Delta robot, we should focus on the dynamic characteristics of the driven arm.

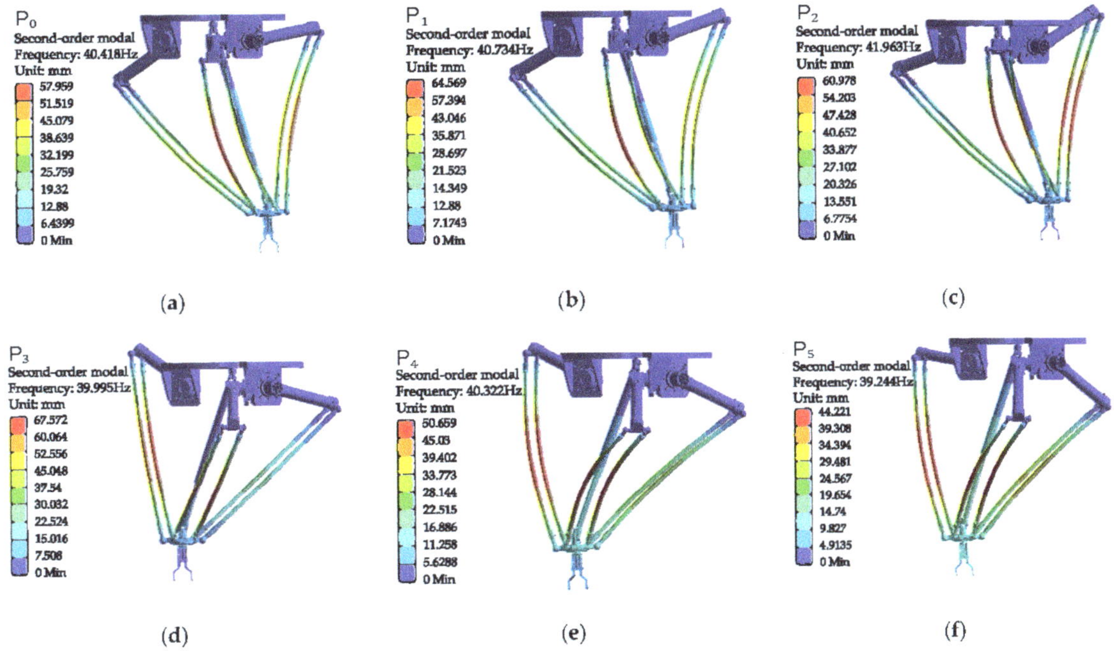

Figure 7. The second-order modal of each position point: (**a**) mode shape contour at point P_0; (**b**) mode shape contour at point P_1; (**c**) mode shape contour at point P_2; (**d**) mode shape contour at point P_3; (**e**) mode shape contour at point P_4; (**f**) mode shape contour at point P_5.

4. Design of the Input Shaping Controller

4.1. Principle of Input Shaping Technology

Input shaping is a vibration suppression algorithm. It convolves a group of pulse signals with the system input signal to obtain the actual input signal of the control system to avoid unnecessary vibration in the system. The input shaping method is based on the posicast principle [22,23], that is, sending out the first pulse produces a dynamic response in the system. After a certain period of time, a second pulse is introduced. If the second pulse has the correct time and amplitude, it can cancel the response produced by the first pulse. The principle is shown in Figures 8 and 9. Input shaping is a typical feed-forward control method, which can be effectively applied to the oscillation suppression of the system. The frame diagram of the input shaper system is shown in Figure 10. Compared with the closed-loop vibration suppression method, input shaping does not require real-time measurement of the vibration deformation of the system [24]. The input shaper only needs to be designed to identify the natural frequency and damping ratio of the system and does not require accurate modeling of the system; it is also suitable for complex and difficult-to-model structures.

However, reference signals for robots typically consist of a reference position, possibly in conjunction with other reference signals such as the reference velocity, reference torque, etc. Thus, it is necessary to modify these reference signals such that they will not introduce vibrations in the system [17].

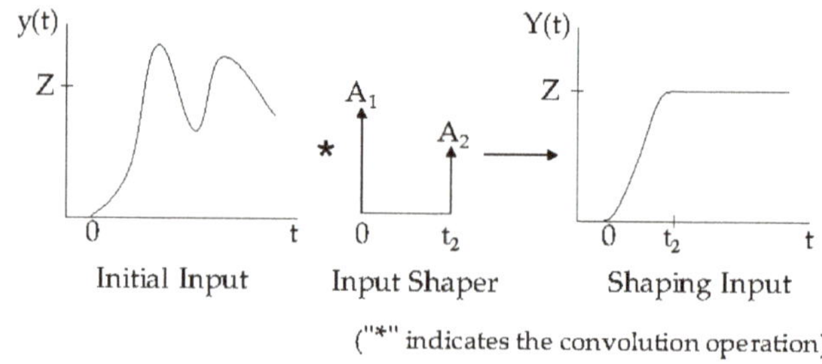

Figure 8. Single-modal input shaping control principle.

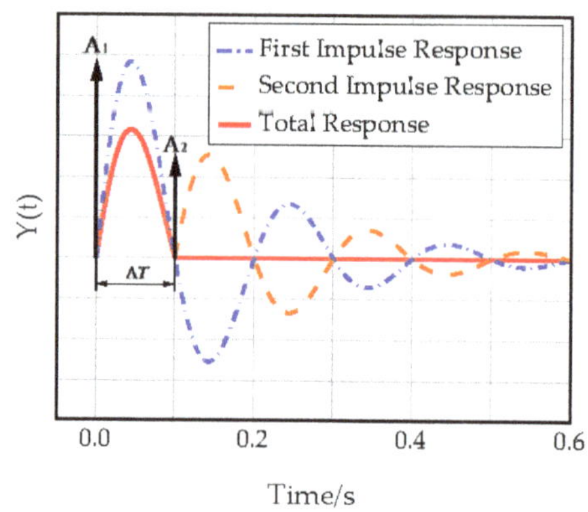

Figure 9. ZV input shaper impulse response.

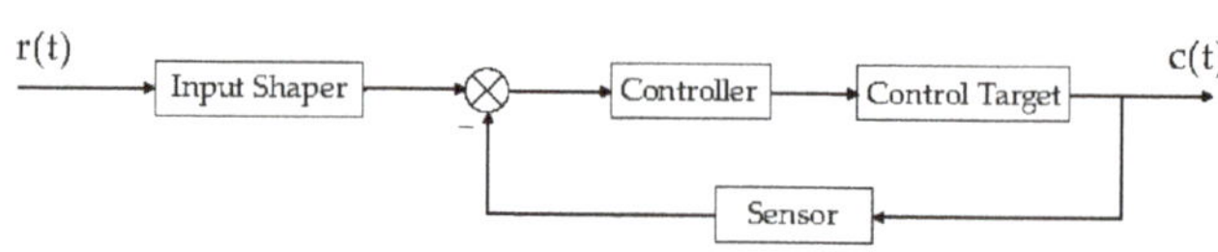

Figure 10. Frame diagram of the input shaping system.

The Delta robot can be modeled as an underdamped second-order system with a transfer function of the load system as

$$G(S) = \frac{{\omega_n}^2}{S^2 + 2\omega_n \zeta S + {\omega_n}^2} \tag{1}$$

where ω_n is the undamped natural frequency, ζ is the damping ratio, and S is the Laplace operator.

The unit impulse response $y(t)$ of an underdamped second-order system is

$$y(t) = \frac{\omega_n}{\sqrt{1-\zeta^2}} e^{-\zeta \omega_n t} \sin(\omega_d t)\ (t \geq 0) \tag{2}$$

where $\omega_d = \omega_n \sqrt{1-\zeta^2}$, ω_d is the damping natural frequency.

The input shaper is a set of pulse trains. For a Delta robot, if there is an input shaper with n pulses, it can be calculated by convolution, which can be expressed in the time domain as

$$f(\mathrm{t}) = \sum_{i=1}^{n} A_i \delta(t - t_i)\ ,\ 0 \leq t_i < t_{i+1},\ A_i > 0 \tag{3}$$

where $\delta(t)$ is the Dirac function (unit impulse function), A_i is the amplitude of the i-th pulse, and t_i is the time of the i-th pulse. After Laplace transformation, the above formula can be expressed in the frequency domain as

$$F(S) = \sum_{i=1}^{n} e^{-t_i s} \tag{4}$$

To ensure that the amplitude of the gain before and after shaping is the same, determine the amplitude as

$$\sum_{i=1}^{n} A_i = 1 \tag{5}$$

When the amplitudes are all positive, that is, $A_i > 0$, it is said that there is no overshoot at this time. Shaping controllers are designed to implant time delays into the controlled system. To improve the response speed, it is necessary to shorten the response time as much as possible so that the first pulse is realized at zero time; at this time, $t_1 = 0$. If input shaping control is to be performed, $t_i > 0$ should be set [19].

The input pulse sequence $f(t)$ shaped in the time domain is convolved with the original unit impulse response $y(t)$, and the model of the obtained response $Y(t)$ after time $t\ (t > t_n)$ is

$$\begin{aligned} Y(t) &= \sum_{i=1}^{n} A_i y(t - t_i) \\ &= \sum_{i=1}^{n} A_i \left[\frac{\omega_n}{\sqrt{1-\zeta^2}} e^{-\zeta\omega_n (t-t_i)} \sin(\omega_d t - \omega_d t_i) \right] \\ &= \frac{\omega_n}{\sqrt{1-\zeta^2}} e^{-\zeta\omega_n t} [A(\omega_n, \zeta) \sin(\omega_d t) - B(\omega_n, \zeta) \cos(\omega_d t)] \\ &= \frac{\omega_n}{\sqrt{1-\zeta^2}} e^{-\zeta\omega_n t} C(\omega_n, \zeta) \sin(\omega_d t + \gamma) \end{aligned} \tag{6}$$

where

$$\begin{aligned} A(\omega_n, \zeta) &= \sum_{i=1}^{n} A_i e^{\zeta\omega_n t_i} \cos(\omega_d t_i) \\ B(\omega_n, \zeta) &= \sum_{i=1}^{n} A_i e^{\zeta\omega_n t_i} \sin(\omega_d t_i) \\ C(\omega_n, \zeta) &= \sqrt{A(\omega_n, \zeta)^2 + B(\omega_n, \zeta)^2} \\ \tan\gamma &= -\frac{B(\omega_n, \zeta)}{A(\omega_n, \zeta)} \end{aligned} \tag{7}$$

The amplitude ratio of the shaped impulse response $Y(t)$ and the original impulse response $y(t)$ after t_n time is used as the performance index of the input shaping, and the ratio is called the residual vibration percentage, which is defined as

$$V(\omega_n, \zeta) = e^{-\zeta\omega_n t_n} C(\omega_n, \zeta) \tag{8}$$

where $e^{-\zeta\omega_n t_n}$ represents the introduction of a time delay t_n in the shaped response. This ratio reflects the suppression effect of the residual vibration, and the design goal of the input shaper is to make $V \approx 0$ [13].

4.2. Construction of a Single-Modal Input Shaper

The types of single-modal input shapers mainly include the ZV (Zero-Vibration) input shaper, ZVD (Zero-Vibration and Derivative) input shaper, and EI (Extra-Insensitive) input shaper.

4.2.1. ZV Input Shaper

The ZV input shaper is also known as a dual-pulse input shaper because it becomes zero residual system vibration when the system's natural frequency and damping ratio are accurately calculated. Take $n = 2$ as the double pulse (ZV) input shaper. The ZV input shaper shapes the pulse signal into two pulse signals with amplitudes A_1 and A_2. After the first pulse signal A_1 generates a dynamic response in the system, the second pulse signal A_2 is sent out at interval ΔT. If the two pulse signals are equal in magnitude and opposite in direction, the signals can be canceled out to achieve vibration suppression. The shaping process of the ZV shaper under the impulse response is shown in Figure 9.

The ZV input shaper can be described in the time domain as

$$F(t) = A_1\delta(t) + A_2\delta(t - t_2) \tag{9}$$

It can be described in the frequency domain as

$$F(S) = \sum_{i=1}^{2} A_i e^{-t_i s} \tag{10}$$

Let Formula (8) be equal to 0, and set the equation system as

$$\begin{cases} A(\omega_n, \zeta) = 0 \\ B(\omega_n, \zeta) = 0 \\ \sum\limits_{i=1}^{2} A_i = 1 \end{cases} \tag{11}$$

The ZV input shaper can be obtained:

$$\mathrm{ZV} = \begin{bmatrix} A_i \\ t_i \end{bmatrix} = \begin{bmatrix} \frac{1}{1+K} & \frac{K}{1+K} \\ 0 & \frac{T}{2} \end{bmatrix}, \tag{12}$$

When T is the vibration period, $T = 2\pi/\omega_d$, $K = e^{-\pi\zeta\omega_n/\omega_d}$.

4.2.2. ZVD Input Shaper

In practice, the ZV input shaper has high requirements for the accuracy of mathematical modeling. In theory, when accurate ω_n and ζ are obtained, the ZV input shaper can completely eliminate the residual vibration. If there are certain uncertain factors in the working process of the robot that lead to changes in system parameters, the pulse input to the shaper cannot be completely offset from the pulse of the original signal, resulting in a poor vibration suppression effect. Then, high-order input shapers can be used, which are robust to modeling errors [25].

ZVD is an input shaper with three pulse signals, which can be described in the frequency domain as

$$F(S) = \sum_{i=1}^{3} A_i e^{-t_i s} \tag{13}$$

Calculate the partial derivative of $A(\omega_n, \zeta)$ and $B(\omega_n, \zeta)$ at $\omega = \omega_n$ and make it equal to 0, and set the equation system as

$$\begin{cases} \frac{\partial A(\omega,\zeta)}{\partial \omega}|\omega_n = 0 \\ \frac{\partial B(\omega,\zeta)}{\partial \omega}|\omega_n = 0 \\ A(\omega_n, \zeta) = 0 \\ B(\omega_n, \zeta) = 0 \\ \sum\limits_{i=1}^{3} A_i = 1 \end{cases} \tag{14}$$

The ZVD shaper can be obtained:

$$\text{ZVD} = \begin{bmatrix} A_i \\ t_i \end{bmatrix} = \begin{bmatrix} \frac{1}{D_1} & \frac{2K}{D_1} & \frac{K^2}{D_1} \\ 0 & \frac{T}{2} & T \end{bmatrix} \tag{15}$$

$$D_1 = 1 + 2K + K^2 \tag{16}$$

ZVD is more robust than the ZV input shaper, but the time delay is half a cycle longer than the ZV input shaper. As with all input shapers, a trade-off must be made between the increased robustness and the signal time lag caused by the increased shaper duration.

4.2.3. ZVDD Input Shaper

To further increase the robustness, this process can be repeated with higher-order derivatives with respect to the frequency, whose second derivative (ZVDD shaper) can be expressed as

$$\text{ZVDD} = \begin{bmatrix} A_i \\ t_i \end{bmatrix} = \begin{bmatrix} \frac{1}{D_2} & \frac{3K}{D_2} & \frac{3K^2}{D_2} & \frac{K^3}{D_2} \\ 0 & \frac{T}{2} & T & \frac{T}{3} \end{bmatrix} \tag{17}$$

$$D_2 = 1 + 3K + 3K^2 + K^3 \tag{18}$$

4.2.4. EI Input Shaper

The input shapers discussed above are all designed under the constraints of the residual vibration percentage $V = 0$. However, in actual situations, the natural frequency and damping ratio of the system cannot be accurately estimated, and it is difficult to obtain an accurate model, which will lead to poor vibration suppression effects, and the use of high-order input shapers will double the signal time delay.

If the residual vibration percentage $V = 0$ is not required but at the system frequency $\omega = \omega_n$, $V = V_{\exp}(V_{\exp} > 0)$, $\frac{\partial V(\omega,\zeta)}{\partial \omega}|\omega_n = 0$. When the system frequency is $\omega \neq \omega_n$, $V = 0$. Therefore, the input shaper can guarantee the residual vibration percentage $V \leq V_{\exp}$ of the system. Such an input shaper that allows the percentage of residual vibration to remain below a certain value is called an EI shaper [26]. The EI shaper has three pulse signals, which can be described in the frequency domain as

$$F(S) = \sum_{i=1}^{3} A_i e^{-t_i s} \tag{19}$$

The EI input shaper can be expressed as

$$\text{EI} = \begin{bmatrix} A_i \\ t_i \end{bmatrix} = \begin{bmatrix} A_1 & A_2 & A_3 \\ 0 & t_2 & T \end{bmatrix} \tag{20}$$

When the damping ratio is $\zeta \neq 0$, the pulse signal A_i and pulse time t_i of the EI input shaper can be expressed as [19]

$$\begin{aligned} A_1 &= 0.2479 + 0.2496V_{\exp} + 0.8001\zeta + 1.233V_{\exp}\zeta + 0.496\zeta^2 + 3.173V_{\exp}\zeta^2 \\ A_3 &= 0.2515 + 0.2147V_{\exp} - 0.8325\zeta + 1.415V_{\exp}\zeta + 0.8518\zeta^2 + 4.9V_{\exp}\zeta^2 \\ A_2 &= 1 - (A_1 + A_3) \end{aligned} \tag{21}$$

$$t_2 = T\left(0.4999 + 0.46159V_{\exp}\zeta + 4.26169V_{\exp}\zeta^2 + 1.75601V_{\exp}\zeta^3 + 8.57843{V_{\exp}}^2\zeta - 108.644{V_{\exp}}^2\zeta^2 + 336.898{V_{\exp}}^2\zeta^3\right) \tag{22}$$

The EI input shaper has the same signal time delay as the ZVD input shaper, but it allows a certain percentage of residual vibration, so it is more robust.

4.3. Simulink Implementation of a Single-Modal Input Shaper

According to the parameters of the required pulse amplitude A_i and pulse time t_i calculated above, the input shapers of ZV, ZVD, ZVDD, and EI are constructed in Simulink in Figure 11.

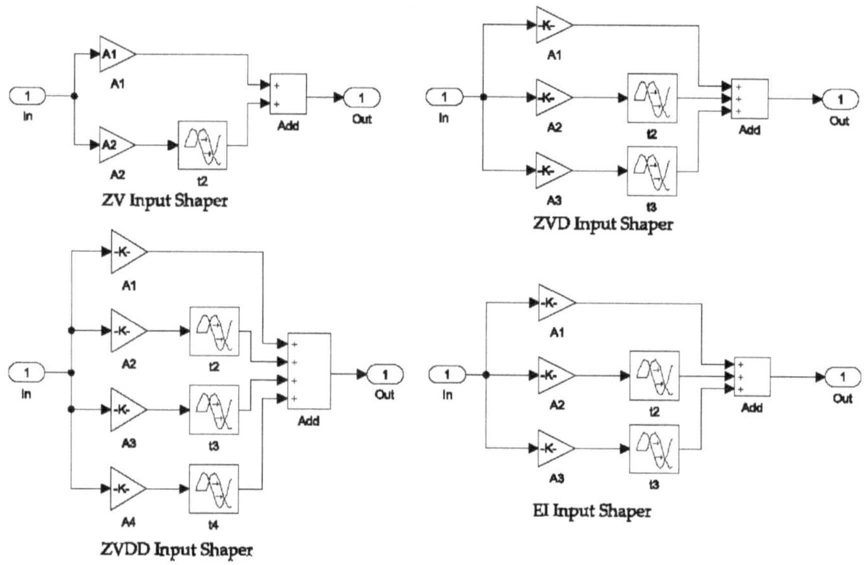

Figure 11. Input shaper module.

4.4. Multimodal Input Shaper

When the high-order modal has a great influence on the residual vibration of the system, the single-modal input shaper cannot effectively suppress the residual vibration of the multi-modal system, and a multi-modal input shaper is required [27]. Multiple single-modal input shapers are convolved to form a multi-modal input shaper, which has the advantage of suppressing the residual vibration generated by multiple modals of the system at the same time and has good vibration suppression for systems with multiple modal effects and robustness. The principle of multimodal input shaping is shown in Figure 12.

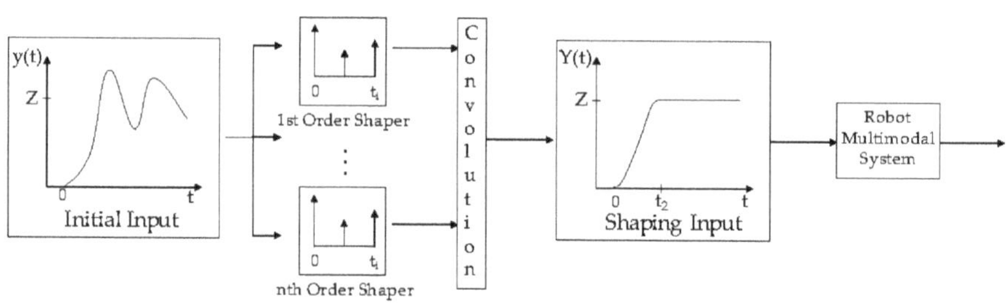

Figure 12. Multimodal input shaping control principle.

4.5. Design of a Dual-Modal Input Shaper

According to the modal analysis in the previous section, the first two order modal vibration shapes of each position point of the Delta robot have a great influence on the residual vibration of its end effector, and the frequency bandwidth and amplitude of the two modes are quite different, so the vibration suppression ability of the single-modal input shaper to the robot is reduced. Therefore, a dual-modal input shaper is designed to overcome these drawbacks. The dual-modal input shaper is built by convolving two single-modal input shapers. For example, the first-order modal of the system is 10 Hz, and the second-order modal is 15 Hz; the input shaper of each order modal can be described as

$$\mathrm{ZV} = \begin{bmatrix} A_i \\ t_i \end{bmatrix} = \begin{bmatrix} A_1 & A_2 \\ 0 & t_2 \end{bmatrix} (10\ \mathrm{Hz}\ \mathit{shaper}) \tag{23}$$

$$\mathrm{ZVD} = \begin{bmatrix} A_i \\ t_i \end{bmatrix} = \begin{bmatrix} A_1 & A_2 & A_3 \\ 0 & t_2 & t_3 \end{bmatrix} (15\ \mathrm{Hz}\ \mathit{shaper}) \tag{24}$$

where the ZV input shaper is used for the first-order modal shaping and the ZVD input shaper is used for the second-order modal shaping so that the ZV-ZVD dual-mode input shaper can be established.

Convolve the single-mode input shaper given by Formulas (23) and (24), which generate a ZV-ZVD shaper, as shown in Formula (25):

$$\mathrm{ZV} - \mathrm{ZVD} = \begin{bmatrix} A_i \\ t_i \end{bmatrix} = \begin{bmatrix} A_1 & A_2 & A_3 & A_4 \\ t_1 & t_2 & t_3 & t_4 \end{bmatrix} \tag{25}$$

According to the above, this paper establishes six dual-mode input shapers for the Delta robot system, namely ZV-ZV, ZVD-ZVD, EI-EI, ZV-ZVD, ZV-EI, and ZVD-EI. Since the ZVDD input shaper response time lag is too high, ZVDD is not used for the design of the dual-modal input shaper, and it is only used as a reference.

5. Simulation and Analysis

5.1. Design of a Simulink Block Diagram

According to the modal analysis in the third section, the average value of the natural frequency of the first two modes at each position point is selected to establish the input shaper; $\omega_{n1} = 39.36$ and $\omega_{n2} = 40.45$ can be obtained, and the damping ratio is $\zeta = 0.05$. Substitute the above parameters into the second-order transfer function of the system (1). Taking the unit step signal as the input signal of the system, the six dual-mode input shapers established in the previous section are simulated and analyzed, and the simulation time is 0.5 s. The block diagram of the dual-mode input shaper is shown in Figure 13. The simulation results are shown in Figure 14.

It can be seen from Figure 14 that the ZV-ZV dual-modal input shaper has the best vibration suppression performance, but the ZV-ZV dual-modal input shaper requires precise system modal parameters. However, the modal parameters of the Delta robot will change continuously during the movement in the workspace and there will be errors between the simulation and the actual system, so the robustness of the dual-modal input shaper is particularly important.

5.2. Robustness Analysis of the Dual-Modal Input Shaper

Robustness is a key index to evaluate the ability of the input shaper to suppress residual vibration, so the designed dual-modal input shaper can be used in practical engineering only if it has good robustness. In order to analyze the robustness of each input shaper, a certain error value is given to the modal parameters in this paper, and then the vibration suppression performance of each input shaper in the actual work of the Delta robot is simulated.

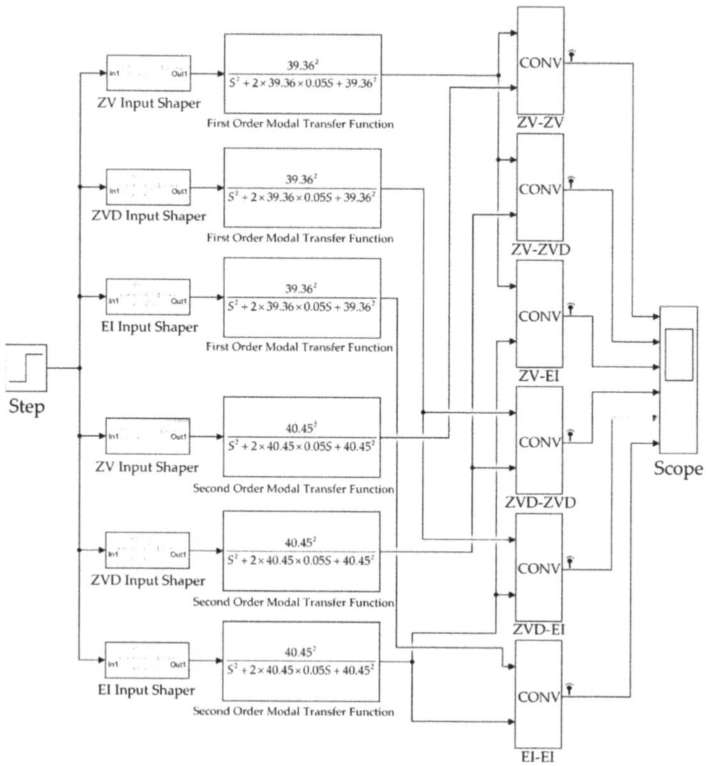

Figure 13. Block diagram construction of dual-modal input shapers.

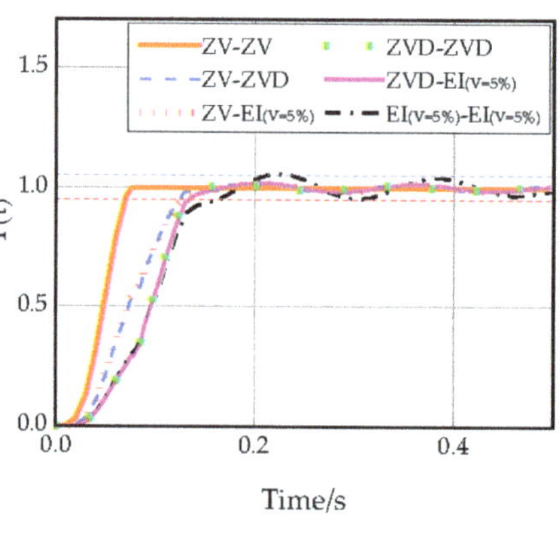

Figure 14. Unit step response of each dual-mode input shaper.

It can be seen from Figure 15 that all types of dual-modal input shapers can effectively suppress the residual vibration of the first two order modals. Among them, the ZV-ZV input shaper has the fastest response speed and no overshoot, but when the system modal parameter error increases, its residual-vibration-suppression effect gradually becomes worse, and its robustness is poor, as shown in Figure 15a–d. The EI-EI input shaper can also have good robustness when the system modal parameter error increases, but at the cost of increasing the adjustment time. Taken together, the ZVD-EI input shaper can significantly reduce the residual vibration of the first two order modals of the Delta robot and it has good robustness and a short adjustment time.

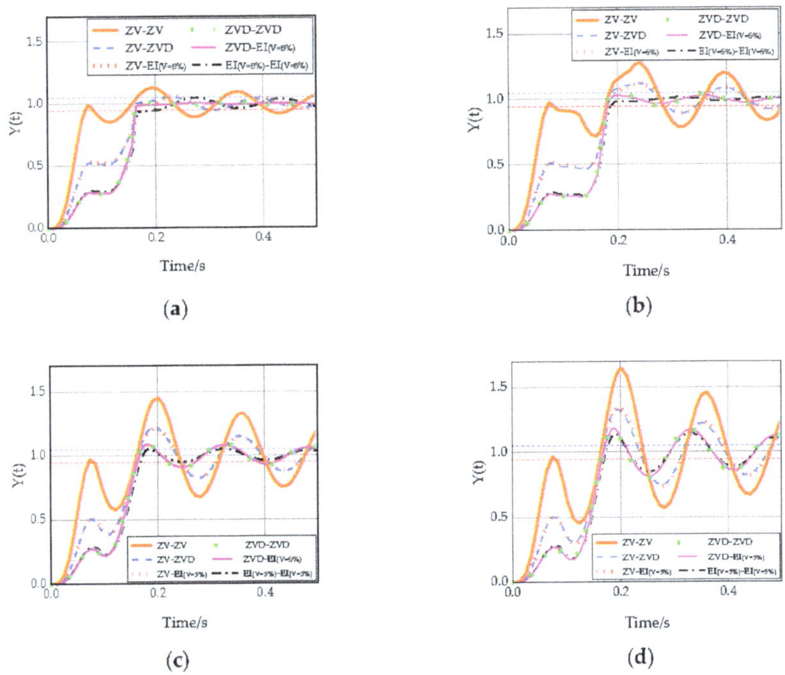

Figure 15. Unit step response of each dual-modal input shaper when the system modal parameters have error values: (**a**) response when the system modal parameter error value is 5%; (**b**) response when the system modal parameter error value is 10%; (**c**) response when the system modal parameter error value is 15%; (**d**) response when the system modal parameter error value is 20%.

5.3. Verification of the ZVD-EI Dual-Modal Input Shaper

The natural frequency parameter of the single-modal input shaper in this section is the average natural frequency of the first-order modal as ω_{n1}= 39.36. According to Figure 5, it can be seen that the natural frequency difference of the first two order modals of the Delta robot at each position point does not exceed 10% of the average value. Therefore, the modal parameter error value is set to 10% for simulation analysis, that is, ω_{n1}= 35.424, ω_{n2}= 36.405. Taking the unit step signal as the input signal of the system and the simulation time is 0.5 s, the simulation results are shown in Table 5. It can be seen from Table 5 that among single-modal input shapers, the ZV input shaper has the largest overshoot, the longest adjustment time, and the worst robustness; the ZVDD input shaper has the smallest overshoot but the longest adjustment time. Among the dual-modal input shapers, the ZVD-ZVD input shaper has the shortest adjustment time but slightly higher overshoot; the ZVD-EI input shaper has the smallest overshoot, and the adjustment time is similar to the ZVD-ZVD input shaper. On the whole, the ZVD-EI input shaper has the advantages of small overshoot and a short adjustment time, and its robustness and vibration suppression ability are the best.

Table 5. Performance of each input shaper at 10% error value.

Input Shaper	Overshoot/%	Adjustment Time/s
ZV	12.66	0.606
ZVD	2.28	0.156
ZVDD	0.04	0.244
$EI_{(V=5\%)}$	2.05	0.166
ZV-ZV	11.33	0.606
ZV-ZVD	11.53	0.682
ZV-$EI_{(V=5\%)}$	12.22	0.687
ZVD-ZVD	1.35	0.156
ZVD-$EI_{(V=5\%)}$	0.16	0.158
$EI_{(V=5\%)}$-$EI_{(V=5\%)}$	3.92	0.170

In order to further verify the superiority of the ZVD-EI dual-modal input shaper, it is compared with the ZVD single-modal input shaper commonly used in industrial robot vibration suppression [19–21], and the robustness and vibration suppression effect of the traditional ZVD single-modal input shaper and the ZVD-EI dual-modal input shaper proposed in this paper are mainly compared. As shown in Figure 16a, when there is no modal parameter error value in the system, compared with the traditional ZVD single-modal input shaper, the adjustment time of the ZVD-EI dual-modal input shaper is reduced by 24.14% and the amplitude is smaller, which has better vibration suppression performance. As shown in Figure 16c, when the modal parameter error value is 10%, the adjustment time of ZVD-EI dual-modal input shaper is 17.2% shorter than that of ZVD input shaper. With a certain modal parameter error value, compared with the traditional ZVD input shaper, the ZVD-EI input shaper still has good robustness and residual vibration suppression performance, as shown in Figure 16b–e.

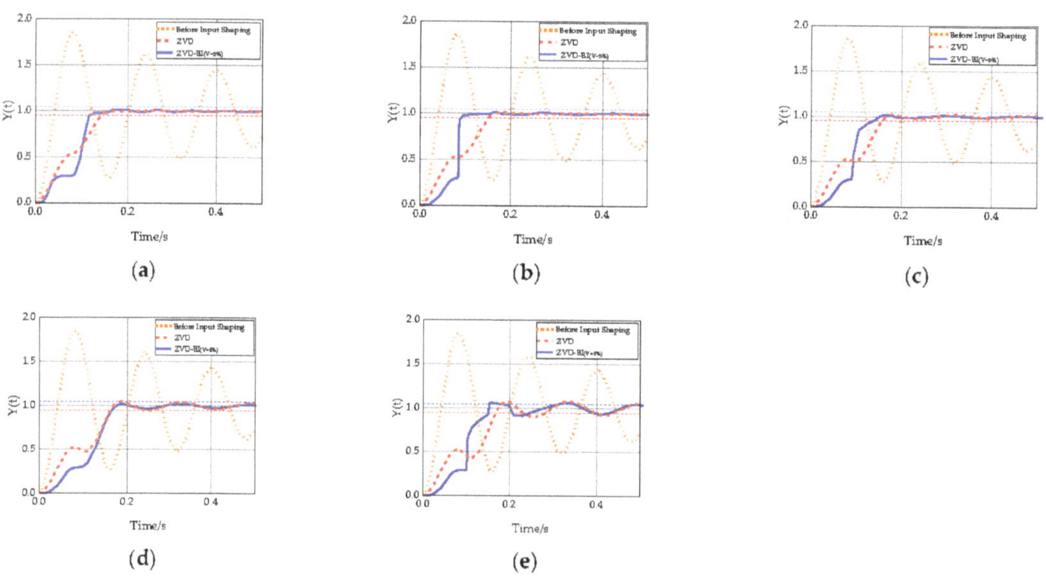

Figure 16. Unit step response of the ZVD-EI dual-modal input shaper and the ZVD single-modal input shaper: (**a**) response when the system modal parameter error value is 0; (**b**) response when the system modal parameter error value is 5%; (**c**) response when the system modal parameter error value is 10%; (**d**) response when the system modal parameter error value is 15%; (**e**) response when the system modal parameter error value is 20%.

5.4. Selection Principles of Dual-Modal Input Shapers

For the six dual-modal input shapers established above, the ZVD-EI dual-modal input shaper is most suitable for application in the Delta robot system established in this paper and has good robustness and residual vibration suppression performance.

The dual-modal input shaper designed in this paper can also be applied to other robots with multiple dominant modal and industrial equipment controlled by controllers and motors. When the vibration frequency band of the first-order modal of the system is narrow and the vibration frequency band of the second-order modal is wide, the ZV-EI or ZV-ZVD dual-modal input shaper can be used for residual vibration suppression. When the vibration frequency band of the first-order modal of the system is wide and the vibration frequency band of the second-order modal is narrow, the EI-ZV or ZVD-ZV dual-modal input shaper can be used for residual vibration suppression. Among the above six input shapers, the ZVD-EI dual-modal input shaper has the best robustness and comprehensive performance. The ZV-ZV dual-modal input shaper has the shortest adjustment time but the worst robustness and is suitable for systems where the vibration frequency bands of each modal of the system are narrow. Therefore, the most suitable dual-modal input shaper can be selected for residual vibration suppression according to the vibration frequency bandwidth of each order modal of the system, the expected vibration amplitude, and the adjustment time.

6. Conclusions

(1) A design method of a dual-modal input shaper was presented to solve the poor effect of the traditional input shaper on residual vibration suppression caused by the change of the dynamic characteristics of the Delta robot in its workspace. Through simulation analysis, it is found that among the six dual-modal input shapers established in this paper, the ZVD-EI dual-modal input shaper has the best robustness and residual vibration suppression performance.

(2) The simulation verification in Simulink shows that compared with the traditional ZVD single-modal input shaper, when the system modal parameters are accurate, the adjustment time of the ZVD-EI dual-mode input shaper is reduced by about 24%; when the parameter error value is 10%, the adjustment time of the ZVD-EI dual-modal input shaper is reduced by about 17%. Compared with other input shapers, the ZVD-EI dual-modal input shaper has better robustness and comprehensive performance and can effectively reduce the residual vibration of the Delta robot.

(3) By analyzing the usage principles of the established dual-modal input shaper, different dual-modal input shapers in different robots and industrial equipment can be selected for residual vibration suppression according to their system characteristics. In the future, the ZVD-EI dual-modal input shaper algorithm will be applied to the Delta robot prototype to further improve its residual vibration suppression performance and make this method apply better to actual engineering.

Author Contributions: Z.G.: conception of the study, proposition of the theory and method, supervision; J.Z.: literature search, figures, data collection, manuscript preparation and writing; P.Z.: programming, testing of existing code components. All authors have read and agreed to the published version of the manuscript.

Funding: This research was funded by the Liaoning Provincial Education Department Project (Grant No. LJKZ0114).

Data Availability Statement: Not applicable.

Conflicts of Interest: The authors declare no conflict of interest.

References

1. Clavel, R. DELTA, a fast robot with parallel geometry. In Proceedings of the 18th International Symposium on Industrial Robots, Lausanne, Switzerland, 26–28 April 1988; pp. 91–100.
2. Carabin, G.; Scalera, L.; Wongratanaphisan, T.; Vidoni, R. An energy-efficient approach for 3D printing with a Linear Delta Robot equipped with optimal springs. *Robot. Comput. Integr. Manuf.* **2021**, *67*, 102045. [CrossRef]
3. Scalera, L.; Boscariol, P.; Carabin, G.; Vidoni, R.; Gasparetto, A. Enhancing energy efficiency of a 4-DOF parallel robot through task-related analysis. *Machines* **2020**, *8*, 10. [CrossRef]
4. Wu, M.; Mei, J.; Zhao, Y.; Niu, W. Vibration reduction of delta robot based on trajectory planning. *Mech. Mach. Theory* **2020**, *153*, 104004. [CrossRef]
5. Cheng, H.; Li, W. Reducing the Frame Vibration of Delta Robot in Pick and Place Application: An Acceleration Profile Optimization Approach. *Shock Vib.* **2018**, *2018*, 2945314. [CrossRef]
6. Dai, Z.; Sheng, X.; Hu, J.; Wang, H.; Zhang, D. Design and implementation of Bézier curve trajectory planning in DELTA parallel robots. In Proceedings of the International Conference on Intelligent Robotics and Applications, Portsmouth, UK, 24–27 August 2015; Springer International Publishing: Cham, Switzerland, 2015; pp. 420–430.
7. Zu, Q.; Liu, Q.; Wu, J. Dynamic Pick and Place Trajectory of Delta Parallel Manipulator. In Proceedings of the 6th International Conference on Human Centered Computing, Virtual Event, 14–15 December 2020; Springer International Publishing: Cham, Switzerland, 2021, pp. 1–11.
8. Ni, J.; Mei, J.; Ding, Y.; Yu, D.; Duan, Y.; Le, Y. A trajectory planning approach for Delta robots considering both motion smoothness and dynamic stress. *J. Mech. Robot.* **2022**, *15*, 041012. [CrossRef]
9. Zhao, R.; Wu, L.; Chen, Y.H. Robust Control for Nonlinear Delta Parallel Robot With Uncertainty: An Online Estimation Approach. *IEEE Access* **2020**, *8*, 97604–97617. [CrossRef]
10. Singer, N.C.; Seering, W.P. Preshaping Command Inputs to Reduce System Vibration. *J. Dyn. Syst. Meas. Control* **1990**, *112*, 76–82. [CrossRef]
11. Chatlatanagulchai, W.; Kijdech, D.; Benjalersyarnon, T.; Damyot, S. Quantitative Feedback Input Shaping for Flexible-Joint Robot Manipulator. *J. Dyn. Syst. Meas. Control* **2016**, *138*, 061006. [CrossRef]
12. Piedrafita, R.; Comín, D.; Beltrán, J.R. Simulink implementation and industrial test of Input Shaping techniques. *Control Eng. Pract.* **2018**, *79*, 1–21. [CrossRef]
13. Yu, Z.; Chen, W.; Te, T.; Tomizuka, M. Zero time delay input shaping for smooth settling of industrial robots. In Proceedings of the 2016 IEEE International Conference on Automation Science and Engineering (CASE), Fort Worth, TX, USA, 21–25 August 2016; pp. 620–625.
14. Li, Z.; Wang, H.; Zhao, H.; Ding, H. Force impact suppression of contact transition state in robot grinding and polishing of industrial blades. *Proc. Inst. Mech. Eng. Part C J. Mech. Eng. Sci.* **2022**, *236*, 7387–7397. [CrossRef]
15. Ha, M.T.; Kang, C.G. Experimental analysis of natural frequency error to residual vibration in ZV, ZVD, and ZVDD shapers. In Proceedings of the 2013 10th International Conference on Ubiquitous Robots and Ambient Intelligence (URAI), Jeju, Republic of Korea, 30 October–2 November 2013; pp. 195–199.
16. Masoud, Z.; Alhazza, K. Frequency-modulation input shaping for multimode systems. *J. Vib. Control* **2014**, *22*, 3439–3451. [CrossRef]
17. Thomsen, D.K.; Søe-Knudsen, R.; Balling, O.; Zhang, X. Vibration control of industrial robot arms by multi-mode time-varying input shaping. *Mech. Mach. Theory* **2021**, *155*, 104072. [CrossRef]
18. Singhose, W.; Crain, E.; Seering, W. Convolved and simultaneous two-mode input shapers. *IEEE Proc. Control Theory Appl.* **1997**, *144*, 515–520. [CrossRef]
19. Zheng, K. Research on intelligent vibration suppression control of high-speed lightweight Delta robot. *J. Vib. Control* **2022**, *28*, 3042–3057. [CrossRef]
20. Shan, X.; Li, Y.; Liu, H.; Huang, T. Residual Vibration Reduction of High-Speed Pick-and-Place Parallel Robot Using Input Shaping. *Chin. J. Mech. Eng.* **2022**, *35*, 16. [CrossRef]
21. Wang, H.; Liu, Q.; Ma, Y.; Liu, S. Residual Vibration Suppression of the Diamond Robot Based on the Input Shaping. *Mach. Des. Res.* **2022**, *38*, 75–80. (In Chinese) [CrossRef]
22. Smith, O.J.M. Posicast Control of Damped Oscillatory Systems. *Proc. IRE* **1957**, *45*, 1249–1255. [CrossRef]
23. Hung, J.Y. Feedback control with Posicast. *IEEE Trans. Ind. Electron.* **2003**, *50*, 94–99. [CrossRef]
24. Li, L. Vibration Suppression of Flexible Load System of Robot Based on Actor-Critic Algorithm. Master's Thesis, University of Chinese Academy of Sciences, Beijing, China, 2020. (In Chinese).
25. Singer, N.; Singhose, W.; Kriikku, E. An input shaping controller enabling cranes to move without sway. In Proceedings of the American Nuclear Society Topical Meeting on Robotics and Remote Systems, Augusta, GA, USA, 27 April–1 May 1997.

26. Singhose, W.E.; Seering, W.P.; Singer, N.C. Shaping inputs to reduce vibration: A vector diagram approach. In Proceedings of the IEEE International Conference on Robotics and Automation, Cincinnati, OH, USA, 13–18 May 1990; Volume 922, pp. 922–927.
27. Li, B.; Xie, L.; Wei, Y.; Zhao, J.; Wang, L. Residual Vibration Suppression of a 3-DOF Parallel Robot Using Multimodal Input Shaping. *Machinery* **2010**, *48*, 21–25. [CrossRef]

actuators

MDPI

Article

Numerical Study on Unbalance Response of Dual-Rotor System Based on Nonlinear Bearing Characteristics of Active Magnetic Bearings

Nianxian Wang [1,2,*], Mingzheng Liu [1,2], Junfu Yao [1,2], Pingping Ge [1,2] and Huachun Wu [3]

1 School of Machinery and Automation, Wuhan University of Science and Technology, Wuhan 430081, China
2 Hubei Key Laboratory of Mechanical Transmission and Manufacturing Engineering, Wuhan University of Science and Technology, No. 947, Heping Venue, Qingshan District, Wuhan 430081, China
3 School of Mechanical and Electronic Engineering, Wuhan University of Technology, Wuhan 430070, China
* Correspondence: wangnianxian@wust.edu.cn

Abstract: The magnetic suspended dual-rotor system (MSDS) has the advantage of a high power density. The system can be used in high-speed rotating machinery. The major purpose of this study is to predict the unbalance response of the MSDS considering the nonlinear bearing characteristics of active magnetic bearings (AMBs). Firstly, the nonlinear bearing model was established by a non-linear magnetic circuit method (NMCM). The model considers magnetic flux leakage, magnetic saturation, and working position flotation accurately. Then, the dynamic model of the system was established by using the finite element method and solved by the Newmark-β method. Finally, the effects of external load, rotational speeds, and control parameters were studied. Axial trajectory diagrams, stability zone diagrams, and waterfall diagrams were employed to analyze the dynamic behaviors of the MSDS. The results indicate that the external load, rotational speeds, and control parameters have a significant impact on the unbalance response of the system. Super harmonics of rotational frequencies and their combined frequencies may be excited by heavy load conditions. Appropriate control parameters can suppress the nonlinear phenomena. The obtained results of this research will contribute to the design and fault diagnosis of MSDSs.

Keywords: magnetic bearings; nonlinear bearing characteristic; unbalance response; dual-rotor system; finite element method

Citation: Wang, N.; Liu, M.; Yao, J.; Ge, P.; Wu, H. Numerical Study on Unbalance Response of Dual-Rotor System Based on Nonlinear Bearing Characteristics of Active Magnetic Bearings. *Actuators* **2023**, *12*, 86. https://doi.org/10.3390/act12020086

Academic Editor: Jun Zheng

Received: 14 December 2022
Revised: 27 January 2023
Accepted: 14 February 2023
Published: 16 February 2023

1. Introduction

Dual-rotor system is the core component of aeroengines and gas turbines. Active magnetic bearings (AMBs) have many advantages over conventional bearings, such as contactless operation, lubrication-free operation, and controllable dynamic bearing properties [1]. The substitution of AMB for mechanical bearings to support dual-rotor systems, namely, the magnetic suspended dual-rotor system (MSDS), can improve the power density of the system [2].

The traditional research on the dynamics of magnetic suspended bearing-rotor system were mostly based on the linearized AMBs bearing model, in which the nonlinear factors such as magnetic leakage and the saturation of AMBs were ignored, and the dynamics of system were studied based on the linearized bearing model near the equilibrium position [1]. However, in the dual rotor system, the load and vibration of the rotor is complex, the operation state of the rotor will no longer satisfy the assumption of ignoring those nonlinear factors, and the accurate response of the rotor system cannot be predicted by using the traditional, linearized AMBs bearing model. Therefore, the dynamics of the AMB rotor system with nonlinear bearing models of AMBs need to be studied.

Recently, the nonlinear bearing characteristics of AMBs have attracted the attention of researchers. C. Yu et al. and Cy A et al. considered the effects of magnetic leakage and magnetic saturation in hybrid magnetic bearings, respectively, using a dynamic magnetic circuit model [3,4]. A new mathematical modelling method of studying the suspension force was proposed by W. Zhang et al. The method is based on an exact segmentation of magnetic fields and enables the accurate calculation of edge flux and leakage coefficients [5]. Wang et al. proposed a non-linear magnetic circuit method (NMCM) to establish an analytical bearing capacity model for AMBs. The model considers the fringe flux, leakage flux, and the magnetic saturation. Experiments indicate that the accuracy of the model is high for different air gap lengths and eccentric distances [6]. Based on Wang et al., Wajnert et al. established an accurate bearing capacity model of hybrid magnetic bearings by using the NMCM and verified the efficiency of the model with experiments [7]. In general, due to magnetic leakage and magnetic saturation, the bearing characteristics of magnetic bearings exhibited significant nonlinearity. Accurate nonlinear bearing models of AMBs can be established by methods such as the NMCM.

Nonlinear factors can lead to more complex rotor dynamics, which is significantly different from the dynamic behaviors predicted using simple linearized models. Lu [8], Wang [9], Zhang [10,11], and Su et al. [12] studied the dynamic response of rotor system with nonlinear disturbances (e.g., shaft cracks, rotor rubbing, fluid exciting, and base shock). The results indicate that the system response is significant in the frequency domain, and many harmonic signals are excited in addition to the rotational frequency. Additionally, considering the nonlinearity of squeeze film damper and rolling bearing, the resonance characteristics of the system response become significantly different, and operating parameters such as rotational speed and load can affect the nonlinear behaviors of the system [13–15]. Compared to mechanical bearings, AMBs have different non-linearities, and the introduction of control laws can lead to richer dynamic behaviors. The nonlinear behaviors with large air gaps [16], time-varying stiffness [17–19], magnetic flux saturation and current saturation [20–22] are discussed, and it is shown that the control parameters can affect the nonlinear dynamic characteristics.

The motivation for this paper is to investigate the unbalance response of an MSDS considering nonlinear bearing characteristics of AMBs. Firstly, a non-linear bearing model for an AMB is established, Further, an MSDS dynamics model with a PID controller is built. Finally, the effects of external loads, rotational speed, and control parameters on the unbalanced response of the MSDS are analyzed.

2. System Model

A finite element model of an MSDS was designed for the study (Figure 1), where the dual-rotor system is discretized into nine nodes, including five nodes of the inner rotor and four nodes of the outer rotor, and the two coaxial rotors interlink through a mechanical inter-shaft bearing, and the rotors are supported by three AMBs.

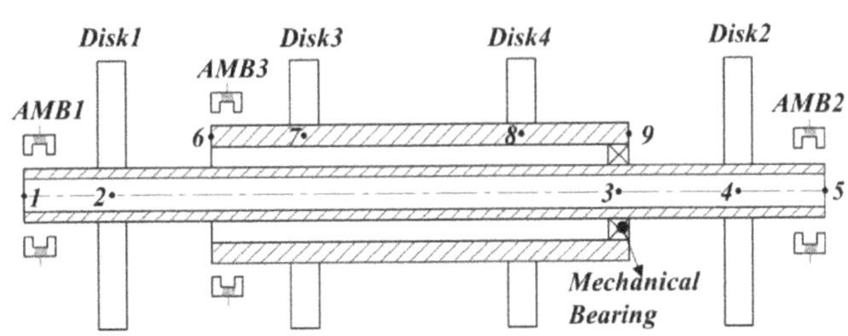

Figure 1. Finite element model of the MSDS.

The Finite Element Method was used in this research to build the mathematical model of the MSDS. In the following research, the MSDS, in which the magnetic flux leakage and magnetic saturation of AMBs have been ignored and the electromagnetic forces are regarded as linear forces, is called the conventional MSDS. The opposite one is called the nonlinear MSDS. The effects of torsional vibration and axial vibration are ignored; only the freedom of radial vibration is considered, and the effects of the rotors' gravity on the system are not considered. Moreover, the inter-shaft bearing is expressed as a linear spring and viscous damping. The main parameters of the structure are shown in Table 1.

Table 1. The main parameters of the MSDS.

Physical Parameter	Value	Physical Parameter	Value
Length of inner rotor (m)	0.706	Outside and inside radius of disk 2 (m)	0.125, 0.0125
Length of outer rotor (m)	0.5011	Outside and inside radius of disk 3 (m)	0.125, 0.02
Outside and inside radius of inner rotor (m)	0.0125, 0.0075	Outside and inside radius of disk 4 (m)	0.125, 0.02
Outside and inside radius of outer rotor (m)	0.02, 0.015	Thickness of disk (m)	0.0273
Density of rotating shaft (kg/m^3)	7850	Density of disk (kg/m^3)	7928.56
Elastic modulus (Pa)	2.1×10^{11}	Eccentric distance of disk 2 and disk 4 (m)	$6 \times 10^{-5}, 5 \times 10^{-5}$
Poisson's ratio	0.3	Inter-shaft bearing stiffness (N/m)	1×10^6
Outside and inside radius of disk 1 (m)	0.125, 0.0125	Inter-shaft bearing damping (N·s/m)	100

2.1. Magnetic Bearing Modeling

In this research, an eight-pole AMB with differential drive mode was used, the basic control loop in the y direction is shown in Figure 2. A PID strategy was applied to the controller, where I_b and I_c are the bias and control currents of the coil, respectively, and g_0 is the air-gap length. The power amplifier A_a = 0.6; the displacement sensor A_s = 800.

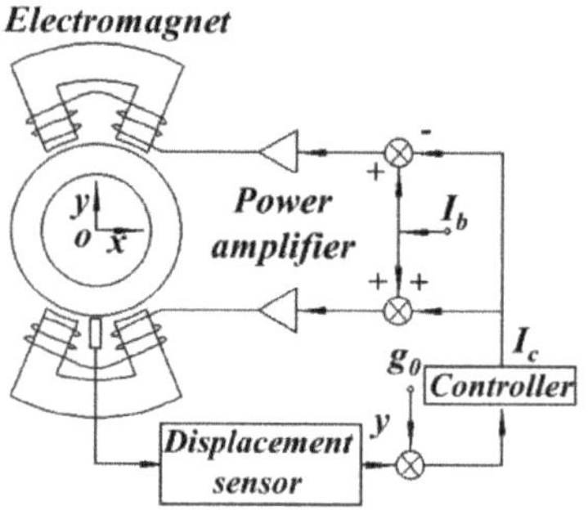

Figure 2. Basic control loop.

To consider the effects of the nonlinear factors on AMBs in this research, electromagnetic forces were applied to the system as external forces. Therefore, the discussion is divided into two cases.

(a) The nonlinear electromagnetic force model: Based on the NMCM, which divided the magnetic field as shown in Figure 3a, the flux areas correspond to the magnetic resistance of the air gap, leakage, yoke, poles, and rotor (corresponding to R_g, R_k, R_y, R_p, and R_r, respectively). Then, the equivalent magnetic circuit (EMC) for one pole pair was obtained, as shown in Figure 3b, and the expression for the electromagnetic force in the x direction was deduced [6]

$$f_x = \frac{\cos\alpha}{\mu_0 A_p}(\Phi_1^2 - \Phi_3^2) \quad (1)$$

$$\begin{cases} \Phi_1 = \dfrac{N(I_b+I_x)R_k}{(R_{g1}+R_{g2}+R_r)(R_k+R_y+2R_p)+R_k(R_y+2R_p)} \\ \Phi_3 = \dfrac{N(I_b-I_x)R_k}{(R_{g3}+R_{g4}+R_r)(R_k+R_y+2R_p)+R_k(R_y+2R_p)} \end{cases} \quad (2)$$

where, μ_0 is the permeability of vacuum; α is the angle between the central lines of the pole and pole pair, $\alpha = \pi/8$; A_p is the sectional area of the stator poles; N is the coil turn; Φ_1 and Φ_3 are the magnetic flux crossing the air gaps between the pole pairs and the rotors on the right and left, respectively.

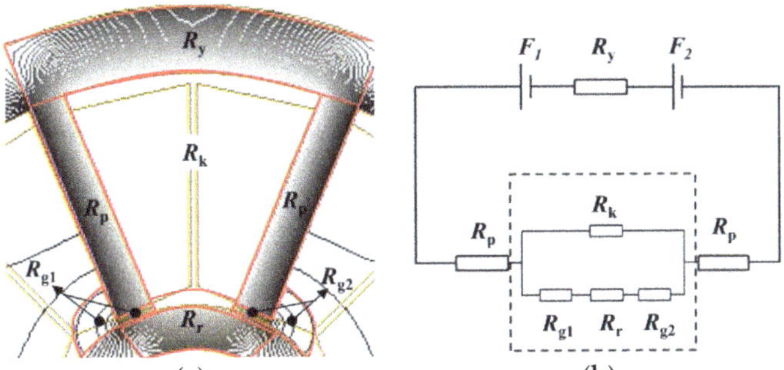

(**a**) (**b**)

Figure 3. The nonlinear electromagnetic force modeling process by using NCMC. (**a**) Magnetic field division. (**b**) EMC for one pole pair based on NMCM.

The electromagnetic force can be obtained after calculating each magnetic resistance value. Therefore, the NMCM was used to model and solve individual reluctances.

The air-gap magnetic resistance model is shown in Figure 4a. According to the magnetic flux distribution, the air-gap magnetic flux can be divided into edge flux tube, a, and main flux tube, b. Using a calculation, the permeability of both of them can be obtained, and R_g can be further calculated. Similarly, the magnetic flux of the leakage magnetic resistance model can be divided into regions 1, 2, and 3, as shown in Figure 4b. After calculating the permeability of each region, R_k can be calculated. In the calculation of R_y, R_p, and R_r, the effects of variation in magnetic permeability of soft magnetic materials are considered by using a power function to fit the B–H curve. The soft magnetic material B–H curve fit is shown in Figure 4c.

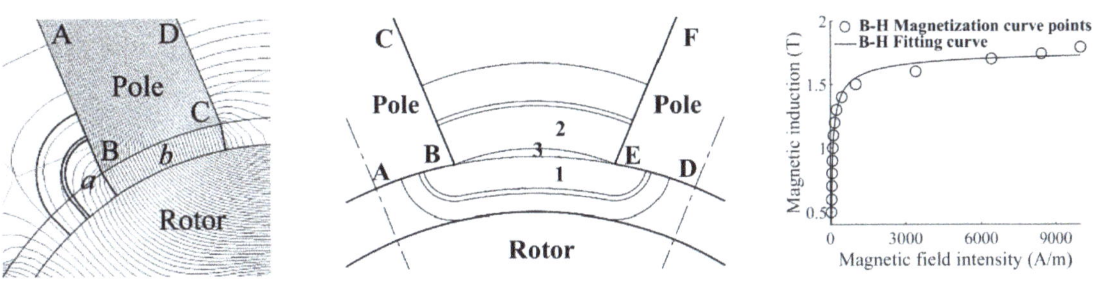

(**a**) (**b**) (**c**)

Figure 4. Each reluctance calculation by using NMCM. (**a**) The air gap reluctance model; (**b**) the leakage reluctance model; (**c**) the magnetization curve points of soft magnetic materials and their fitting curve.

The same method can be used to obtain the electromagnetic force in the y direction.

(b) The conventional electromagnetic force model: Factors such as magnetic flux leakage and magnetic saturation on the bearing characteristics of AMBs were not considered. Therefore, in Figure 4, R_y, R_p, R_r, and R_k are ignored, and only the air-gap reluctances exist. Additionally, the AMB worked in a linear range, and the linearized electromagnetic force in a certain direction (such as x direction) is given by [23].

$$f_x = k_x x + k_i i \tag{3}$$

$$\begin{cases} k_x = -\frac{\mu_0 A_p N^2 I_b^2 \cos\alpha}{g_0^3} \\ k_i = \frac{\mu_0 A_p N^2 I_b \cos\alpha}{g_0^2} \end{cases} \tag{4}$$

where, k_x and k_i are the displacement stiffness and current stiffness of the AMBs, respectively.

Based on Equations (1) and (3), the electromagnetic force of the AMB under conventional and nonlinear conditions can be obtained. For g_0 = 2.0 mm, the bearing capacity curves calculated by the two electromagnetic force models and finite element method (FEM) are shown in Figure 5.

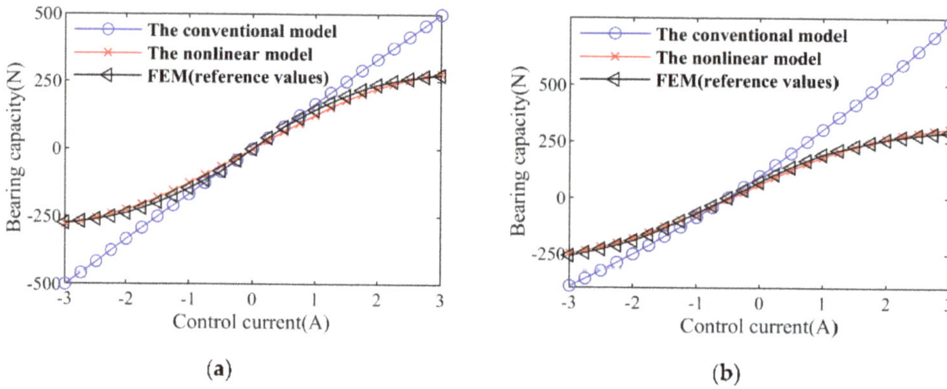

Figure 5. Bearing capacity curves. (**a**) e = 0; (**b**) e = 20%.

In Figure 5, e is the ratio of radial eccentricity to normal air-gap length. The results of the nonlinear electromagnetic force model and FEM are in good agreement, and these two results are obviously different from those of the linear electromagnetic force model, especially when the control current was large.

2.2. Dynamical Equations of the MSDS

In Figure 1, the shaft element is described by the Euler–Bernoulli beam element, which has four degrees of freedom at each node, including two translations and two rotations. The generalized displacement vectors of the inner rotor are

$$\begin{cases} q_{1i} = \{x_1\ \theta_{y1}\ x_2\ \theta_{y2} \cdots x_5\ \theta_{y5}\}^T \\ q_{2i} = \{y_1 - \theta_{x1}\ y_2 - \theta_{x2} \cdots y_5 - \theta_{x5}\}^T \end{cases} \quad (5)$$

where x_k and y_k (k = 1, 2, ..., 5) denote translations of notes 1–5, while θ_{yk} and θ_{xk} (k = 1, 2, ..., 5) denote the rotations of nodes 1–5, respectively.

Then, the motion equations of the inner rotor can be expressed as

$$\begin{cases} M_i\ddot{q}_{1i} + \omega_i J_i \dot{q}_{2i} + K_i q_{1i} = 0 \\ M_i\ddot{q}_{2i} - \omega_i J_i \dot{q}_{1i} + K_i q_{2i} = 0 \end{cases} \quad (6)$$

where M_I, J_i, and K_i are the mass, polar rotational inertial, and stiffness matrices of the inner rotor system, respectively, and ω_i is the rotation speed of the inner rotor.

Similarly, the motion equations of the outer rotor are

$$\begin{cases} M_o\ddot{q}_{1o} + \omega_o J_o \dot{q}_{2o} + K_o q_{1o} = 0 \\ M_o\ddot{q}_{2o} - \omega_o J_o \dot{q}_{1o} + K_o q_{2o} = 0 \end{cases} \quad (7)$$

where M_o, J_o, and K_o are the mass, polar rotational inertial, and stiffness matrices of the outer rotor system, respectively. ω_o is the rotation speed of the outer rotor.

Based on the above analysis, the generalized displacement of the dual-rotor system can be further express as

$$q = \{q_{1i}q_{1o}q_{2i}q_{2o}\}^T \quad (8)$$

Therefore, the motion equations of the system can be expressed as

$$M\ddot{q} + C\dot{q} + Kq = F \tag{9}$$

where $\boldsymbol{M}$, $\boldsymbol{C}$, and $\boldsymbol{K}$ are the total mass, damping, and stiffness matrices of the system, respectively. $\boldsymbol{F}$ is the external incentive force vector.

$$F = f_a + f_q \tag{10}$$

where f_a is the electromagnetic force vector; f_q is the other external incentives vector. In addition, the stiffness and damping of the inner-shaft bearing should be added to the corresponding nodes.

Secondly, the electromagnetic forces of the AMBs should be added to the corresponding nodes.

$$F = F_0 + \begin{bmatrix} \cdots \\ f_{x,k} \\ \cdots \\ f_{y,k} \\ \cdots \end{bmatrix} \tag{11}$$

where $\boldsymbol{F}$ is the generalized external force vector after considering the AMBs, F_0 is the generalized external force vector without considering the AMBs, and $f_{x,k}$ and $f_{y,k}$ are the electromagnetic forces in the x and y directions of the kth node, respectively.

3. Model Validation

Based on the published results in [24], the established dynamic model was verified. When the AMBs are worked in a linear interval, the nonlinear factors have little influence on the bearing characteristics, which can be ignored, and the dynamic responses of the two systems are basically consistent in this moment [17]. A comparison between the numerical simulation and experimental results in [24] was conducted.

The finite element model of the dual-rotor system in reference is shown in Figure 6.

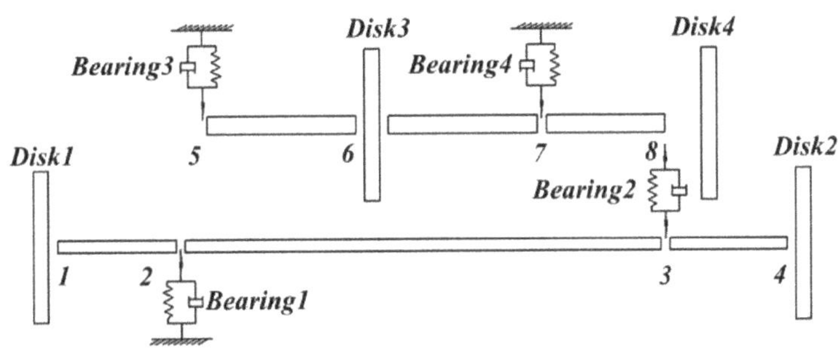

Figure 6. Yang's finite element model of the dual-rotor system [24].

According to the dual-rotor structure in the literature [24], mechanical bearings 1, 3, and 4 were replaced by AMBs, but the other structural parameters of the rotor system remain unchanged. The replaced dual-rotor system is an MSDS, and the structure parameters of the AMBs are shown in Table 2.

To ensure that the bearing characteristics of the system remain unchanged after replacement, the control parameters are $K_P = 50$, $K_I = 0.001$, and $K_D = 0.027$. At $\omega_i = 252.6$ rad/s and $\omega_o = 301.2$ rad/s, the unbalanced response of disk 1 in the vertical direction is shown in Figure 7. By comparing the results in Figure 7, it can be seen that the numerical results are in good agreement with the experimental results in [18].

Table 2. Structure parameters of AMBs.

Name	Parameters	Name	Parameters
Theoretical value of maximum bearing capacity F_{max} (N)	500	Inner diameter of the stator D (mm)	50
Theoretical value of saturated magnetic flux density B_{pmax} (T)	1.3	The width of the pole W_P (mm)	10
g_0 (mm)	0.5	The height of the pole h (mm)	23
Coil turn N	172	The width of the stator yoke W_y (mm)	12
I_b (A)	3	Axial length of the stator L_a (mm)	42

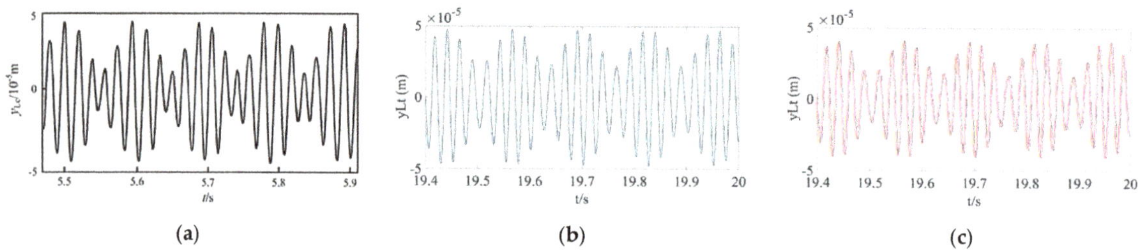

Figure 7. Displacement response of the dual-rotor system (ω_i = 252.6 rad/s; ω_o = 301.2 rad/s). (**a**) Yang's experimental result; (**b**) numerical result of the conventional MSDS; (**c**) numerical result of the nonlinear MSDS.

At ω_i = 184.1 rad/s and ω_o = 276.1 rad/s, a frequency diagram of the unbalance response of disk 1 in the vertical direction is shown in Figure 8. It can be seen that the numerical calculated frequency components have a good consistent with the experimental results [18], except for the amplitude of frequency components, which are mainly related to the position of unbalance and the initial conditions.

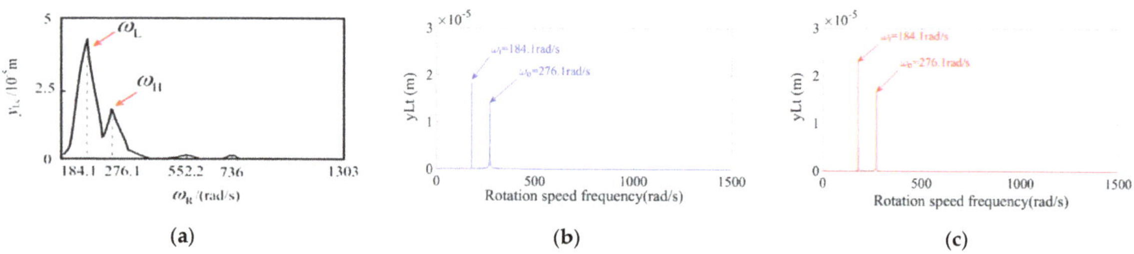

Figure 8. Displacement response of the dual-rotor system (ω_i = 184.1 rad/s; ω_o = 276.1 rad/s). (**a**) Yang's experimental result; (**b**) numerical result of the conventional MSDS; (**c**) numerical result of the nonlinear MSDS.

From the above comparison results, the numerical results have a good agreement with the experimental results, and the dynamic models of the MSDS established in this research are correct.

4. Unbalance Response Analysis

According to [14], the load has a great influence on the bearing characteristics of AMBs; when the rotor system is subjected to a large external load, the AMBs will be deviated from the linear working interval and present strong nonlinear features, then the unbalance response of the rotor system will also tend to be complex. In this research, the dynamic characteristics of the nonlinear MSDS in Figure 1 are discussed by applying loads to the inner and outer rotors' disks, respectively.

4.1. The Effect of Load on the System

To explore the effect of inner rotor's loaded on unbalance response of the nonlinear MSDS with the control parameters and operation parameters $K_P = 40$, $K_I = 100$, $K_D = 0.05$, $\omega_i = 200$ rad/s, $r_s = 1.2$, and $g_0 = 2.0$ mm, the stable unbalance response was analyzed by the Newmark-β method.

When the disk 1 was non-loaded, the time-domain response, shaft center trajectory, and frequency response of the two systems at disk 1 were calculated and are shown in Figure 9.

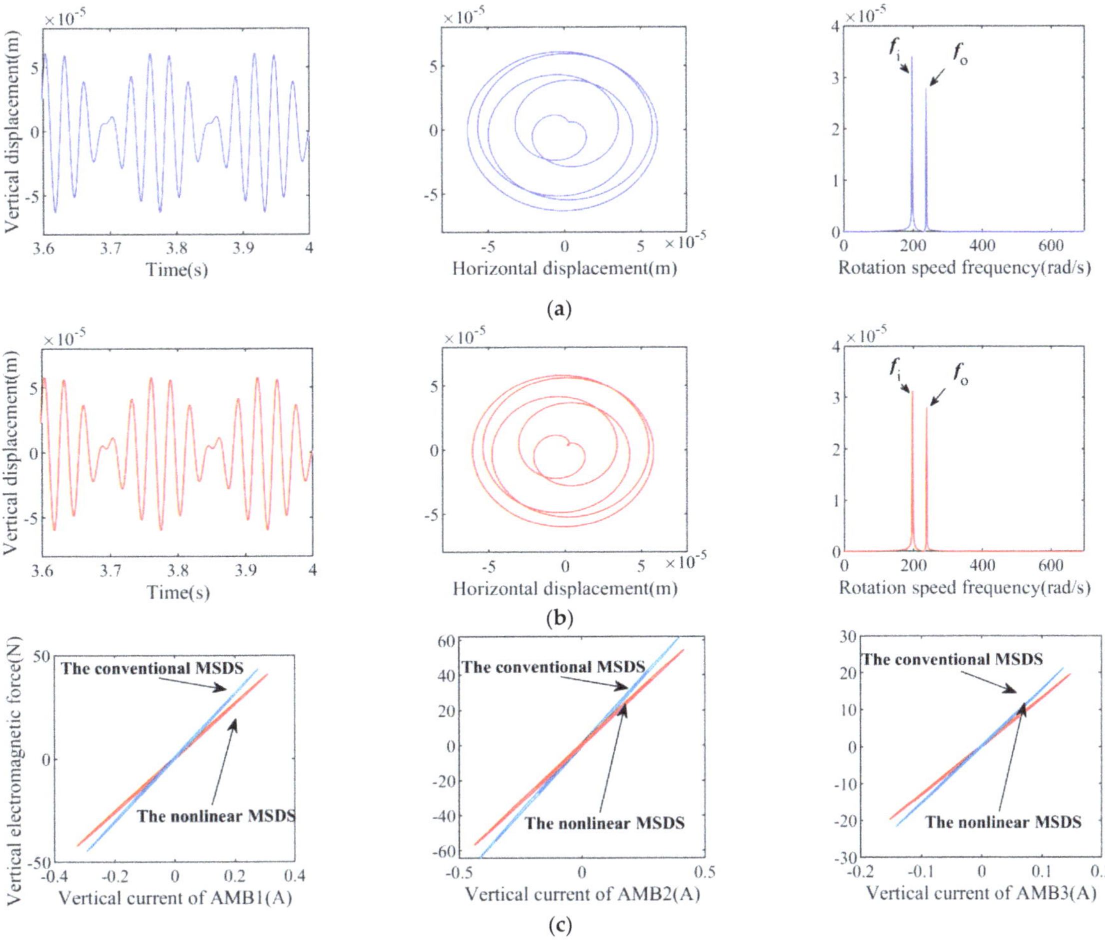

Figure 9. Unbalanced responses of the disk 1 ($\omega_i = 200$ rad/s, $F_L = 0$ N). (**a**) Unbalanced response of the conventional MSDS. (**b**) Unbalanced response of the nonlinear MSDS. (**c**) Bearing capacity curves of AMB 1, AMB 2, and AMB 3.

It can be seen from Figure 9a,b that the unbalance response of those two systems are basically identical, as there are only rotation frequencies f_i and f_o in the spectrum diagram. In Figure 9c, the bearing force of each AMB in the nonlinear MSDS is smaller than that of the conventional MSDS under the same current.

When disk 1 was loaded with 200 N in the vertical direction, the unbalance responses of the two systems were calculated and are shown in Figure 10.

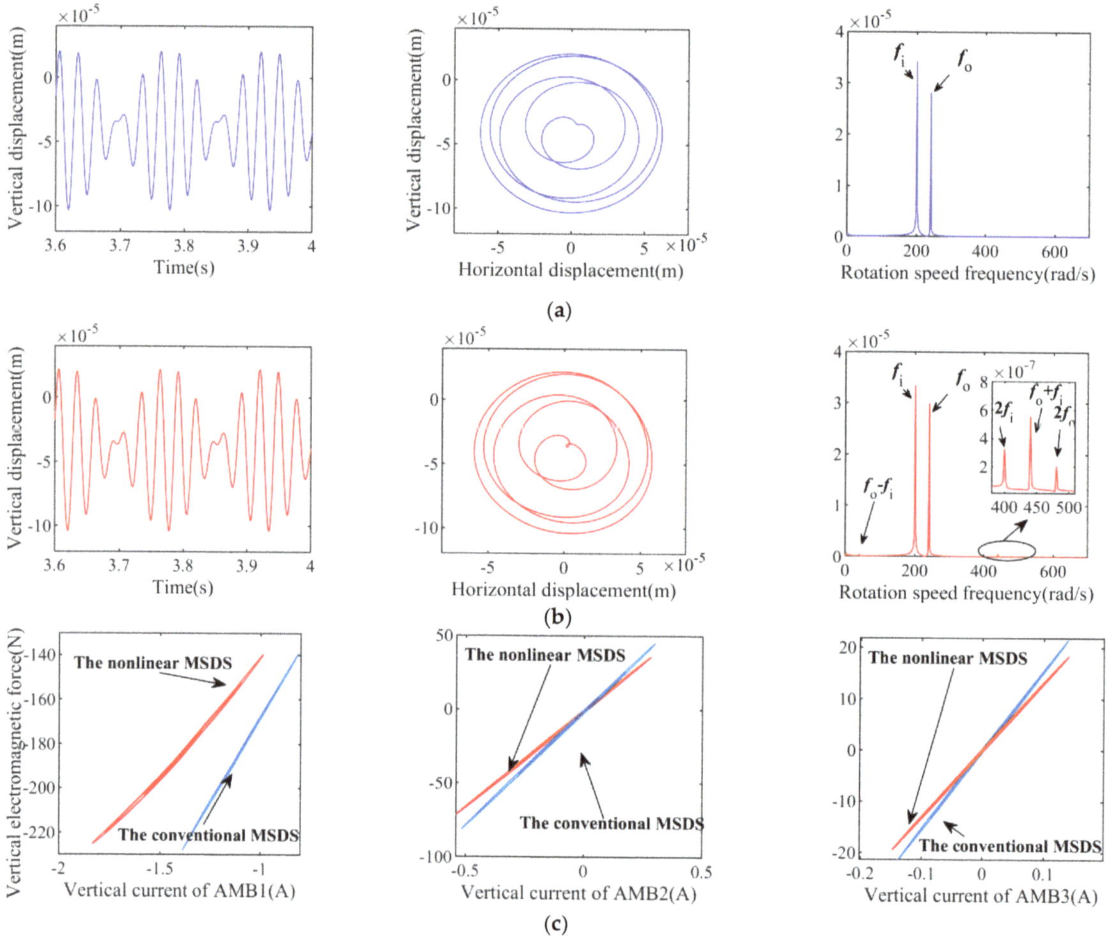

Figure 10. Unbalanced responses of the disk 1 (ω_i = 200 rad/s, F_L = 200 N). (**a**) Unbalanced response of the conventional MSDS. (**b**) Unbalanced response of the nonlinear MSDS. (**c**) Bearing capacity curves of AMB 1, AMB 2, and AMB 3.

It can be seen from Figure 10 that the unbalance response of the conventional MSDS is basically unchanged after the load is applied, and only the rotor is not fully returned to the balance position due to the flexibility of the rotor and the overall stiffness of the system. Additionally, in the nonlinear MSDS, the vertical current and electromagnetic forces of AMB 1 increased significantly, and a nonlinear relationship between the electromagnetic force and the movement of the rotor presented, which lead to significance changes in the unbalance response of the system. In Figure 10b, it can be seen that not only frequencies f_i and f_o appear, but also more combined frequency components appear in the spectrum diagram of the nonlinear MSDS, whose frequency values can be expressed as $mf_o \pm nf_i$ (m, n = 0, 1, 2, ...).

Figure 11 shows the shaft center trajectory of the inner rotor in the two systems. It can be seen that the shaft center trajectory at each node of two systems is also different, especially near the AMB 1. This phenomenon is also caused by the nonlinear bearing characteristics of the AMBs.

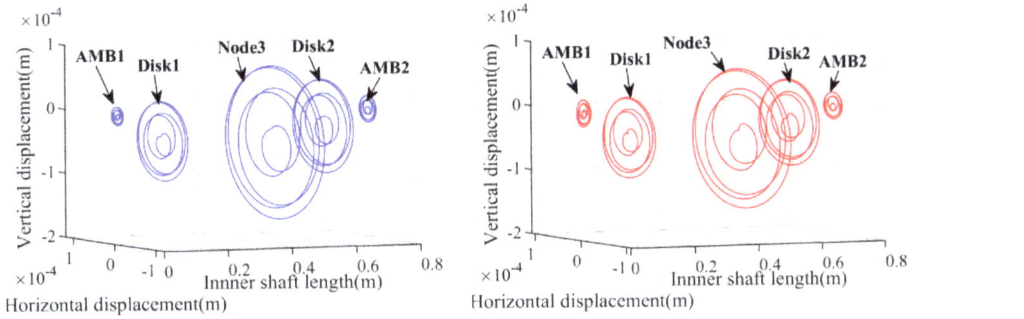

(**a**) (**b**)

Figure 11. Shaft center trajectory diagrams of the inner rotor (ω_i = 200 rad/s; F_L = 200 N). (**a**) The conventional MSDS, (**b**) the nonlinear MSDS.

In addition, Figure 12 shows the current responses of AMB 1 in the vertical direction when disk 1 was loaded with 200 N. It can be seen that the maximum current value of the nonlinear MSDS is closer to the system's designed limit (−3A). Moreover, the amplitude of current in the nonlinear MSDS is larger than that in the conventional MSDS.

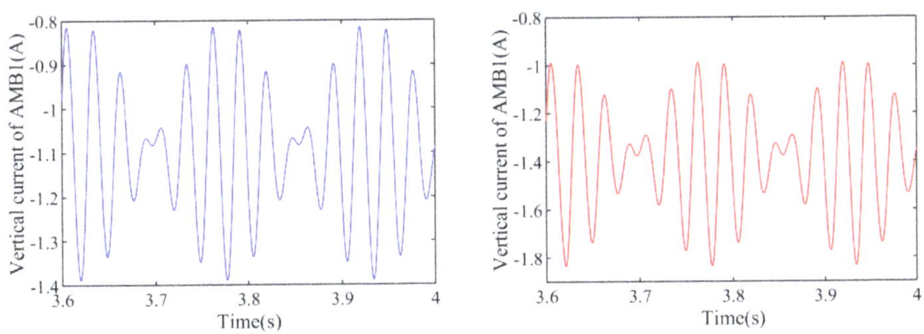

(**a**) (**b**)

Figure 12. Current in the vertical direction of the AMB 1 (ω_i = 200 rad/s; F_L = 200 N). (**a**) The conventional MSDS. (**b**) The nonlinear MSDS.

The waterfall plots of frequency responses of the two systems under different loads are shown in Figure 13.

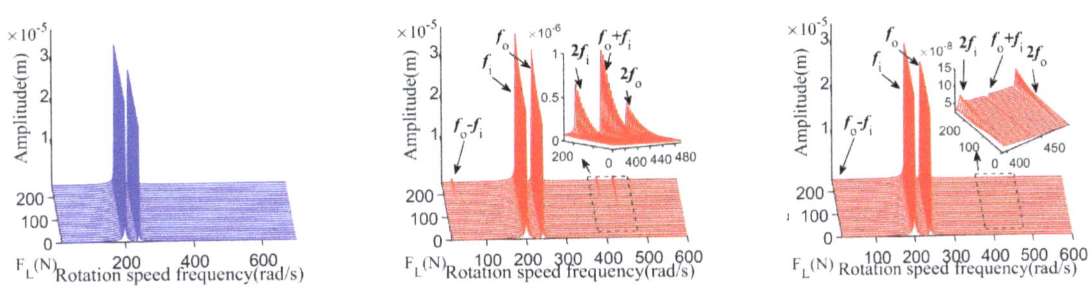

(**a**) (**b**) (**c**)

Figure 13. Waterfall plots of frequency responses under different load. (**a**) At disk 1 of the conventional MSDS. (**b**) At disk 1 of the nonlinear MSDS. (**c**) At disk 4 of the nonlinear MSDS.

In Figure 13a, only rotation frequencies f_{i} and f_{o} are found with the increase in the load in the conventional MSDS. However, in the nonlinear MSDS, as the load increases, the nonlinear bearing characteristics of the AMBs become more severe, and some combined frequency components appear in the system, such as $f_{\mathrm{o}}-f_{\mathrm{i}}$, $f_{\mathrm{o}}+f_{\mathrm{i}}$, $2f_{\mathrm{i}}$, and $2f_{\mathrm{o}}$. Additionally, the larger the load is, the more combined frequency components that will appear. It can be seen in Figure 13c that the same situation occurs for disk 4: with the increase in the load, more combined frequency components appear. The unbalance responses of other disks also show the same phenomenon as disk 1 does, which are not shown here.

From the above analysis, in the nonlinear MSDS, when the system is attached to a large external load, the AMBs show clear nonlinear bearing characteristics, and the response of the dual-rotor system appears to be obviously nonlinear, and the frequency response components of the system become more complex.

When the outer rotor's disks are loaded, the responses of the system are similar, which are not shown in here. Because the response laws of the four disks are similar, only the response at disk 1 was analyzed in the following research.

4.2. The Effect of Operation Parameters

With the control parameters K_{P} = 40, K_{I} = 100, and K_{D} = 0.05, when r_{s} = 1.2 and F_{L} = 200 N, the waterfall plots of frequency responses at disk 1 of the two systems under different rotation speeds were calculated and are shown in Figure 14.

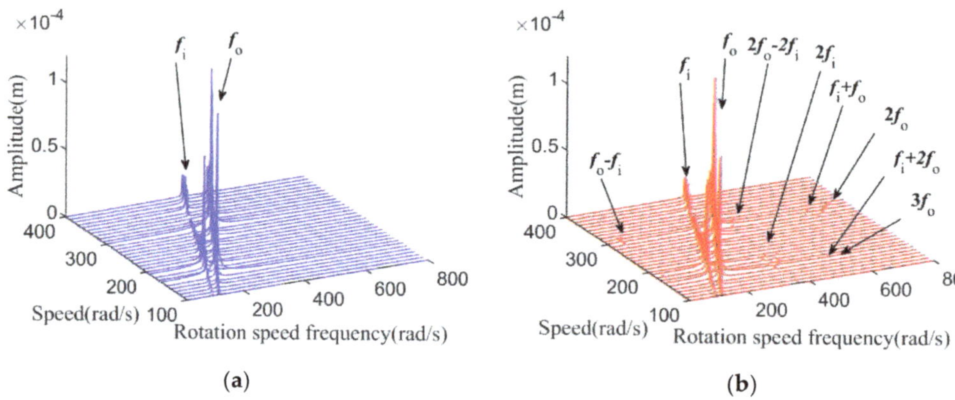

Figure 14. Waterfall plots of frequency responses of the disk 1 under different rotation speeds. (**a**) The conventional MSDS. (**b**) The nonlinear MSDS.

In the conventional MSDS, only rotation frequencies f_{i} and f_{o} appear in the waterfall diagram under different rotation speed. However, it can be seen from Figure 14b that other combined frequency components also appear in the nonlinear MSDS, but the combined frequency components will not always be excited by the load. When the rotation speed is in a range from [150, 200] rad/s to [270, 320] rad/s, the vibration of the system increases and the nonlinear bearing characteristics of the AMBs in this range become more severe; more combined frequency components appear in the waterfall diagram. The results mean that the rotation speed and external load have a large influence on the responses of the MSDSs, and the effect is reflected not only in the frequency components of rotor steady-state responses, but also in the stability of the system.

With the same control parameters as Figure 14, when r_{s} = 1.2, Figure 15 shows the stable operation area of the two systems when ω_{i} is [100, 400] rad/s.

It can be seen in Figure 15, the maximum load capacity of the system varies with the rotation speed, which indicates that the stable operation area of the system is related to the rotation speed. Moreover, the stable area of the nonlinear MSDS is smaller than that of the conventional MSDS under the same operating conditions, which means the stability of the nonlinear MSDS is worse.

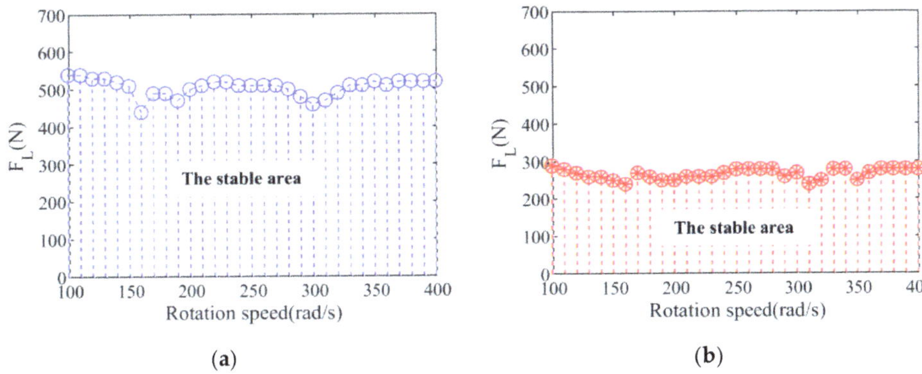

Figure 15. The stable operation area of the two systems under load conditions. (**a**) The conventional MSDS. (**b**) The nonlinear MSDS.

From above analysis, the nonlinear responses of the MSDS are determined by the external load and rotation speed, and only when the amplitude of the vibration responses is large will the combined frequency components appear. Additionally, the stability of the nonlinear MSDS is worse than that of the conventional MSDS under the same operating conditions.

4.3. The Effect of Control Parameters

When the structural parameters and operating conditions of the MSDS are determined, the system response can be changed by adjusting the control parameters. With heavy load conditions, the effect of the PID control parameters on the system nonlinear unbalance response is investigated in this section

With the same operating parameters as Figure 14, when $K_I = 100$ and $\omega_i = 200$ rad/s, the frequency response waterfall plots of the system at disk 1 at different K_P or K_D values are shown in Figure 16, respectively.

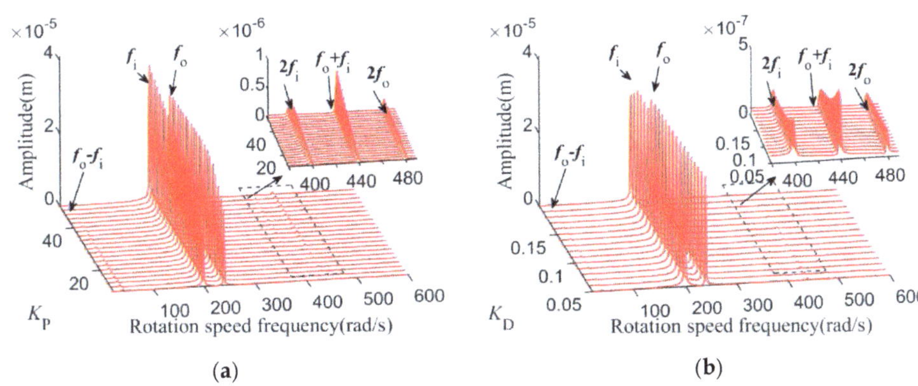

Figure 16. Waterfall plots of frequency responses of disk1 under different control parameters. (**a**) KP; (**b**) KD.

In Figure 16a, the $f_o - f_i$ component of the vibration responses is obvious when K_P is small; with the increase in K_P, the amplitude of the system increases, more frequency combined components appear, and the $2f_i$, $f_o + f_i$ and $2f_o$ components of the vibration responses become larger.

It can be seen from Figure 16b that when K_D is small, the vibration suppression ability of the system is limited and the nonlinear bearing characteristics of AMBs become serious; the rotation frequencies $2f_i$ and $f_o + f_i$ are obvious in the waterfall diagram. The response amplitudes of these frequency components decrease with the increase in K_D, meanwhile, the nonlinear bearing characteristics are minimized.

Therefore, control parameters K_P and K_D can affect the dramatic level of the nonlinear response characteristics of the system under heavy loads. A larger KP and a smaller KD induce more combined frequency components and increase the amplitude of the components, making the unbalance response of the system more complex.

The effect of K_I on the system response is mainly reflected in the low-frequency segment, and it can be ignored in the high-frequency segment; the research results are not shown here.

5. Conclusions

In this paper, the unbalance response of an MSDS based on the nonlinear bearing characteristics of AMBs was investigated. A finite element model of the system was built, and the Newmark-β method was employed to obtain the stable unbalance response of the system. Accordingly, the time-domain response, shaft center trajectory, and frequency response of the system with different operation parameters and control parameters were studied. The results indicated that the external load, speed, and control parameters have influences on the unbalance response of the system. The specific conclusions are as follows:

(1) Combination frequencies ($mf_o \pm nf_i$) may exist in the responses of the nonlinear MSDS under heavy loading conditions, which could excite the nonlinear bearing characteristics of the AMBs, and the effects of the inner and outer rotors' disks that are loaded are basically identical. Moreover, the stability of the nonlinear MSDS is worse than that of the conventional MSDS under the same operating conditions.

(2) K_P has an influence on the amplitude of the nonlinear system; when the amplitude of the system becomes larger, the unbalance response of the system becomes more complex, and a small K_D will lead to more combination frequencies. The results obtained in this paper may contribute to the construction of models based on the nonlinear bearing characteristics of AMBs and restraining the negative effect of nonlinear bearing characteristics on the system.

Author Contributions: N.W.: conception of the study, propose theory and method, supervisor; M.L.: literature search and writing; J.Y.: software, drawing, data collection and processing, writing—original draft; P.G.: software, drawing; H.W.: data interpretation and analysis. All authors have read and agreed to the published version of the manuscript.

Funding: This research was funded by National Natural Science Foundation (Grant No. 51975427) and Shenzhen Science and Technology Plan Project (Grant No. JCYJ20190809150603586).

Data Availability Statement: The data used to support the findings of this study are available from the corresponding author upon request.

Conflicts of Interest: The authors declare that there are no conflict of interest regarding the publication of this paper.

References

1. Maslen, E.H.; Schweitzer, G.; Bleuler, H.; Cole, M.; Traxler, A. *Magnetic Bearings—Theory, Design, and Application to Rotating Machinery*; Springer: Berlin/Heidelberg, Germany, 2009.
2. Wang, D.X.; Wang, N.X.; Chen, K.S.; Ye, C. Dynamic Characteristics of Magnetic Suspended Dual-Rotor System by Riccati Transfer Matrix Method. *Shock. Vib.* **2019**, *2019*, 9843732. [CrossRef]
3. Yu, C.M.; Deng, Z.Q.; Mei, L.; Peng, C.; Cao, X.; Chen, S.S.; Ding, Q. Evaluation Criteria of Material Selection on 3-DOF Hybrid Magnetic Bearing. *IEEE Trans. Ind. Appl.* **2021**, *57*, 4733–4744. [CrossRef]
4. Yu, C.; Deng, Z.; Mei, L.; Peng, C.; Cao, X. Multiobjective optimization of 3-DOF magnetic bearing considering eddy current effects and saturation. *Mech. Syst. Signal Process.* **2023**, *182*, 109538. [CrossRef]

5. Zhang, W.Y.; Yang, H.K.; Cheng, L.; Zhu, H.Q. Modeling Based on Exact Segmentation of Magnetic Field for a Centripetal Force Type-Magnetic Bearing. *IEEE Trans. Ind. Electron.* **2020**, *67*, 7691–7701. [CrossRef]
6. Wang, D.X.; Wang, N.X.; Ye, C.; Chen, K.S. Research on analytical bearing capacity model of active magnetic bearings based on magnetic saturation. *IET Electr. Power Appl.* **2017**, *11*, 1548–1557. [CrossRef]
7. Wajnert, D.; Tomczuk, B. Nonlinear magnetic equivalent circuit of the hybrid magnetic bearing. *Compel-Int. J. Comput. Math. Electr. Electron. Eng.* **2019**, *38*, 1190–1203. [CrossRef]
8. Lu, Z.Y.; Hou, L.; Chen, Y.S.; Sun, C.Z. Nonlinear response analysis for a dual-rotor system with a breathing transverse crack in the hollow shaft. *Nonlinear Dyn.* **2016**, *83*, 169–185. [CrossRef]
9. Wang, D.; Wang, N.; Chen, K. Fixed-point Rubbing Characteristics of Magnetic Suspended Dual-rotor Systems. *China Mech. Eng.* **2021**, *32*, 1686–1699.
10. Zhang, S.Y.; Zhou, J.; Wu, H.T.; Zhang, Y. Dynamic Analysis of Active Magnetic Bearing Rotor System Considering Alford Force. *J. Vib. Eng. Technol.* **2021**, *9*, 1147–1154. [CrossRef]
11. Zhang, S.Y.; Zhou, J.; Han, X.M.; Ma, Y.C. Analysis of Nonlinear Characteristics and the Factors Affecting the Operation of the Active Magnetic Bearings Rotor System Considering Alford Force. *Appl. Comput. Electromagn. Soc.* **2022**, *37*, 253–261. [CrossRef]
12. Su, Y.X.; Gu, Y.P.; Keogh, P.S.; Yu, S.Y.; Ren, G.X. Nonlinear dynamic simulation and parametric analysis of a rotor-AMB-TDB system experiencing strong base shock excitations. *Mech. Mach. Theory* **2021**, *155*, 104071. [CrossRef]
13. Yi, H.; Hou, L.; Gao, P.; Chen, Y. Nonlinear resonance characteristics of a dual-rotor system with a local defect on the inner ring of the inter-shaft bearing. *Chin. J. Aeronaut.* **2021**, *34*, 110–124. [CrossRef]
14. Gao, P.; Chen, Y.S.; Hou, L. Nonlinear thermal behaviors of the inter-shaft bearing in a dual-rotor system subjected to the dynamic load. *Nonlinear Dyn.* **2020**, *101*, 191–209. [CrossRef]
15. Wang, F.; Luo, G.H.; Yan, S.; Cui, H.T. A Comparison Study on Co- and Counterrotating Dual-Rotor System with Squeeze Film Dampers and Intermediate Bearing. *Shock. Vib.* **2017**, *2017*, 5493763. [CrossRef]
16. Ji, J.C.; Hansen, C.H. Non-Linear Oscillations of a Rotor in Active Magnetic Bearings. *J. Sound Vib.* **2001**, *240*, 599–612. [CrossRef]
17. Zhang, G.; Xi, G. Vibration Control of a Time-Delayed Rotor-Active Magnetic Bearing System by Time-Varying Stiffness. *Int. J. Appl. Mech.* **2022**, *14*, 2250007. [CrossRef]
18. Zhang, W.; Wu, R.Q.; Siriguleng, B. Nonlinear Vibrations of a Rotor-Active Magnetic Bearing System with 16-Pole Legs and Two Degrees of Freedom. *Shock. Vib.* **2020**, *2020*, 5282904. [CrossRef]
19. Wu, R.Q.; Zhang, W.; Yao, M.H. Nonlinear dynamics near resonances of a rotor-active magnetic bearings system with 16-pole legs and time varying stiffness. *Mech. Syst. Signal Process.* **2018**, *100*, 113–134. [CrossRef]
20. Zhang, X.S.; Fan, T.P.; Sun, Z.; Zhao, L.; Yan, X.S.; Zhao, J.L.; Shi, Z.G. Nonlinear Analysis of Rotor-AMB System with Current Saturation Effect. *Appl. Comput. Electromagn. Soc. J.* **2019**, *34*, 557–566.
21. Sun, Z.; Zhang, X.; Fan, T.; Yan, X.; Zhao, J.; Zhao, L.; Shi, Z. Nonlinear dynamic characteristics analysis of active magnetic bearing system based on cell mapping method with a case study. *Mech. Syst. Signal Process.* **2019**, *117*, 116–137. [CrossRef]
22. Kang, K.; Palazzolo, A. Homopolar Magnetic Bearing Saturation Effects on Rotating Machinery Vibration. *IEEE Trans. Magn.* **2012**, *48*, 1984–1994. [CrossRef]
23. Soni, T.; Dutt, J.K.; Das, A.S. Dynamic behavior and stability of energy efficient electro-magnetic suspension of rotors involving time delay. *Energy* **2021**, *231*, 120906. [CrossRef]
24. Yang, Y.; Cao, D.; Yu, T.; Wang, D.; Li, C. Prediction of dynamic characteristics of a dual-rotor system with fixed point rubbing—Theoretical analysis and experimental study. *Int. J. Mech. Sci.* **2016**, *115*, 253–261. [CrossRef]

MDPI AG
Grosspeteranlage 5
4052 Basel
Switzerland
Tel.: +41 61 683 77 34

Actuators Editorial Office
E-mail: actuators@mdpi.com
www.mdpi.com/journal/actuators

www.ingramcontent.com/pod-product-compliance
Lightning Source LLC
Chambersburg PA
CBHW061513210926
51850CB00008B/1903
9783725817535